Sergey Pulinets
Kirill Boyarchuk
Ionospheric Precursors of Earthquakes

Sergey Pulinets
Kirill Boyarchuk

Ionospheric Precursors of Earthquakes

With 182 Figures, 3 in colour

Springer

Professor Sergey Pulinets
Instituto de Geofísica
Universidad Nacional Autónoma de México
Ciudad Universitária, Delegación de Coyoacán
Código 04510,
México D.F.

México

E-mail: pulse@geofisica.unam.mx

Dr. Kirill Boyarchuk
Institute of Terrestrial Magnetism
Ionosphere and Radiowave Propagation
Russian Academy of Sciences
Troitsk, Moscow Region, 142190

Russia

E-mail: boyar@izmiran.rssi.ru

Library of Congress Control Number: 2004107505

ISBN 3-540-20839-9 **Springer Berlin Heidelberg New York**

Springer is a part of Springer Science+Business Media
springeronline.com

Printed in Germany

Cover design: E. Kirchner, Heidelberg
Production: A. Oelschläger
Typesetting: G. Oesterlein
Printing: Mercedes-Druck, Berlin
Binding: Stein + Lehmann, Berlin

Printed on acid-free paper 32/2132/AO 5 4 3 2 1 0

Preface

Using the kind permission given to me by my co-author, this short preface will be written in my name.

I want to devote this book to San Juan city in Argentina. It is not only due to the fact that the city was twice completely destroyed after the devastating earthquakes in 1941 and 1977, but also because my stay there completely changed my life. Changes included changing my career from the field of space plasma physics to Earth sciences and geophysics, and changes in my personal life giving me happiness and compliance in my present family.

Going back to the subject of the book, it should be noted that the history of the question asked by the book is very complicated and intricate. Starting in the 1930s from the observation of seismogenic electric fields, the area of seismo-ionospheric coupling became an area of fighting and conflicts, hopes and frustrations. Speculation and misunderstanding on the interdisciplinary borders made this field for many years (even up to now) taboo for so-called "serious scientists". But due to the courageous efforts of several groups in Russia and the former USSR states such as Kazakhstan and Uzbekistan, Japan, later France and Taiwan, Greece and Italy the situation started to improve. From the late 80-th — early 90-th regular workshops started to be organized both as separate events (like the Chofu workshops in Japan), and within the framework of the World's main scientific Assemblies and conferences such as URSI, IAGA, EGS and COSPAR General Assemblies, Wroclaw International Symposium on Electromagnetic Compatibility, PIERS, WPGM (Western Pacific Geophysical Meeting) etc. This intensive work led to two major results. First, the field of seismo-ionospheric coupling and the electromagnetic precursors of earthquakes started to be acknowledged and internationally coordinated. The common working group of URSI commissions E, G, H on Lithosphere-Atmosphere-Ionosphere Coupling was organized coordinating this activity within URSI, and in 2002 the IAGA-IASPEI-IAVCEI Inter-Association Working Group on "Electromagnetic Studies on Earthquakes and Volcanoes (EMSEV)" was organized.

The second very important result is that governmental bodies in different countries started to be involved in the practical application of the results of seismo-ionospheric coupling studies which led to the financing of several projects of ground based observations (FRONTIER projects in Japan, iSTEP project in Taiwan, PREVENTION project in Mexico), and satellite based studies of the ionospheric precursors of earthquakes: DEMETER (France), Warning and Variant (Ukraine), COMPASS and Vulkan (Russia), and the ESPERIA mission (Italy).

The lithosphere-atmosphere-ionosphere coupling is a very complicated subject involving a lot of physical effects and interactions on all levels starting from underground up to the Earth's magnetosphere. The volume of knowledge stored up to now is so large that it is impossible even in a short time to research all available directions. So we decided to limit ourselves to the following. There is the natural divide in the subject reflected in the existence of two groups of scientists. One of them is involved in studies of seismo-electromagnetic phenomena expressed mainly in electromagnetic emissions and fields associated with the seismic activity, and another one — dealing with ionospheric and space plasma variations observed before the approaching earthquakes. Our book will concentrate on the second of these two areas of research. This can be explained in two ways. Firstly, it is very natural as the authors belong to the second group, and the second reason is that the number of publications on seismo-electromagnetic noises is essentially larger than publications on ionospheric phenomena. So we decided to fill this gap by describing the underlying physical mechanism and systemizing the information on ionospheric precursors of earthquakes.

The book consists of 7 chapters. Taking into account the multidisciplinary character of the described subject the first chapter is devoted to main composite constituents or fields of science involved in seismo-ionospheric coupling. It provides elementary knowledge on the main terms of seismology, ionosphere, radiative plasmachemistry and atmospheric electricity.

The second chapter gives an idea of how look the ionospheric precursors, the techniques of its registration.

The model description is concentrated in Chapters 3 and 4. Chapter 3 describes the physical mechanism of anomalous electric field generation and presents the new conception of near ground atmospheric plasma. Chapter 4 is devoted to the physical mechanism of the electric field penetration into the ionosphere and its effects in the ionosphere and magnetosphere.

To be able to distinguish the ionospheric precursors from the other kinds of ionospheric variability it is necessary to know their main morphological features. Chapter 5 is devoted to their description.

Working with something that has the name "precursor", it is impossible to avoid the questions of practical application. Some of our approaches to practical application of the ionospheric precursors are posed in Chapter 6.

Seismo-ionospheric coupling is a science which is developing now, in our eyes, so there still exist issues which are not well established and proved. The seventh chapter is devoted to issues not as yet clarified but which have a relation to seismo-ionospheric coupling, and describes some very new results obtained during the writing of the book. It also deals with the future directions of our studies.

Such multidisciplinary study is impossible without involving a large scientific community, experimental and theoretical work. So we would like to thank all the people who helped us in our studies. First of all, our colleagues from the IZMIRAN Ionospheric Department, which I headed for many years. And personally I would like to thank Valery Hegai whose important contribution to this work, especially Chapter 4, is difficult to overestimate. His kind but firm support and valuable comments helped very much. I also want to thank our colleagues Anna

Legen'ka, Victor Depuev, Tamara Gaivoronskaya, Alexei Lomonosov, who are our co-authors in many publications devoted to seismo-ionospheric coupling.

My appreciation also to my brother-in-arms, Prof. J. Y. Liu, now Director of the Institute of Space Science, National Central University, Taiwan. A lot of results from Taiwan contributed to our model substantiation, and my fruitful multiple stays in Taiwan always gave inspiration for future work.

Many thanks to our colleagues from different institutions who are working in the field of electromagnetics, ionospheric precursors, and seismology: Hiroshi Kikuchi, Seiya Uyeda, Masashi Hayakawa, Oleg Molchanov, Michel Parrot, Friedemann Freund, Dimitar Ouzounov, Pier Francesco Biagi, Manuel Hernandez-Pajares, Ramesh Singh, Mikhael Gokhberg, Vladimir Kostoglodov, Yuri Taran, Amando Leyva Contreras, Sinna Lomnitz, and many-many others who contributed to our work by fruitful discussions and comments.

I should thank P. Hollis-Watts whose completely new results on the pulsed emission before the earthquakes in Western Australia were obtained as a result of his enthusiastic and voluntary work.

It should be mentioned also that without the friendly and creative atmosphere surrounding me at the Institute of Geophysics of UNAM this book would never have been finished.

We also want to thank our opponents who helped to sharpen our arguments and convince us of our rightness.

Chapters 1, 2, 4, 5, were written by S. A. Pulinets, Chapter 3 — by K. A. Boyarchuk, and Chapter 6 and 7 — by S.A. Pulinets and K.A. Boyarchuk.

Sergey Pulinets 28 of August 2003 Mexico City

Table of contents

Chapter 1 – The Basic Components of Seismo-Ionospheric Coupling

1.1 Introduction

During the last few years in the course of discussions with our colleagues on different scientific forums, and discussions with the referees of our journal papers we realized one very simple problem, which complicates the understanding of seismo-ionospheric coupling – it is lack of knowledge. The subject itself is multidisciplinary, ranging from pure mechanics to space plasma physics, demanding simultaneous knowledge of many special disciplines such as plate tectonics, seismology, atmosphere and ionosphere physics and chemistry, and atmospheric electricity. Of course, the fields of seismology and ionosphere physics are very far one from another and discussions between representatives of these groups remind us of a discussion between the blind and the deaf. This entails a lot of unfounded conflicts and complications. The only answer we see is to familiarize those who really want to understand the problem at least with the basic ideas of the disciplines mentioned above. Of course, the seismologist who picks up our book probably should not read the paragraph devoted to seismology but we strongly recommend he looks through the other paragraphs, especially the ionospheric one.

We do not intend to cover all the principles of modern seismology, as well as other disciplines to be described in the first chapter. Our wish is to give new comers the most important ideas which will be used in the following text discussing the problems of seismo-ionospheric coupling.

From the perspective of a discussion of the physical mechanisms of seismo-ionospheric coupling, we decided to include in the first chapter the following disciplines:

- Seismology
- Ionosphere physics and morphology
- Radiative troposphere plasmachemistry
- Atmospheric electricity

1.2 Seismology

Earthquake physics is a very complex and broad topic. It involves many scales of the Earth's crustal structure starting from tectonic plates and finishing with the microscopic processes involved in the friction, generation of electric charge and chemical reactions.

Earthquake occurrence is connected with the Earth' crustal dynamics. The Earth's crust is the rigid external shell of our planet consisting of the continental and oceanic crust. The thickness of the continental crust is of the order of 40 km and the oceanic one – is respectively near 6 km. The crust and the upper layer of the mantle form the lithosphere consisting of semi rigid plates of different sizes. The slow movement of these plates over the asthenosphere and ocean floor extension is called plate tectonics (Oreskes 2003). Collision of these plates leads to the diving of one plate edge under another (convergent or subduction boundaries). Subduction leads to the formation of different geological structures starting from the oceanic trench in the zone of plates contact, formation of new mountains and zones of volcanic activity. Apart from the subduction process there are other kinds of plate relative motion. Plate separation (divergent boundaries) generating the oceanic crust, and transform boundaries when plates move one along another as occurs at the San-Andreas fault in California. A simplified diagram of plate boundary configuration is presented in Fig. 1.1. All these movements lead to a strain accumulation within the Earth's crust, to mechanical deformations and crust rupture when the deformation exceeds the limit of mechanical strength. The process of rupture is the earthquake itself. It is natural to expect a difference in earthquake characteristics for every type of tectonic plate contact.

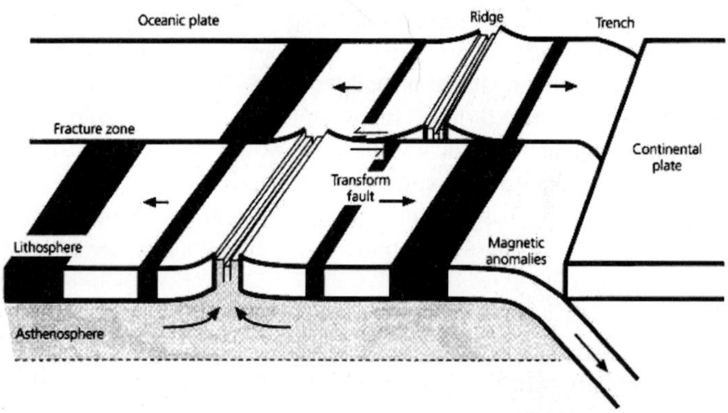

Fig. 1.1 Schematic presentation of the different types of tectonic plate borders

Before we move on to the physical aspects of earthquakes, some introductory information is necessary. Those who are interested in more extended information can find this in recent monographs starting from elementary books (Bolt 1993;

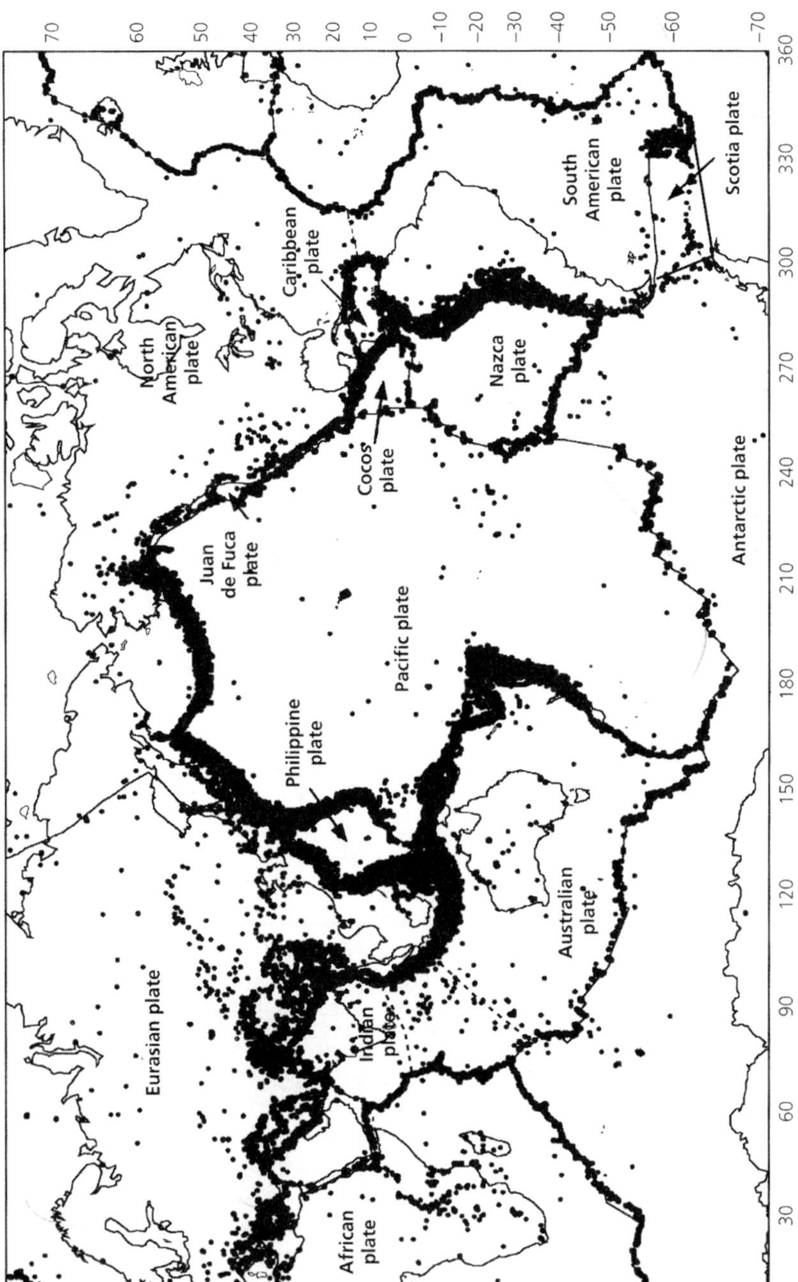

Fig. 1.2 Global earthquake seismicity in relation to the tectonic plate boundaries

Brumbaugh 1999) up to more advanced manuals (Shearer 1999; Stein and Wysession 2003). We strongly recommend the last one. Taking into account that earthquakes are the result of the tectonic plates contact, their global distribution is not uniform and is concentrated close to the tectonic plate's boundaries, which is demonstrated in Fig. 1.2. Except for the Himalayan region where the plate border is not obvious looking at the distribution of seismic activity, the other regions demonstrate a clear concentration of earthquakes on the tectonic plate borders.

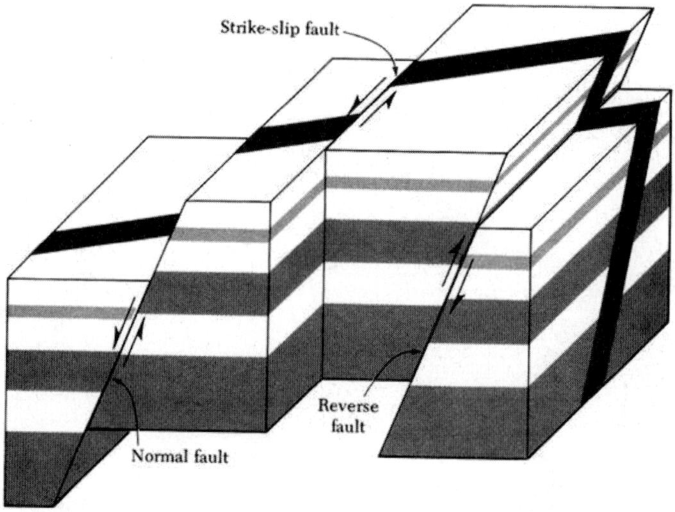

Fig. 1.3 The main types of tectonic fault geometry. Their combination is observed also

Fracturing of the Earth's crust has different scale sizes and in the more regional scale, it is associated with so-called tectonic faults (this is now generally excepted). The conception of a fault was introduced by Anderson (1905) and he was the first who tried to explain the physics of faulting (Anderson 1942) by brittle fracture and applying the Coulomb criterion to this problem. Not all faults geometry could be explained by Anderson's theory, and it was developed and accomplished later by other investigators, but it is not our task to go deeply into faults mechanics – the only important conclusion we have is that the earthquakes occur on the fault – an existing discontinuity of the Earth's crust formed by the previous crustal movements. Additional information can be obtained in the monograph by Scholz (1990). From the mechanical point of view, the earthquake takes place when the applied stress exceeds the dynamic friction. The rupture criterion was introduced first by Griffith (1920). After the earthquake the dynamic friction becomes more than the stress on the rupture zone. If one takes two bricks and tries to move one in relation to another, you may find different types of relative movement. These relative movements of the fault walls (the main ones) are presented in Fig. 1.3.

The faults have also their hierarchy. Their length varies from thousands of kilometers (Anatolian and San Andreas Faults) up to local discontinuities of several kilometers long. Usually after the earthquake a scarp is formed causing damage of mechanical constructions as one can see in Fig. 1.4 demonstrating damage along the fault of the recent catastrophic Chi-Chi earthquake in Taiwan on 21 September 1999 (M_W=7.6)

Fig. 1.4 Top panel: the bridge damage and scarp of 8 m height formed along the fault of the Chi-Chi earthquake on the river floor. Bottom panel: the fault scarp passing across the racetrack of one of the schools in Puli, Taiwan

The earthquake's intensity is estimated from the oscillation created by different kinds of seismic waves (usually, the surface and body waves). The seismic waves are registered by seismographs and the largest vertical displacement A of the seismograph arrow was selected by Charles Richter in the 1930s to characterize the earthquake intensity by a parameter called the "local magnitude" M_L (Shearer 1999):

$$M_L = \log_{10} A + 2.56 \log_{10} \Delta - 1.67 \quad (1.1)$$

where A_0– is the amplitude of known reference event, Δ – distance to the source of seismic wave. Here A is expressed in millimeters, and Δ – in kilometers. After the Richter table providing A_0 for different distances Bullen and Bolt (1985) proposed the empirical formula for M_L:

$$M_L = \log_{10} A(\Delta) - \log_{10} A_0(\Delta) \quad (1.2)$$

But this determination was very local because it depended on a specific place and was proposed for a specific instrument (Wood-Anderson seismograph), so development of a more universal parameter was necessary. Two other determinations of magnitude were proposed for two different types of seismic waves: the body wave named in seismology the P-wave, and the surface Rayleigh wave, named the S-wave. These waves have different propagation velocity, and different periods (the body wave period is smaller by order of magnitude in comparison with the surface wave). Taking into account the amplitude-frequency characteristic of the seismographs, one should expect that these magnitude definitions would be different which is the case. The body wave magnitude is expressed as:

$$m_b = \log_{10}(A/T) + Q(h, \Delta) \quad (1.3)$$

where T is the dominant period of the body waves, and Q again the empirical function depending on the range and the depth of the event. The surface wave magnitude will be expressed as:

$$M_s = \log_{10}(A/T) + 1.66 \log_{10} \Delta + 3.3 \quad (1.4)$$

Here again T is the dominant period but now of the surface wave which is usually of the order of 20 s.

In addition to the frequency dependence of the seismographs, m_b and M_L are exposed to the saturation effect for the strong earthquakes. In addition it was necessary to elaborate a more physical estimation of earthquake intensity. That's why Kanamori (1977) proposed the magnitude definition based on the seismic moment of the earthquake, which is a more physical estimation of the earthquake energy. Generally speaking, the seismic moment is a tensor but for simplified estimations one can use the scalar definition of the seismic moment:

$$M_o = \mu D A \quad (1.5)$$

where μ is the shear modulus, D is the fault displacement, and A is the area of the fault. The Kanamori's definition of the moment magnitude looks like:

$$M_w = 2/3 \log_{10} M_0 - 10.7 \tag{1.6}$$

The variety of magnitude definition methods leads to the fact that one can find different magnitude values for the same earthquake in the literature. This is demonstrated in Table 1.1 (after Shearer 1999):

Table 1.1. Some Big Earthquakes (M_0 in 10^{20} Nm)

Date	*Region*	m_b	M_S	M_w	M_0
1992 June 28	Southern California	6.2	7.3	7.5	2
1906 April 18	San Francisco		8.2	7.9	10
1989 May 23	Macquarie Ridge	6.4	8.2	8.2	20
1994 June 9	Bolivia	7.0		8.2	26
1977 August 19	Indonesia		8.1	8.3	30
1957 March 9	Aleutian Islands		8.2	9.1	585
1964 March 28	Alaska		8.4	9.2	820
1960 May 22	Chile		8.3	9.5	2000

To give an idea how much energy is released during strong earthquakes and to imagine the relationship between the number and energy released by small, moderate, and strong earthquakes, it is worth taking a look at Fig. 1.5 and Table 1.2 (Stein and Wysession 2003). One can see that the most destructive and dangerous earthquakes have a magnitude higher than 7.

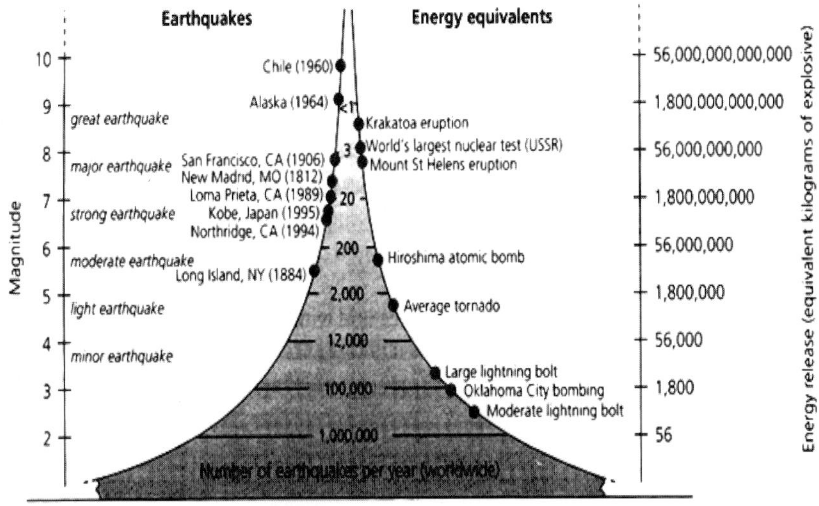

Fig. 1.5 The number of events per year (horizontal axis), magnitude (*left axis*) and energy in trinitrotoluol equivalent (*right axis*) released yearly by Earth's seismic activity

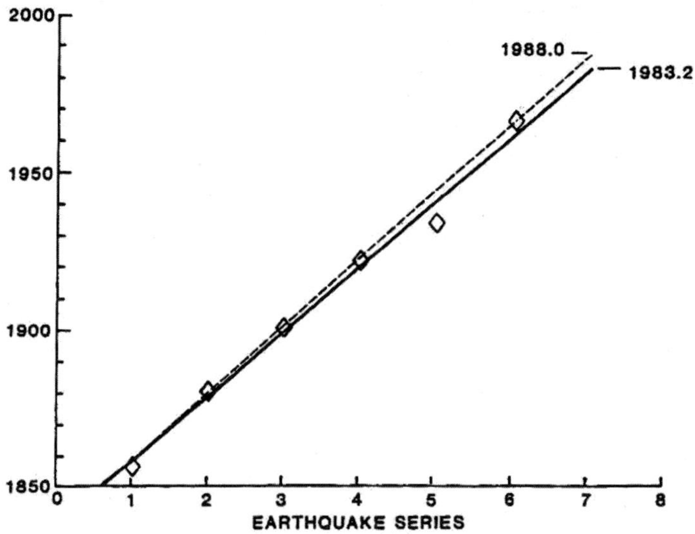

Fig. 1.6 Regular periodicity of earthquakes with M≥6 observed at Parkfield on San Andreas Fault

Table 1.2. Earthquakes number and energy released per year

Earthquake magnitude (M_s)	Number per year	Energy released (10^{15} J/yr)
≥ 8.0	0-1	0-1,000
7-7.9	12	100
6-6.9	110	30
5-5.9	1,400	5
4-4.9	13,500	1
3-3.9	>100,000	0.2

1.2.1 Physical Background of Earthquake Prediction

The classical approach to seismic prediction was based on the so-called seismic cycle implying the periodical storing and release of the seismic stress taking into account the continuous tectonic plate movement. This concept is usually connected with H. F. Reid who summarizing the mechanism of the California earthquake in 1906 published the theory of elastic rebound (Reid 1910). Turcotte and Spense (1974) made the calculations of periodicity for a homogeneous medium based on the elasticity theory. Of course, it is difficult to expect from the Earth's crust the ideal homogeneous structure but sometimes the earthquake sequence has a strikingly regular periodicity which is demonstrated in Fig. 1.6 for the Parkfield earthquakes series. These comparatively simple conceptions were recently reconsidered from the point of view of the fast developing theory of fractals and chaos (Bak et al. 1988). The conception of self-organizing complex systems was put forward and described in the monograph by Rundle et al. (2000). But this approach, using pattern recognition techniques and self-organizing system theory, disregards the physical and chemical changes happening inside the fault and the seismic source. Our goal is just to consider the variations of different parameters observed before the earthquake.

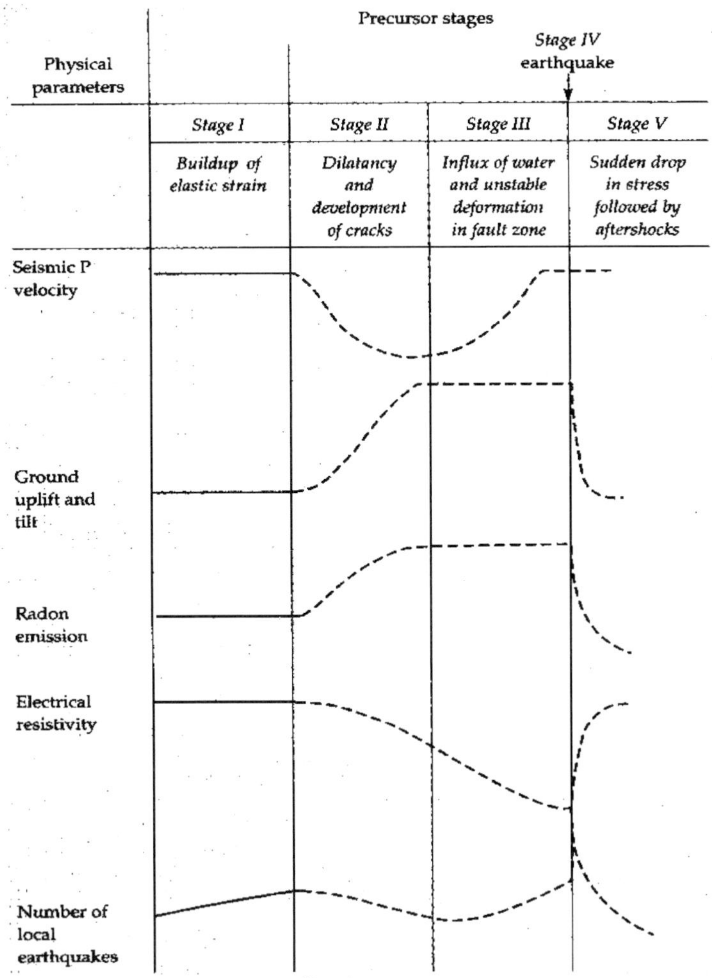

Fig. 1.7 The precursory variations of different parameters within the interval of earthquake preparation

Abstracting from the earthquake forecast as an end in itself, we will pay more attention to the physics of the earthquake preparation process, especially its final stage. We will base this on the Scholz et al. approach (1973) regarding the earthquake preparation process and associated physical and chemical transformations within the Earth's crust in the zone of earthquake preparation. So we will take only one period of the seismic cycle to look how the fault transformation develops and what parameters manifest these changes. To have an idea of the seismic cycle stages we will use the picture from Bolt (1993), which is a slightly modified presentation of the Scholz picture, is used by many authors and presented here (Fig. 1.7). The basis of this approach relies on dilatation theory regarding crack formation within the Earth's crust and finally the formation of the main fault in the process of shallow earthquakes preparation. Actually, there were two models developed differing in the role of the underground water (Mjachkin et al. 1975). The process of earthquake preparation within one period of the seismic cycle was divided into five stages starting from the moment of the previous earthquake. Different physical parameters are traced. The first one (which actually put forward the dilatation theory) is the change of the P-wave velocity. The build of dilatation is detected by the velocity diminution in the second stage. Other precursors are ground uplift and tilt, radon emanation, electric resistivity and a number of small earthquakes within the area of earthquake preparation. Basing on present knowledge, we can say that these are not the only precursors registered. Most likely they are representatives of different groups of precursors, namely: mechanical deformation, geochemical and hydrological precursors, electromagnetic precursors, and naturally, the seismic ones. The stages of the cycle show the transition from the elastic deformation to a nonlinear process up to the rupture. The duration of the stages is variable depending on many factors that we will not discuss here. Our interest will concentrate mainly on stages III and IV, i.e. just before and during the earthquake. Some indirect data show that the final stage when the process becomes irreversible and the earthquake will take place, lasts at least several days.

We will not describe all types of precursors, for example, the water level changes or geoelectric potentials. One can find such descriptions in the literature (Rikitake 1976; Mogi 1985, Turcotte 1991; Bolt 1993; Lomnitz 1994; Uyeda et al. 2000). We concentrate mainly on the processes participating in the seismo-ionospheric coupling. It does not mean that some precursors are more important than others; simply the book's limits put restrictions on the items to be considered.

1.2.2 Radon Emanation as a Precursory Phenomenon

Radon monitoring is one of the widely used techniques in geological studies. As is known, radon is the product of radium decay, which in turn appears after ^{238}U decay. Radon is an ideal indicator in geological research because it is generated continuously in any geological structure. Its concentration loss due to decay (period of semi-decay is 3.825 days) and due to migration into the atmosphere is always compensated by new production. So the radon concentration in the given geological structure is more or less constant and is proportional to the uranium (radium) content in the given massif. In the presence of cracks and upflowing gas streams the convectional transport of the radon could be from the depths up to 200 m. There are no problems with the radon registration due to its high radioactivity. It is surely registered under 30 – 50 decays per second within one cubic meter, i.e. its activity is 3 – 50 Bk/m^3

Radon migration is determined by the macroscopic diffusion coefficient which depends on the mode of geological structure deformation. It is obvious that under compression the diffusion coefficient decreases, and under unloading – increases. The loading-unloading process during earthquake preparation is the reason of the radon concentration variations before the earthquakes. Many authors reported radon variations before the earthquake and used them as a short-term earthquake precursor (Holub and Bready 1981; Segovia 1989; King et al. 1993; Utkin and Yurkov 1998; Garavaglia et al. 2000; Belyaev 2001). One can find a perfect review on the geochemical precursors of earthquakes in Toutain and Baurbon (1998). The example of the radon anomaly before the Kobe earthquake is presented in Fig. 1.8.

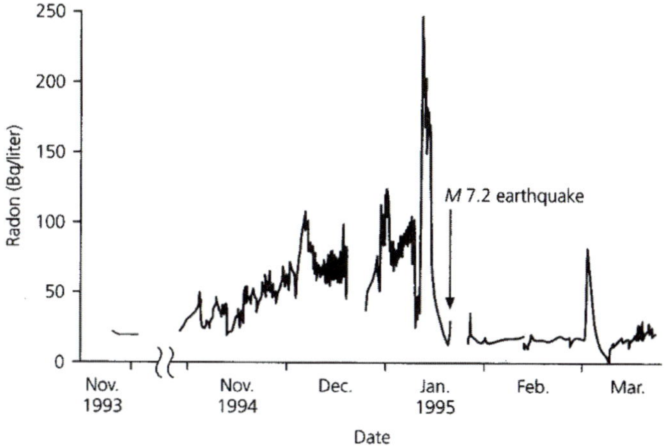

Fig. 1.8 Radon in groundwater before and after the 16 January 1995, Kobe earthquake in Japan (Igarashi et al. 1995)

King et al. (1993) measuring the radon concentration in the area of tectonic faults in California have detected that the increase in radon concentration is observed not directly over the fault but some distance away. It coincides with the conception of Utkin and Yurkov (1998) who analyzed many cases of radon monitoring by different experimenters and came to the conclusion that radon variations have some regularities connected with the earthquake source. The epicenter of the future earthquake is, as a rule, inside the compression zone where radon concentration decreases before the earthquake (they called it the "near zone"). This near-zone radius for the earthquake with magnitude 5 is near 24±15 km. The size of the near-zone increases with the earthquake magnitude. The compression zone is surrounded by the stretching zone where the radon concentration growth occurs before the earthquake. They named it the “far-zone” and determined its size as 110±40 km for an earthquake with magnitude 5. The problem of the distance where radon anomalies associated with earthquakes are measured will be discussed below.

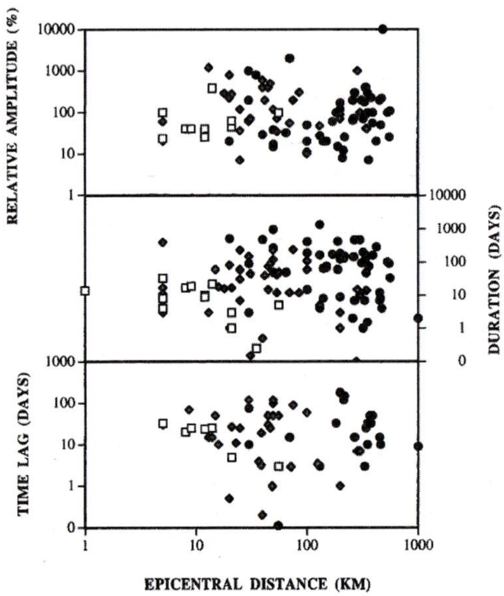

Fig. 1.9 Plot of relative amplitudes (log scale), time lag (time between the onset of the anomaly and the related earthquake, in days) and duration (in days) of radon anomalies listed in Toutain and Baurbon (1999), review as a function of epicentral distances for selected magnitudes. *Dots, squares* and *lozenges* display earthquakes with magnitudes 0–3, 3–6 and 6–9, respectively. A general increase of maximum time lag and duration is seen with increasing epicentral distance

The radon concentration dynamic indicates the earthquake preparation process 3–4 months before the seismic shock with maximum dynamics registered 1–2 weeks before the event. The distance, time of advance and duration of radon anomaly growth with earthquake magnitude is shown in Fig. 1.9 (after Toutain and Baurbon 1999).

Besides its precursory value, we should attract the attention of the reader to the ionizing abilities of radon. It allows the creation of so-called “atmospheric plasma”. The atmospheric plasma conception will be discussed in the following paragraphs. Practically, radon is the first link in the chain of the physical processes providing the seismo-ionospheric coupling.

1.2.3 Other Geochemical Precursors

Earth degassing is more and more attracting the attention of scientists (King 1986; Wakita 1988; Toutain and Baurbon 1999; Heinicke et al. 2000; Taran et al. 2001). Furthermore, some recent ideas suggest that gas emanation is not the result of strength storing in the Earth's crust but is its reason (Iudin et al. 2002). These ideas develop the conception of gas convection (not diffusion) within the crust and formation of giant gas bubbles (Korovkin et al. 2002). Irwin and Barnes (1980) describe the carbon dioxide discharges connected with the seismic activity. They claim that CO_2 discharges may indicate the high pore pressure in the crust and potential seismic regions. Such discharges can initiate acoustic gravity wave generation. The CO_2 discharges have regular character at tectonic faults as a result of thermal crustal anomalies but sometimes they are not observed. In this case, helium-enriched nitrogen is registered in seismic faults (Toutain and Baurbon 1999). The large flows of gas can carry with them other substances that are formed during the morphology transformation process before the earthquake in the form of aerosols. Their presence in the volcano fumaroles is well established (Taran et al. 2001) but aerosol measurements in the seismically active zones are more complicated due to the mosaic character of the gas emanation in the seismic zones and the uncertainty of aerosol origin in gas probes.

According to Toutain and Baurbon (1999), the helium emanating from the crust has a deeper origin in comparison with radon and can indicate the uprising of magma. The $^3He/^4He$ isotope ratio permits us to distinguish between atmospheric, crustal and mantle origin of helium.

The other gases such as H_2, CH_4,, aromatic hydrocarbons are also registered at tectonic faults.

The emanation of volatile metals such as Hg, As, Sb at tectonic faults reported by some authors (Sugisaki et al. 1980; King et al. 1993; Klusman 1993) is also of interest because the metallic ions play a specific role in the developed mechanism of seismo-ionospheric coupling. The emanation of metallic aerosols in seismically active areas was reported also by Alekseev et al. (1995).

1.2.4 Anomalous Electric Field

Earthquake preparation is usually accompanied by electromagnetic phenomena in different frequency bands starting from DC up to VHF radio emissions (Hayakawa 1999). Due to the ability of electromagnetic emissions to propagate in the ambient environment, these emissions are registered at different distances from the earthquake epicenter. But one phenomenon, namely the anomalous vertical constant electric field is registered just within the area of the earthquake preparation. This phenomenon was first registered in 1924 by Prof. Chernyavsky (1955). In April 1924 their expedition arrived at Dzhelal-Abad in Kirgizia to study atmospheric electricity. Prof. Chernyavski describes this in the following way: "During the day, when the behavior of our device struck us, the sky was clear. But the equipment with all due evidence demonstrated – electric storm got up in the atmosphere with an extremely high potential. What exactly, we were not able to detect because the device arrow went out of the scale limit immediately. And four hours later the Earth yawned. I thought at that time, maybe, the earthquake was the reason of the anomalous state of the atmospheric electric field". The epicentral distance was 120 km, and the earthquake magnitude was 4.5.

We can mention also the early measurements of anomalous atmospheric electric field before earthquakes by Bonchkovsky (1954) and Kondo (1968). The more recent measurements (Hao 1988; Nikiforova and Michnovski 1995; Vershinin et al. 1999; Hao et al. 2000) with modern techniques confirming the early results. To discuss them more definitely, let's look at the examples of anomalous electric field registration presented in Fig. 1.10. Usually, in fair weather conditions the atmospheric electric field is of the order of 100 V/m and directed down to the ground (the atmospheric electricity physics will be discussed below), which is clearly seen in Fig. 1.10 (top left panel, continuous bold line). Some daily variations within the limit of 50 V/m can be observed. But one day before the strong earthquake in the Carpathian mountains (30.08.86, M=6.9) the picture changed drastically (Nikiforova and Michnovski 1995). One can observe the increase of the electric field up to the value of 770 V/m, and even a flip of the natural electric field direction at 04 LT to –280 V/m. The peak-to-peak variation of the electric field exceeds 1 kV/m. It should be mentioned also that the electric field measurements were conducted in Poland, 700 km apart from the earthquake epicenter what implies the large size of the earthquake preparation zone, and consequently, the large earthquake magnitude.

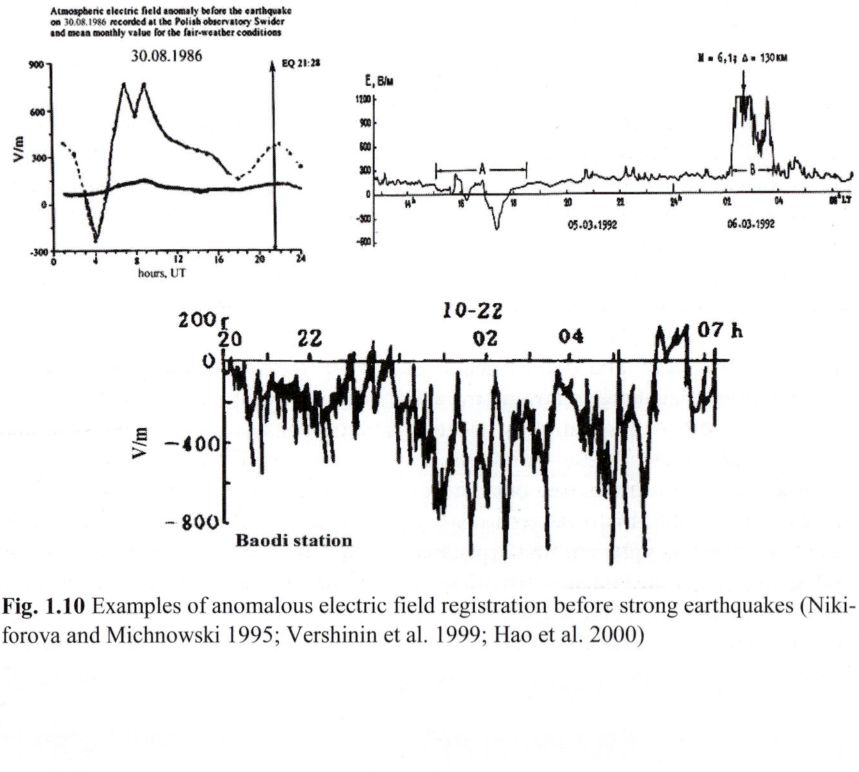

Fig. 1.10 Examples of anomalous electric field registration before strong earthquakes (Nikiforova and Michnowski 1995; Vershinin et al. 1999; Hao et al. 2000)

A similar picture is observed in Kamchatka records (Vershinin et al. 1999, Fig. 1.10 top right panel) for the earthquake with magnitude 6.1 at the distance 130 km from the epicenter. Here one can see again the increase of electric field in the natural direction up to the device saturation over 1 kV/m, and an electric field flip with the negative value near –500 V/m with the peak-to-peak span over 1.5 kV/m. A slightly different picture is observed in Chinese records (Hao et al. 2000, Fig. 1.10, bottom panel). The prevailing negative electric field is observed in close to epicenter records (38 and 130 km), and increased electric field of natural direction at the far measurement point (320 km) for the earthquake with magnitude 4.7. And again, the peak-to-peak value of the anomalous electric field exceeds 1 kV/m.

Summarizing the aforesaid, we can state that before the strong earthquakes in fair weather conditions the anomalies of atmospheric electric field have been observed in the form of electric field increase up to values exceeding sometimes 1 kV/m, electric field overturn with negative values of electric field also reaching sometimes – 1 kV/m. The peak-to-peak difference between the extremes of the positive and negative values of electric field usually exceeds 1 kV/m. The area occupied by the anomalous electric field is very large and for earthquakes with magnitude close to 7 is at least 1,400 km in diameter. According to Hao et al. (2000) the time in advance of the anomalous electric field appearance could be more than one month (the maximum value indicated in Table 1 of their paper is 38 days).

1.2.5 Earthquake Preparation Zone

The conception of the earthquake preparation zone was developed by different authors (Dobrovolsky et al. 1979; Keilis-Borok and Kossobokov 1990; Bowman et al. 1998). In general words, this is an area, where the local deformations connected with the source of the future earthquake are observed. Naturally, the deformations imply the changes of the crust properties, which could be measured by different techniques. First of all the deformations are accompanied by the strain storing, so the mechanical properties could be measured by strain meters, dephormographs and tilt meters (Yoshihisa et al. 2002), as well as by a recently developed GPS technique (Kostoglodov et al. 2003). According to dilatation theory (Scholz et al. 1973; Myachkin et al. 1975) formation of the cracks happens within the preparation zone and will be accompanied by changes in the seismic waves velocity, density, electric resistivity, ground water level, and geochemical precursors. All these changes can be monitored experimentally. All these physical and chemical changes within the earthquake preparation zone create the physical basis for the earthquake prediction (Rikitake 1976; Mogi 1985; Sobolev 1993). Regardless of the fact that dilatation theory is not more widely accepted and self-organized criticality became a more common approach of the seismologists, the earthquake preparation zone (or earthquake activation zone in other publications) remains commonly accepted with the same scale parameters as for dilatation (Kossobokov et al. 2000).

To determine the size of the earthquake preparation zone some authors use pure seismic precursors like foreshocks and the distribution of deformations, whereas other authors – the whole complex of physical parameters. There is no doubt that deformations are accompanied by other kinds of changes in crustal properties, so in the following discussion we will use the second, more physical approach. Dobrovolsky (1984) gave an approximate (but sufficiently accurate) solution of the problem of an inclusion in an elastic homogeneous isotropic half-space. After integration of the volume parameters the dimensions of the disturbed volume could be obtained, the radius of the spherical inclusion (Dobrovolsky et al. 1989):

$$a = 10^{0.414M - 1.696} \text{ km} \tag{1.7}$$

Taking into account that the inclusion has the spherical shape, the formula is valid for the deep earthquakes as well. For ground surface projection one may use the formula (Dobrovolsky et al. 1979):

$$\rho = 10^{0.43M} \text{ km} \tag{1.8}$$

Keilis-Borok and Kossobokov (1990) obtained the value of the preparation zone diameter (in grades of arc):

$$l(M_0) = \exp(M_0 - c) + 2\varepsilon \tag{1.9}$$

where M_0 is the seismic moment of the earthquake ε – the possible error of 0.5°, and c is calculated from the data-fitted $l(8)$=12°. This estimation is in good agreement with the Eq. (1.8).

For completeness we should mention one more determination of the radius of the earthquake preparation zone given by Bowman et al. (1998):

$$\xi = 10\, A_E^{1/2} \tag{1.10}$$

where A_E is the area of the rupture zone. The great value of this formula is that it uses the size of the rupture zone, the final result of the earthquake. That's why it could be used with the equal applicability for shallow and deep earthquakes. But the dependence on magnitude in this case is determined from the graph presented in the cited paper. That's why in the following estimations and discussion we will use the Dobrovolsky Eq. (1.8) giving the direct relationship between the radius of the preparation zone and earthquake magnitude. Table 1.3 gives an idea of the earthquake preparation zone's size for different magnitudes:

Table 1.3 Earthquake preparation zone radius for different magnitudes according to Dobrovolsky Eq. (1.8)

Magnitude	3	4	5	6	7	8	9
Earthquake preparation zone radius ρ (km)	19.5	52.5	141	380	1022	2754	7413

Different tectonic zones (subduction, spreading, intraplate, etc.), different tectonic faults and correspondent focal mechanisms (thrust, strike slip, etc.), different rock structure and properties in the seismically active area do not permit us to obtain universal formulas for the earthquake preparation zone estimation. Also, the complex of precursory phenomena will be different in different zones. This means that for the prediction procedure special studies for the every given region will be necessary. Nevertheless, some common features for all earthquake zones exist which proves the comparison of the graphs for all types of precursors (Fig. 1.11a, after Dobrovolsky et al. 1979) and for geochemical precursors only (Fig. 1.11b, after Toutain and Baurbon 1998).

From the discussion above the most important conclusion is that the area, where precursory radon emanations are observed, is of the same order of magnitude as the area of the earthquake preparation zone; we will use this information in the following construction of the seismo-ionospheric coupling model.

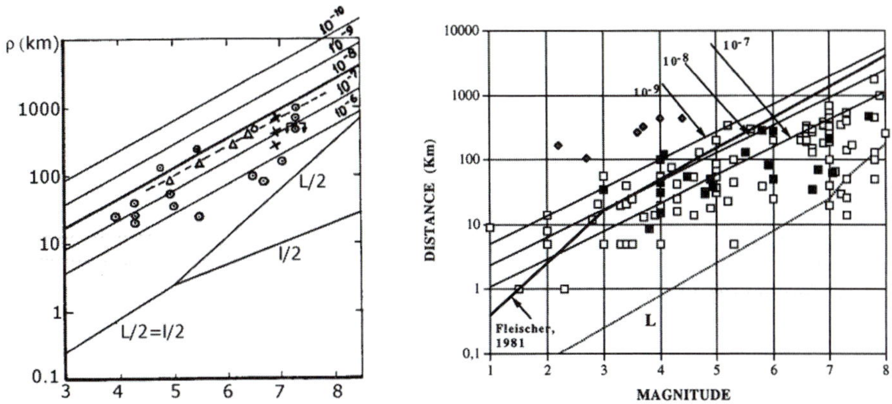

Fig. 1.11a,b a The precursor epicentral distance versus the magnitude of an earthquake. ⊕ geochemical precursors; □□□ resistivity; **x** electric field; Δ telluric current; ∏-radon; 1 light effects. **b** The epicentral distances versus magnitudes for geochemical data listed in Table 1 of Toutain and Baurbon (1998). *Open* and *filled squares* represent Rn and other gas anomalies, respectively. *Plain lines* characterize the relationship between strain radius and magnitudes for strains ranging from 10^{-7} to 10^{-9} using the empirical relationship proposed by Dobrovolsky et al. (1979). The *single bold line* characterizes the empirical relationship of Fleischer (1981) who calibrated the maximum distance of a radon anomaly for a given magnitude on the basis of a shear dislocation of an earthquake. *Dashed line L* characterizes typical rupture length of active faults as a function of magnitude by using the empirical law of Aki and Richards (1980)

1.2.6 Concluding Remarks

Here we summarize the physical precursory phenomena which will be important for the further development of the seismo-ionosphere coupling model. The first important conception is the *earthquake preparation zone* – the area on the ground surface where precursory phenomena associated with the approaching earthquake are observed. The size of this zone can be estimated by Eq. (1.8) but the more precise value could be obtained only by concrete measurements in the given area where the differences in geological structure, tectonic zone parameters and focal mechanism of the earthquake will be taken into account. Earthquake preparation is accompanied by a set of precursory phenomena connected with the transformations within the Earth's crust. The *gaseous emanation* from the Earth's crust in the earthquake preparation zone is one of the important earthquake precursors. Among different species emanating from the crust we will pick out the *radon* as a highly radioactive gas able to produce air ionization. The regular gas discharges, especially of CO_2 can carry with them the products of geomorphology in the form of aerosols including high metal content aerosols. The metals could also be emanated from the crust in the form of volatile metals such as Hg, As, Sb. These *metallic aerosols* could be easily charged and therefore able to change the electrical properties of the near ground air. It is important to note that the area of geochemical precursors is equal to the area of the earthquake preparation zone. And the last important phenomenon associated with earthquake preparation is the *anomalously large vertical electric field* registered also within the earthquake preparation zone several days or hours before the seismic shock.

1.3 Ionosphere Physics and Morphology

The ionosphere is a part of the upper atmosphere where there are enough electrons and ions to effectively interact with electromagnetic fields. As a conducting media, it plays an important role in the global electric circuit, and as partially ionized gas, it is subjected to the laws of plasma physics. Due to its electromagnetic properties the ionospheric plasma interacts with the Earth's magnetic field and to a considerable extent is controlled by it. Therefore, we refer to the ionosphere as magnetoactive plasma.

Presently, the altitude borders of the ionosphere are determined at between 50 and 1,000 km. Some authors propose higher values for the upper boundary, but usually the higher altitudes are regarded as the Earth's plasmasphere, because at these altitudes the Earth's atmosphere becomes fully ionized plasma (Bilitza 2001; Ondoh and Marubashi 2001). Usually, it is not underlined but we would attract the attention of the reader to the fact that the concentration of ionized part of the Earth's atmosphere even at the height of ionosphere maximum concentration, which is between 250 and 400 km depending on geophysical conditions, is by three orders of magnitude lower than the neutral one which is demonstrated in Table 1.4. This means that the neutral component plays a very important role in ionosphere formation and dynamics. The physical and chemical processes responsible for ionosphere formation are very different at different altitude levels what provides the layered structure of the ionosphere. That's why the practice which is adopted is to divide the ionosphere into different layers described and studied separately. In the following chapters our attention will concentrate mainly on the F2 layer of the ionosphere where the peak of the electron concentration is situated but for tutorial reasons let us familiarize ourselves some general information on the ionosphere as a whole. Additional information on the ionosphere can be found in several monographs (Fatkullin 1982; Brunelli and Namgaladze 1988; Kelley 1989; Davies 1990; Schunk and Nagy 2000) and in reference books (Dieminger et al. 1996; Ondoh and Marubashi 2001).

Table 1.4 Electron N and neutral density n comparison for upper ionosphere

z, km	N, m^{-3}	n, m^{-3}
200	$4\cdot10^{11}$ (day)	$8\cdot10^{15}$
300	$2\cdot10^{12}$	$5\cdot10^{15}$
400	$1{,}5\cdot10^{12}$	$3\cdot10^{14}$
500	$1\cdot10^{12}$	$1\cdot10^{14}$
600	$7\cdot10^{11}$	$1\cdot10^{13}$
800	$(0{,}8\div2)\cdot10^{11}$	$2\cdot10^{12}$
1000	$(0{,}4\div1)\cdot10^{11}$	$4\cdot10^{11}$
1500	$(2\div6)\cdot10^{10}$	$1\cdot10^{10}$
2000	$(1\div4)\cdot10^{10}$	$3\cdot10^{9}$

1.3.1 Ionosphere Formation

Ionization in the ionosphere is produced mainly by the solar ultraviolet radiation with a wavelength shorter than 102.7 nm which is effectively absorbed by atmosphere molecules and atoms. There are other sources of ionization: solar X-rays, solar cosmic rays, energetic particles precipitating at high latitudes but they have a sporadic character connected with the solar activity. Figure 1.12 shows the ionosphere height structure and the main source of the ionization for every layer of the ionosphere. Sometimes (depending on the time of day and geophysical conditions) ionospheric layers are subdivided into two or more layers, that's why Dieminger et al. (1996) propose we distinguish between the ionospheric region and layer which is reflected in Table 1.5

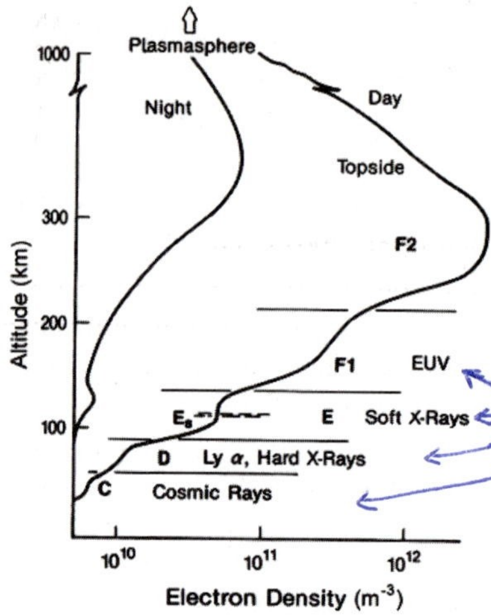

Fig. 1.12 Ionosphere structure on a summer day and night in a middle latitude and the main ionization regions (after Davies 1990)

Table 1.3. Ionospheric regions and layers

Region	Approximate Height Range (km)	Layer	Approximate Height Range (km)	Approximate Daytime Electron Density (m^{-3})
C			65	10^8
D	50-90			
		D	70-90	10^9
		El	100-110	10^{11}
E	90-140	E2	115	10^{11}
		E_s	110-120	variable
		F1,F1$^{1/2}$	140-200	$2x10^{11}$
F	140-350	F2	200-350	10^{12}
topside	350-1000			10^{10}-10^{11}
plasmasphere	> 1000			10^9

1.3.2 D-Region

The lower boundary for penetration of solar radiation shorter than 102.5 nm is 90 km but Lyman-α radiation (λ=121.6 nm) can ionize the minor component of the neutral atmosphere NO having very low ionizing potential and forming NO^+

ions at the heights of 60 – 90 km. The low concentration of NO is compensated by the high intensity of Lyman-α. The hard solar and galactic X-rays (λ<1 nm) together with energetic precipitating particles ionize the main molecular components O_2 and N_2, and solar ultraviolet radiation in the interval 102.8<λ<111.8 nm ionize the excited oxygen molecules $O_2(^1\Delta_g)$. The heights of the D-region are very difficult for study. The main source of information is in-situ measurements made by rockets (Danilov 2000; Friedrich et al. 2001). This region is characterized by the most complex ion composition. It contains the large ion clusters such as $H^+(H_2O)_n$, $NO^+(H_2O)_n$, $NO^+(CO_2)_n$ etc. It is the only region of the ionosphere where the concentration of the positive ions is not equal to the electron concentration due to formation of the negative ions. At altitudes lower than 75 km the negative ion concentration prevails over the electron concentration. Due to the NO reaction extinction in winter the electron concentration grows, that's why the electron density is higher in the D-region during winter than in the summer for the same solar zenith angles. The diurnal variations of electron concentration in the D-region (difference between the noon and midnight values) can reach 1.5 orders of magnitude for low solar activity and 2.5 orders for the periods of high solar activity.

Within the D-region the C-layer can form (see Fig. 1.12) controlled by the cosmic rays.

In relation to atmospheric regions, D-region heights correspond to the mesosphere where temperature inversion is observed from 260 K to 180 K at corresponding heights from 50 to 85 km, and then the temperature increases steadily.

1.3.3 E-Region

The E-region of the ionosphere is formed under action of the solar ultraviolet radiation within the band 80<λ<102.8 nm. The most important among them are the lines L_β=102.6 nm and CIII=97.7 nm ionizing O_2, as well as soft X-rays 1<λ<10 nm. The main ionized components are O_2 and N_2, and main ions O_2^+ and NO^+, which are formed from O_2^+ and N_2^+ as a result of ion-molecular reactions. The main mechanism of the charged particles loss is the dissociative recombination of the molecular ions with the electrons.

During the daytime the behavior of the E-region is very regular and controlled by the Sun's zenith angle. It can be well described within the frames of photochemical approximation on the basis of formation and loss of the molecular ions without taking into account of their transport.

During dusk and nighttime the behavior of the E-region is more complex, the ionization sources could be scattered by the geocorona, the solar radiation in hydrogen lines 121.6 nm, 102.6 nm and helium 58.4 nm, helium ion 30.4 nm.

Recent material on the E-region aeronomy one can find in Antonova et al. 1996.

Inside the E-region very thin-patched layers could be formed. This formation is called the sporadic E layer and designated as E_S. The formation of these layers is due to convergence of the vertical flux of long living metallic ions. Experimentally the following ions were registered: Na^+ (23^+), Mg^+ (24^+), Mg^+ (26^+), Mg^+ (27^+), Al^+ (27^+), Si^+ (29^+), K^+ (39^+), Ca^+ (40^+), K^+ (41^+), Ca^+ (44^+), Ti^+ (48^+), Fe^+ (54^+), Mn^+ (55^+), Fe^+ (56^+), Fe^+ (57^+), Fe^+ (58^+), Ni^+ (58^+), Co^+ (59^+), Ni^+ (60^+), Ni^+ (61^+), Ni+ (62^+), Ni^+ (64^+), Xn^+ (66^+), oxides and hydrates of metallic ions: AlO^+, SiO^+, MgO^+, $SiOH^+$, $Na^+(H_2O)$, $Na^+(H_2O_2)$, NaO^+, $NaOH^+$ (Fatkullin 1982). It is believed that these ions are of meteor origin but in a later discussion we will try to show that transport of metallic ions is possible from the ground level too. For the E_s formation the ion transport processes are very important, they are connected with winds, tides, gravitational waves and electric fields. Widely accepted is the wind shear mechanism of the formation of E_s put forward by Dungey (1959) and then developed by Whitehead (1970). It was shown also that the sporadic E layer could be formed by the anomalous vertical electric field of seismic origin penetrating into the ionosphere (Kim et al. 1993). This question will be discussed in detail in the following chapters.

The direct rocket in-situ measurements have shown that the thickness of sporadic E-layer changes from several hundred meters up to a few kilometers with a very strong density gradient on the edges, which is approximately an order of two higher than the mean vertical gradients of the regular ionospheric layers (Fatkullin 1982). The layers are very irregular in both vertical and horizontal directions. The vertical structure demonstrates multilayer construction, and in the horizontal direction the layers are patchy, sometimes periodical, implying the wave instabilities participating in their formation. The electron density of the sporadic E-layer can be very high, sometimes higher than the peak density of the ionosphere screening the F-layer from the ionosonde sounding by radiowaves. This will be demonstrated in the following chapters when discussing the effects of E_s connected with earthquakes.

1.3.4 F-Region

The F-region of the ionosphere also consists of two layers F1 and F2. The F1 layer appears as a ledge or bending point on the vertical profile of electron concentration between the E and F2 layers at the height 160 – 200 km. It rarely develops in the distinct maximum appearing more often near 180 km. Its formation owes to solar ultraviolet radiation within the band $10 < \lambda < 90$ nm. The main ionized components are N_2 and O, and to a lesser extent O_2. The prevailing ions appearing as a result of ion-molecular reactions are the same as in the E-region, i.e. NO^+ and O_2^+. The processes in both regions are very similar. Here, photochemical reactions prevail over the transport processes, at least during daytime. The new feature appearing at the altitudes of the F1 layer is electron-ion temperature anisotropy – difference between the electron and ion temperatures.

The F1 layer is more likely to appear during summer daytime conditions, and during nighttime it is the cavity between the E and F2 layers named in ionospheric physics the valley.

1.3.5 F2-Layer

The F2 layer is the most dynamical and most dense (from the point of view of plasma density) layer of the ionosphere. Here the main ionosphere maximum is located which, depending on the geophysical conditions, may be at an altitude of between 250 and 500 km. It includes the small part of the bottom ionosphere from 200 km up to peak density, and the whole topside ionosphere from the peak up to 1,000 – 2,000 km. It determines the propagation of the decametric radiowaves, and makes the main contribution in all errors (delay, phase distortion, scintillations) of navigational systems, such as GPS and GLONASS.

Fig. 1.13 Ionospheric ion composition according to International Reference Ionosphere model (IRI) for noon and midnight conditions (solar maximum)

It is the most simple in photochemical consideration but the most complex in dynamic and thermodynamic behavior. It is formed under action of the same band

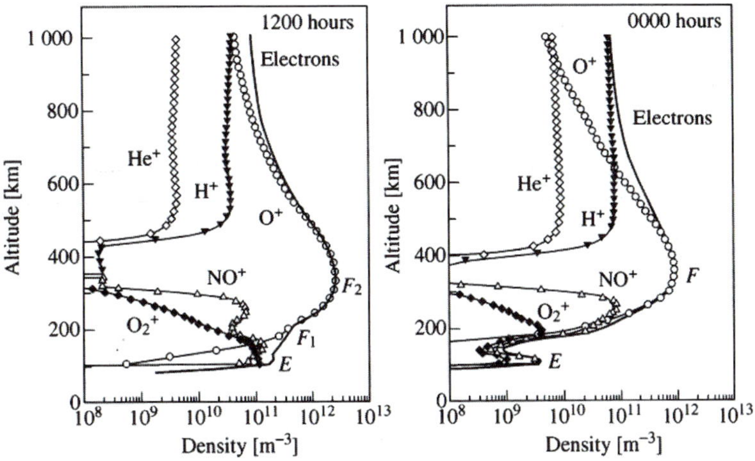

of the solar emission, as the F1-layer, i.e. $10 < \lambda < 90$ nm. The main ionized species is the atomic oxygen, but N_2 and O_2 also play important roles in the atom-ion interchange process leading to the loss of electrons by dissociative recombination. The prevailing ion in the F2 layer is O^+ and only at altitudes higher than 700 km do the light ions He^+ and H^+ start to exceed the O^+ concentration (Fig. 1.13). The main difference between the F2-layer and other ionospheric layers is that the ionization loss velocity is proportional to the ion concentration when in other layers it is proportional to the square of the concentration. The lifetime of the O^+ ions is comparable and at the peak is higher than characteristic transport time which leads to the ionospheric maximum formation. The layer is formed during the daytime, and during the nighttime rapidly decreases, but the plasma density is supported by the ambipolar diffusion which conserves the shape and the maximum peak altitude, but with decreased density.

1.3.6 Vertical Ionospheric Sounding

Before we discuss the problems of ionosphere dynamics and morphology we need to familiarize ourselves with the techniques of ionospheric parameter measurement. If the local probe measurements are intuitively understandable for any scientist, the remote sounding of the ionosphere by radiowaves needs special explanation. The ionosphere has the ability to reflect radiowaves (Appleton and Barnet 1925; Breit and Tuve 1926). It is well known from the plasma physics field that the plasma reflects the radiowaves, the frequency of which is lower or equal to the plasma frequency ω_p (Krall and Trivelpiece 1973):

$$\omega_p = (\frac{4\pi ne^2}{m})^{1/2} \text{ rad/s} \tag{1.11}$$

where n – electron concentration, e – electron electric charge, m – electron mass which for the linear frequency will be approximately:

$$f_O \approx 8.98\sqrt{n} \text{ Hz} \tag{1.12}$$

Here electron concentration is expressed in m^{-3}. Usually, in ionospheric physics, near plasma frequency has the symbol ***O*** which means the "ordinary wave" – the wave of the left circular polarization which propagates not 'filling" the magnetic field. As we mentioned above, the ionospheric plasma is magnetoactive. Due to this fact two additional propagating modes appear in the decametric wave band, namely, the ***X***-mode and ***Z***-mode:

$$f_X = \frac{\sqrt{4f_O^2 + f_H^2}}{2} + \frac{f_H}{2} \text{ - right-hand polarized fast extraordinary mode} \tag{1.13}$$

$$f_Z = \frac{\sqrt{4f_O^2 + f_H^2}}{2} - \frac{f_H}{2} \text{ - left-hand polarized slow extraordinary mode} \tag{1.14}$$

Now let us imagine what happens if we would radiate ionosphere by radiowaves of different frequencies taking into account the shape of the electron concentrations vertical profile (Fig. 1.12). Let us start from the lower frequencies. Until the frequency is lower than the local plasma frequency, it will be reflected from the ionosphere. When the frequency exceeds the local plasma frequency at the given altitude, it will be reflected from the upper level where the density (and consequently, the plasma frequency) is higher. So, changing the sounding frequency from lower to higher, we will obtain reflections from higher and higher levels of the vertical profile of the ionosphere until we reach the peak altitude designated as h_mF2. Here, the frequencies higher than the peak frequency designated as f_OF2 and named the "critical frequency" will simply penetrate through the ionosphere and propagate into outer space; we will not receive reflections from the ionosphere. If we illuminated the ionosphere not by continuous radiowaves but by short pulses, and we measured the time taken for the pulse to travel from the ground to the ionosphere and back, we would be able to measure the height from which the pulse was reflected. In this way we will obtain the height-frequency characteristic of the ionosphere named the "ionogram". The pulsed radiofrequency radar for ionosphere studies is named an ionospheric station or ionosonde. Usually, the ionosondes work in the frequency band 1–20 MHz.

Fig. 1.14 The schematic presentation of the topside (*upper panel*) and bottomside (*bottom panel*) ionograms (after Ondoh and Marubashi 2001)

But using sounding from the ground we will be able to obtain information only

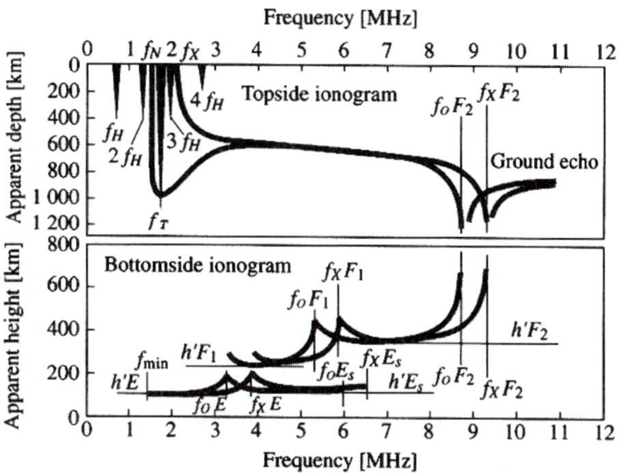

on the bottom part of the ionospheric profile. The upper (topside) part will be screened from us by the peak of the electron density. Fortunately, the space era gives us the opportunity to put the ionospheric station onboard the orbiting satellite and to make the same sounding from space. The composite topside- bottom side ionogram is shown in Fig. 1.14. On the vertical axis we have the delay time for pulse propagation (upper part from the height of the satellite orbit to the reflection point, bottom part- from the ground – to the ionosphere). On the horizontal axis the sounding frequency is shown which could be conversed into the electron concentration with the help of Eq. (1.12).

We should mention that in the discussion above we talk only about one frequency mode (***O***-mode). The same will relate to the other wave modes, but they will be reflected from the levels corresponding to ***X*** and ***Z*** cut-off frequencies. The radiowave group velocity in magnetoactive plasma depends on the electron density, magnetic field and direction of propagation. This means, that to obtain the real values of the ionospheric layers height, the apparent height in ionograms where the vacuum light speed velocity was used, should be reduced into real height which needs the special iterative procedure named the N(h) profile calculation. The technique of the bottom and topside profiles calculation one can find in the literature (Jackson 1969; Titheridge 1985, 1988; Huang and Reinisch 1996; Denisenko et al. 1998). The history of the ionosondes and the modern state of the ground based and topside sounding one can find in the work by Pulinets (1989); Reinisch (1996); Bibl (1998); and Pulinets and Benson (1999).

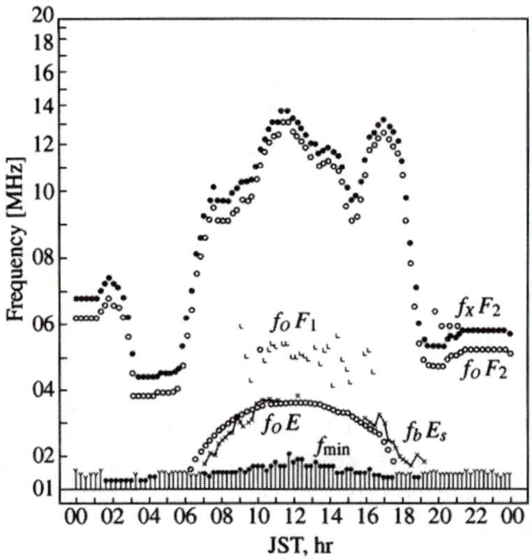

Fig. 1.15 Typical daily variations of the critical frequency $f_O F2$ and other characteristic frequencies scaled from the ionogram

Usually, the standard sounding by the ground based ionosonde is carried every 15 minutes. Taking into account the described higher daily variation of ionospheric parameters (maximum during daytime and minimum during nighttime), we will obtain the tooth-like characteristic for the daily variations presented in Fig. 1.15. To obtain the average ionosphere behavior from daily variations, the monthly median is used (Piggott and Rawer 1972a, 1972b) which is representative for the given month. This characteristic is used in ionospheric modeling. This characteristic is used also to determine the level of ionospheric variations during different kinds of disturbances. We will also use the monthly median in our estimations of ionospheric variations caused by the seismic activity.

Another quickly developing technique of ionosphere monitoring is GPS TEC measurements. This is based on the ionosphere effect on the propagation of the signal transmitted from the Global Positioning System (USA) and GLONASS (Russia) satellites. The space segment of the GPS consists of a constellation of 24 operating satellites in six circular orbits 20,200 km above the Earth at an inclination angle of 55° with a 12-hr period. The satellite transmits two frequencies of signals (f_1=1,575.42 MHz and f_2=1,227.60 MHz). Since the ionosphere is a dispersive medium, scientists are able to evaluate the ionospheric effect with measurement of the modulations on the carrier (codes) and the carrier phases recorded by dual-frequency receivers (Liu et al. 1996; Jakowski 1996; Hernandez-Pajares et al. 1997). The received signals contain many effects, in addition to the ionospheric one, the troposphere and hardware errors. With the help of techniques described in cited above papers only the ionospheric effect is selected from the signal, is transformed into the slant TEC along the ray between the satellite and the receiver, and then into the vertical TEC as a final product. TEC – is the abbreviation of the Total Electron Content and is the integrated number of the electrons within the slab between the satellite and the receiver, or finally, in the vertical direction over the receiver. In many cases the TEC dynamics are very similar to the f_OF2 dynamics because the main contribution to TEC is from the F2 layer. In other cases (for example during geomagnetic storms) their behavior is counterphase because of the larger contribution to TEC from the plasmasphere. The network of permanent GPS receivers is growing explosively, which permits in addition to one point measurements the realization of regional and global ionospheric mapping.

1.3.7 The Global Structure of the Earth's Ionosphere

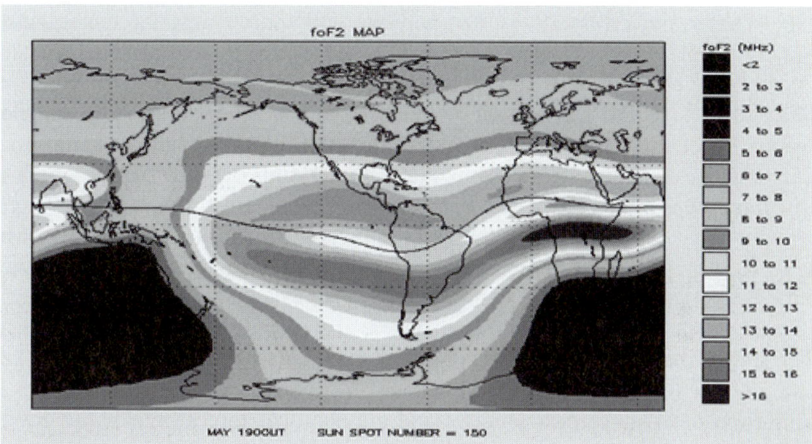

Fig. 1.16 The model global distribution of the critical frequency within the ionosphere according to IRI (Bilitza 2001)

At the heights higher than h_mF2 the ionosphere is essentially controlled by the geomagnetic field. The particle diffusion is possible only along the geomagnetic field lines. It means that every horizontal movement of the neutral part of the atmosphere provides the vertical component to the ionospheric plasma movement. If electric fields appear in the ionosphere, the particle movement will be not along the electric field vector but perpendicular to it (so called ***ExB*** drift, Krall and Trivelpilse 1973). Due to non-coincidence of the geomagnetic and geodetic pole positions, the additional forces appear in the ionosphere changing the global distribution of the ionospheric plasma. The presence of the South Atlantic geomagnetic anomaly leads to a strong deviation of the geomagnetic equator from the geodetic one, bringing one more irregularity in the ionospheric plasma global distribution. In addition, the Earth's geomagnetic field has a different configuration at different latitudes. If at middle latitudes it could be approximated by dipole, at high latitudes the lines of the geomagnetic field go into deep magnetosphere (so called, open field lines) which creates a completely different plasma convection pattern on high, middle and low latitudes.

Generally speaking, the Earth's ionosphere by its global properties can be divided into several parts: polar ionosphere, high latitude (or auroral) ionosphere, middle latitude ionosphere, low latitude and equatorial ionosphere. The ionosphere has specific global scale irregularities, which are characterized by their own dynamics and morphology. These are the main ionospheric trough centered near 60 ° geomagnetic latitude, and the equatorial anomaly – two peaks of enhanced electron density from both sides of the geomagnetic equator (10° – 15° depart from it). Both structures have complex behavior, which is out the scope of the present consideration. The global distribution of the critical frequency according to the IRI model is presented in Fig. 1.16. In the following discussion we will not consider questions connected with the main trough, so we will not discuss here the mechanisms of its formation. One can find this in the literature (Karpachev et al. 1996; Werner and Prölss 1997). But here we will shortly describe the equatorial anomaly formation mechanism which will be necessary in our further discussions of the seismic effects in the equatorial ionosphere.

From Fig. 1.16 one can see two undulation belts not far from the geomagnetic equator. This large-scale ionospheric structure is called the equatorial anomaly. Its formation one can understand from Fig. 1.17. During daytime according to experimental measurements the eastward electric field is always present on the geomagnetic equator which creates the steady upward drift of the ionospheric plasma due to ***ExB*** drift in crossed electric and directed to North horizontal geomagnetic field.

Plasma rises up to higher altitudes where the recombination loss is lowering. After sunset the electric field increases which increases the uplift effect. At the same time the lower altitude density decays quickly. But at the higher altitudes the equilibrium between electromagnetic and gravitational forces breaks, the gravitation starts to dominate and plasma slides down along the geomagnetic force lines to the tropical latitudes creating the enhanced ionization on both sides of the geomagnetic equator and forming the equatorial anomaly. A global model for the equatorial anomaly does not exist yet, but at regional scale it is possible to model the effects within the equatorial anomaly.

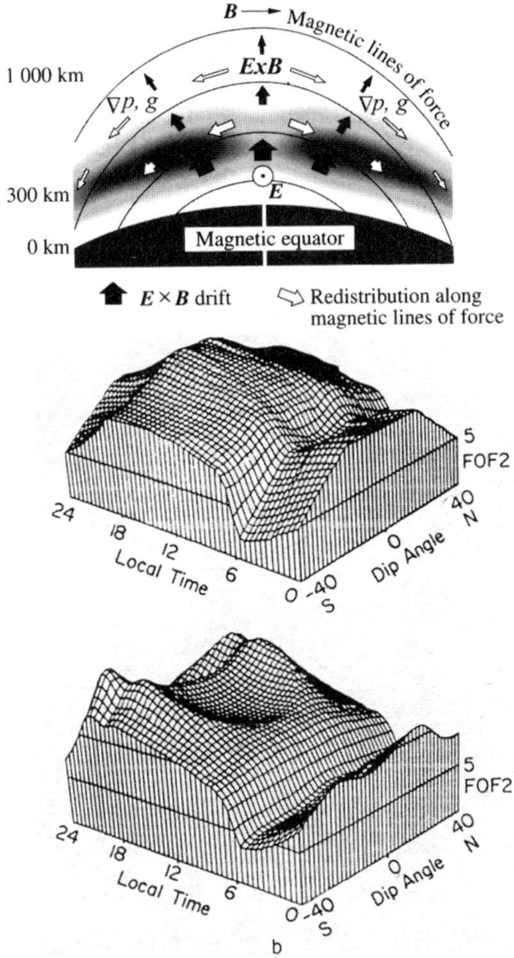

Fig. 1.17 *Upper panel* – fountain effect over geomagnetic equator due to the ***ExB*** drift of the ionospheric plasma (after Ondoh and Marubashi 2001). Contour plots of $f_O F2$ distribution near geomagnetic equator in the absence of eastward electric field – *middle panel*, and in the presence of eastward electric field *bottom panel* (after Kelley 1989)

1.3.8 Ionospheric Regular Variations

We concentrate here mainly on the variations of the F2 layer due to the fact that in the following chapters mainly the F2 layer seismo-ionospheric effects will be discussed. Ionospheric variability can be subdivided on several temporal scales. Leaving out of the scope of our consideration the different kinds of long-term trends in the ionosphere, the largest temporal scale of the ionosphere variability is connected with the cycle of the solar activity. It is a well known 11-year cycle observed visually in the number of dark spots on the solar disk, but having inside a lot of different manifestations of the solar activity in the form of solar flares, coronal mass ejections (CME), solar proton events, etc. (Lang 1995; Ondoh and Marubashi 2001). The ionosphere "feels" the solar activity in the form of optical and electromagnetic emissions, and in the form of severe changes of plasma convection in the magnetosphere after the encounter of the magnetosphere with the disturbed solar wind carrying the fluxes of energetic particles, changing solar magnetic field, and streams of the thermal plasma. The latter transient events, known as geomagnetic storms, will be discussed below, and now we start with the overall long-term dependence of ionosphere behavior on the solar activity.

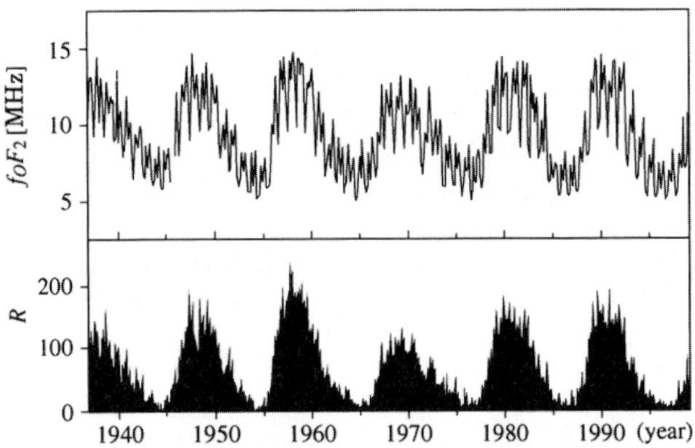

Fig. 1.18 The solar cycle variations of the critical frequency at Tokyo (*upper panel*) and solar 10.7 radio flux (*bottom panel*) for 60 years of the solar activity(after Ondoh and Marubashi 2001)

The solar activity in geophysics is usually described by the Wolf number of the solar spots W or by the solar radio flux with the wavelength 10.7 cm, R. Both services are now available from the Internet (http://sidc...; http://spidr...; http://swdcdb...). As one can see from Fig. 1.18, the critical frequency (upper panel) closely correlates with the solar activity. This means that the absolute daytime value of the critical frequency grows with the growth of solar activity. It reflects the fact of changing within the solar cycle the EUV flux as a main ionizing agent for the atmospheric gases. Besides the absolute value changes of the electron density, the seasonal effect connected with the level of solar activity are present too. The peak-to peak values in seasonal variations diminish in the solar cycle minimum, as well as the amplitude of the daily variations. Also observed are latitudinal changes with the solar cycle. According to Intercosmos-19 and Cosmos 1809 topside sounding satellites data, the equatorial anomaly persists during nighttime in high solar activity conditions while during the low solar activity it disappears (Pulinets et al. 1989).

The next temporal scale in the ionospheric variability is the seasonal variations. Actually, it is difficult to separate the seasonal and solar cycle variations, which is clearly seen from Fig. 1.19. The noon and midnight values of the electron concentration for 4 consecutive years of the 22nd solar cycle are plotted in the figure for two ionospheric stations situated in the same longitudinal sector. One of them (Tokyo) is in the Northern hemisphere, and another one (Canberra), respectively, in the Southern hemisphere. To make the seasonal variations clear the running 30-day mean is superimposed on the daily plot. The sporadic character of the daily plots reflects the effects of the geomagnetic activity and so called day-to-day variability of the ionosphere which will be discussed below. Now let's look at the continuous plots. For the noon values the clear seasonal effect is expressed in the maxima around the equinoxes, and the minimum in the summer. But the amplitude of the seasonal effect decreases with the declining of solar activity. The difference between the Northern and Southern hemispheres exist: in the solar cycle minimum the fall equinox maximum disappears in the Southern hemisphere. The midnight seasonal behavior is simpler but with a very interesting feature – a flip of the seasonal dependence within the solar cycle. If for years of maximum solar activity we have the spring maximum of the midnight values, for years of the solar minimum we observe the summer minimum and two equinox maxima similar to the noon values.

In our further discussion we will need only the parameters of the overall seasonal behavior, so we will not discuss the physics of the observed seasonal variations within the ionosphere and their solar cycle modulation.

And the smallest temporal scale of the ionosphere regular variations is the daily variations. The typical curve of the daily variations for undisturbed conditions is shown in Fig. 1.15. Daily variations of the critical frequency may be different in different locations. They have their own specific features (it could be one or two maxima during the daytime, some features like the pre-sunrise minimum may appear or not, etc.) but the main evident wave-like variation is the day-to night oscillation. The amplitude of this oscillation as one can understand from the discussion above will be modulated by the solar cycle, season, and other factors to be discussed below. In addition to well-understood modulation factors of daily variations of the ionosphere like solar and geomagnetic activity, season, latitude and longitude, there are variations that are still out of the scope of our understanding. One of the most intriguing features is day-to-day ionosphere variability, when even the essential difference is observed without any evident reason.

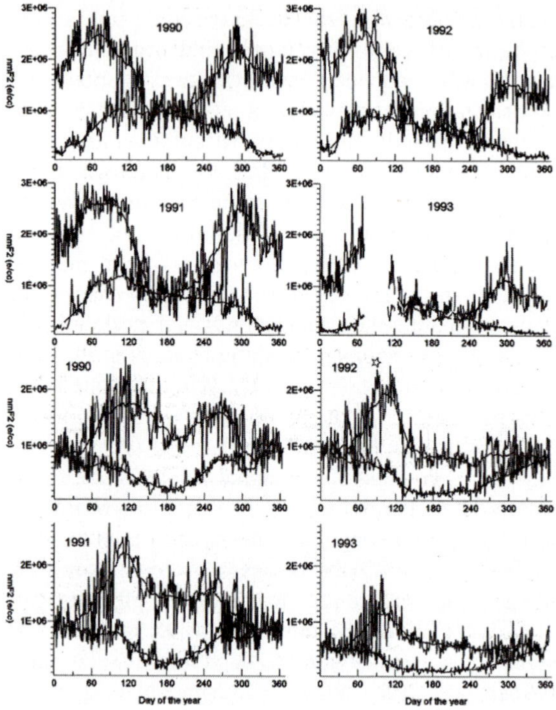

Fig. 1.19 The noon (*upper curves* on every plot) and midnight (*bottom curves* on every plot) values of the electron concentration for 4 years of the 22-th cycle of the solar activity for Tokyo (Northern hemisphere, *upper two panels*) and Canberra (Southern hemisphere, two bottom panels). A 30-day running mean is superposed on the daily meanings of every curve (After Wilkinson et al. 1996)

1.3.9 Effects of Magnetic Storms in the Ionosphere

Let us look again at Fig. 1.19. Why do the daily variations look like a fringe? The most deep and prolonged are the result of the magnetic storms, and the smaller and shorter ones – effects of the smaller magnetic disturbances and other kinds of the ionospheric variability to be discussed below.

Before we discuss the effects of magnetic storms on the ionosphere, let us define the magnetic storm itself. The main source of the magnetic storms is severe irregularities in the solar wind. Regardless of their origins (solar mass ejections, proton events, solar flares, etc.) their final product in the vicinity of the Earth's magnetosphere is strong changes in the interplanetary magnetic field interacting with the magnetosphere. According to Gonzalez et al. (1994) "the geomagnetic storm is an interval of time when a sufficiently intense and long lasting interplanetary convection electric field leads, through a substantial energization in the magnetosphere-ionosphere system, to an intensified ring current sufficiently strong to exceed some key threshold of the quantifying storm time D_{st} index". Leaving the reader to familiarize himself independently with the solar wind-magnetosphere interactions, we will only provide some information on the D_{st} index definition and quantification. This index is a quantitative measure of the ring current forming around the Earth during the geomagnetic storm and supported by the ions with energies of several tens of keV and electrons with energy near 10 keV. It is commonly accepted that the magnetic storm is triggered by the large dawn-to-dusk electric field produced by the southward turn of the interplanetary magnetic field B_z. The electric field leads to the earthward plasma convection, which permits high-energy particles to flow into near-earth space (Ondoh and Marubashi 2001). The ring current, and other currents of modified magnetospheric convection during the magnetic storm induce the variations of magnetic field on the ground surface that could be measured by magnetometers. The network of the reference magnetometers equally distributed along longitude within the latitudinal band 21° – 33° give the quantitative contribution to calculate the D_{st} index. It is commonly agreed that the magnetic storm can develop when the D_{st} index exceeds the threshold –50 nT (which corresponds to the interplanetary B_z component – 5 nT) and stay over this threshold at least 2 hours (Gonzalez et al. 1994). There are other indices to describe the geomagnetic activity (K_p, A_p, a_p, etc.) but all of them have a contribution from the high latitude activity and can show the increase without development of the magnetic storm, so called substorms, which are out of the scope of the present book. The definition of these indices one can find in Dieminger et al. (1996).

The ionospheric effects of the magnetic storm is a very complex phenomenon of increased dissipation of the solar wind energy, probably one of the most complex in ionospheric physics involving a lot of changes in the ion composition, substantial changes in plasma temperature and concentration, etc. Practically all ionospheric parameters are affected by the magnetic storm. One can find the perfect review of the magnetic storms effects in the F-region of the ionosphere in the work by Prölss (1995). The attempt to describe the global picture of the ionospheric storm effects using the data provided by a ground based ionospheric station network is made in the work by Szuszczewicz et al. (1998), and using the topside sounding data from the Intercosmos-19 satellite is described in Karpachev et al. (1995). The empirical model of the magnetic storm now included in the IRI describing the variations of the ionosphere parameters during the storm is published in the work by Araujo-Pradere (2002), and effects of the magnetic storm in all regions of the ionosphere and thermosphere – in Danilov and Lastovicka (2002).

To have an idea how the geomagnetic storm effects the ionosphere let us consider the critical frequency variations during geomagnetic storms (Fig. 1.20). Two consecutive storms in September 1989 resulted in a deep negative depression of the critical frequency on mid-latitude ionospheric stations. The storm depression lasts more than one day. A slight positive deviation can be marked in all stations at the beginning of the first storm. These deviations are called the positive and negative phase of the ionospheric storm respectively.

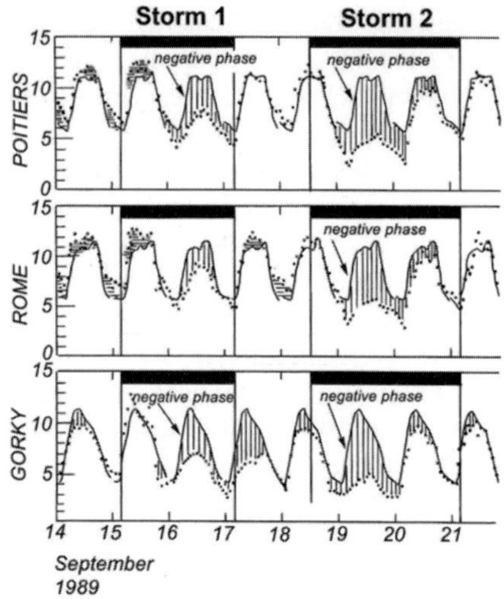

Fig. 1.20 Comparison of daily variations *foF2* (*points*) with the diurnal quiet *foF2* (*thin solid curve*) formed by the average of quiet days September at several European ionospheric stations during two consecutive magnetic storms: Storm 1 – Sept. 15 –16; Storm 2 – Sept.18 – 20

One can find an explanation of the physical mechanisms of the positive and negative phases in the literature cited above. We would only like to mention that the widely accepted explanation of the negative phase is the decrease of the $[O]/[N_2]$ relation and the increased recombination due to increased temperature (Mikhailov and Förster 1997). To summarize the effects of the magnetic storm in the F-region we use the results of the Szuszczewicz et al. (1998) discussion:

- Disturbances in *low* latitudes are observed, basically, during night-time with positive and negative phases noticed, first of all in a period between sunset and sunrise
- The corotation with the Earth is characteristic of all phases of a storm
- In middle and low latitudes, the negative phases of a storm always began in a period between sunset and sunrise and then corotated in a day time hemisphere
- The hmF2 value during a storm exceeded 60 km with hmF2 in quiet conditions not less than 250 km
- As a result of the development of a storm the average molecular mass at heights of the F-layer was increased owing to an increase of the mass relation $([N_2] + [O_2]) / [O]$.
- The ionospheric disturbance represents a continuous process, developing within 40–48 hours
- The ionospheric variations during a storm have global character

1.3.10 Day-to-Day Ionosphere Variability

The problem of day-to-day ionosphere variability is one of the most actual and complex in modern ionospheric physics. This is due to the fact that it is impossible to determine the main cause of this effect. A lot of different factors make a contribution to this phenomenon making it unpredictable in a deterministic sense. In addition, there is no commonly accepted way of quantitative description of this ionosphere variability. It exists more in an intuitive rather than in a well-determined manner. One of the attempts to regulate the terminology and find some approaches to its quantitative description was the special Session on "Ionospheric variability, modeling and predictions" held within the frame of the EGS XXVI General Assembly at Nice in March 2001.

Nevertheless, let's try to find a way through this jungle of sources on the day-to-day variability. As a first step, we can divide the sources into two large categories: the effects on the ionosphere from above and from below. Naturally that the main source of the effects from above is the solar variability. This can also be divided into optical (electromagnetic emission) effects and the effects of the particle precipitation in the auroral region. The chromospheric flares on the Sun lead to the enhancement of the ultraviolet and X-ray emission causing an increase of the electron density in the ionosphere. These short and sudden density variations in the ionosphere are called Sudden Ionospheric Disturbances (SIDs) (Davies 1990). For a recent discussion of this event one can reference the work by Afraimovish et al. (2002). The day-to-day variability associated with the particle precipitation is usually observed on high latitude and middle-high latitude ionospheric stations (Pulinets et al. 1996). The NOAA/TIROS particle power index was proposed as a better estimation of the following ionospheric variability than the K_p and a_p geomagnetic indices.

Estimations show that the thermal energy of solar radiation dissipated in the atmosphere is much larger than the energy dissipated during the geomagnetic storm in the Earth's magnetosphere. This means that the part of this energy in the form of movement of the neutral components of the atmosphere will be introduced into the ionosphere in the form of involving the ionized particles in the neutral movements. They are atmospheric tides and winds, acoustic gravity waves, and other kinds of regular and irregular air movements. How it is reflected in the ionosphere on can find in the recent review by Kazimirovski (2002). We can classify all these effects as meteorological effects on the ionosphere. Attempts to classify the terminology of ionospheric variability one can find in the review by Bradley and Cander (2002). Usually the parameter variability is expressed as a deviation (in %) from the mean or median value. One can find the quantitative estimations of the ionosphere variability in the works by Forbes et al. (2000), Rishbeth and Mendillo (2001), and Mendillo et al. (2002) showing that day-to day variability of the critical frequency *foF2* lies within the limits 10–30%. The effect on the ionosphere from below is regarded as a main source of the day-to-day variability.

Besides the meteorological influence on the ionosphere from below we can mention at least two other connected sources: this is the atmospheric electricity and radioactivity. The ionosphere is a part of the global electric circuit but in the papers on the global electric circuit it is always present simply as a capacitor plate, that's all. At the same time, any local or global changes in the atmosphere electric properties in one or other way are reflected in the ionosphere (Pulinets et al. 1998a). The atmosphere's radioactive pollution changes the electrical properties of the air which is immediately reflected onto the ionosphere which is also discussed in the paper cited above.

1.3.11 Concluding Remarks

It was a very difficult task to summarize in a few pages such a rich subject as ionosphere morphology. We hope that the readers have got an impression of how extended and complex this subject is and how variable is the ionosphere in different spatial and temporal scales. And in this regard we should clearly demonstrate not only the existence of the ionospheric variability connected with seismic activity but to find a way surely to separate it from all other kinds of ionospheric variability. Let God bless us to do this in the following discussion.

1.4 Radioactive Troposphere Plasmachemistry

Natural radioactivity plays a very important role in atmosphere morphology, as well as in the determination of the global electric circuit parameters. This fact is neglected by many researchers or at least underestimated. But before we start with the radiation effects on the atmosphere, let us remind ourselves of the basic facts concerning the Earth's atmosphere. The atmosphere is the gaseous shell of the Earth. Its weight (in view of the volume occupied by the continents above sea level) is equal to $5.157 \cdot 10^{15}$ tons, i.e. makes a little bit less than one million weight of the Earth ($5.98 \cdot 10^{21}$ tons). Approximately up to a height of 200 km the air shell fits the Earth as a thin rather identical everywhere layer. But higher than 200 km the temperature and density of air undergo strong temporal and spatial variations, and the atmosphere becomes pulsating and of irregular shape. Atmosphere is made out of several layers differing by their properties (Fig. 1.21). The *troposphere* spreads up to an altitude near 17 km within the latitude band approximately 42° N and 24° S, i.e. over 67% of the Earth's surface, and up to 11 km outside of this band. The temperature in the troposphere drops with height in such a way that may stimulate the instabilities, strong vertical movements and air mixing. It is usually the lower *planetary boundary layer* marked out which is of 1.0 – 1.5 km depth (it can be thicker in tropical areas). The wind may be slowed down by friction in this layer. The active exchange by the moment, heat and water vapor between the atmosphere and surface of the Earth and oceans happens in the boundary layer. The lowest part of the boundary layer up to 50 m with large vertical gradients of temperature, wind velocity and humidity is called the *near-ground atmospheric layer*. Over the troposphere transient layer *tropopause* the *stratosphere* is situated which is characterized by a very low vertical temperature gradient in the lower part, and a sharp increase of the temperature from the heights 34 – 36 km up to the level of the *stratopause* near 50 km. The stratosphere here is almost as warm as the Earth's surface (270 K in average). Such temperature distribution is unfavorable for the vertical movements and instabilities development. The air humidity within the stratosphere is low and clouds are formed very rarely. The stratosphere's composition differs from troposphere composition only by the ozone admixture. Next by height in the atmospheric layer is the *mesosphere* where temperature lowers again reaching 163 K at its upper part. Noctilucent clouds maybe observed sometimes here. Their shape serves as evidence of the wave and turbulent activity in the mesosphere. The transient layer of the *mesopause* at an altitude near 82 km separates the mesosphere from the above lying thermosphere where temperature sharply increases and at the altitude of 200–250 km reaches 1,800 K in high solar activity periods. Actually the thermosphere height corresponds to the ionospheric height. That's why their thermodynamics are usually considered together. At the higher altitudes we do not observe essential temperature changes except in auroral latitudes and during magnetic storms when higher temperatures up to 3,400 K could be observed.

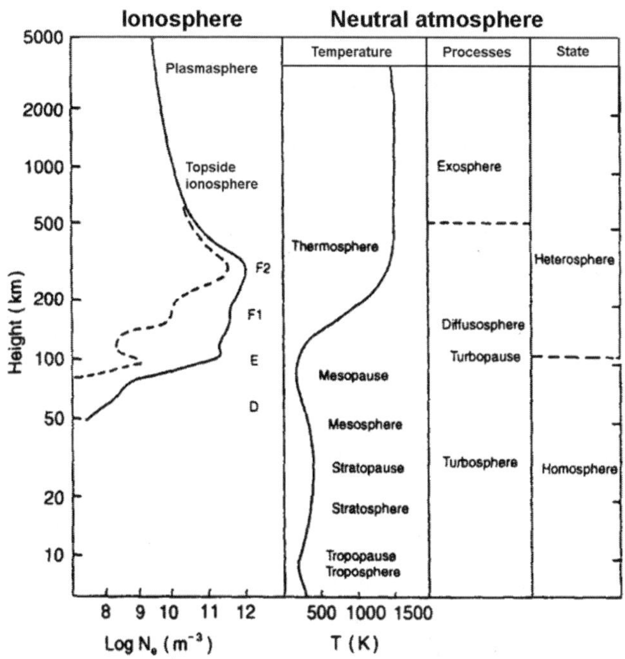

Fig. 1.21 Atmospheric layers classification and their height comparison with the ionosphere (after Deminger et al. 1996)

Up to a height near 106 km atmospheric air is well mixed due to flow currents and winds which provides the same air composition, but higher then this level is the *turbopause* where air composition essentially changes: diffuse separation of gases starts to prevail, the atomic oxygen appears in large amounts, air ionization leads to other changes in air composition due to chemical and photochemical reactions. Actually these are the ionosphere heights, which were discussed earlier.

The atmosphere's composition depends on five groups of substances:

1. The main gases are nitrogen (N_2), oxygen (O_2) and argon (Ar). They are continuously present up to the heights of the turbopause. The water vapor (H_2O) can be related to this group too while its content in the air strongly changes depending on time and place.
2. Small stable constituents – the gases which are chemically stable but present in small amounts in the air. They are CO_2, carbon oxide (CO), and methane (CH_4) etc. The troposphere and lower stratosphere ozone can also be considered as relatively stable minor air constituent.
3. The unsaturated and unstable molecules named "free radicals" present as small but very active constituents which are formed and destroyed very fast usually with the participation of the elements of groups 1 and 2 (CH_3OOH, CH_2O, NO, HO_2, OH etc.). The upper atmosphere ozone can be also attributed to this group.
4. Aerosols – the smallest solid and liquid particles of the various substances, floating in air. Their size usually varies from submicron up to several tens of microns.
5. The structure of an atmosphere also includes molecular and atomic ions of oxygen, nitrogen and other gases (O_2^+, O_2^-, N_2^+, O^+ etc.) and free electrons – the content of both is very variable.

1.4.1 Ionizing Radiation Effect on the Atmosphere

The mixture of atmospheric gases prone to the action of a hard ionizer (high energy electrons or the heavy charged particles being products of radioactive decay, and also γ- and X-ray radiation), is the low temperature recombinationally-nonequilibrium plasma in which plenty of the various charged and neutral components are formed. Their concentration is not characteristic of atmospheric air in usual conditions.

Atmospheric ions are the charged particles in an atmosphere of sizes ranging from molecular ions up to atmospheric aerosol particles. To exclude ambiguity in the classification of ions let us keep the following definitions:

1. Elementary, or molecular ions are formed during the first moment after ionization of a molecule, possessing a mobility around several $cm^2/(V{\cdot}s)$. Average minimal value of their lifetime in normal atmospheric conditions is 10–7 - 10^{-8} s.
2. Complex, or light ions are the ions formed after attachment of molecular ions to the molecules usually having the dipole moment. In the Earth's atmosphere they are the molecules of water vapor. Their mobility is 0.5 –

5 $cm^2/(V \cdot s)$. The lifetime of such ions can be significant and under certain conditions reach tens of minutes.

3. Heavy ions are a compound of complex ions and aerosol particles with a predominantly covalent bond. Their mobility's range from 0.01 $cm^2/(V \cdot s)$ up to infinitely small. The lifetime of such ions is determined basically by such factors as the dust content of the atmosphere and the intensity of external impact.

The concept of intermediate ions is sometimes used – the compounds from several light ions where the bond has both electrostatic (Coulomb) and valence character. Their mobility is 0.5 - 5 $cm^2/(V \cdot s)$ 0.5 – 5 $cm^2/(V \cdot s)$.

1.4.2 Sources of Ionization of the Earth's Atmosphere

Radioactive emission of the nucleus of radioactive elements is represented by electromagnetic radiation (γ-; X-rays) or by a flux of fast charged and neutral particles (α-, β-particles, protons, neutrons). Passing through the air environment, radiation provides the energy for ionization and excitation of molecules.

The basic sources of ionization in the lower atmosphere, besides anthropogenic sources, are the natural radioactivity of ground and air, and also cosmic rays. And more than 60% of the general productivity of natural sources of ionization are given with radiation of ground so, for example, in soils, characteristic for central Russia, all three basic radiators are found: α, β and γ.

The basic source of the α-particles representing a nucleus of helium with a charge *2e*, in air is ^{222}Rn (radon), ^{220}Rn (thoron) and ^{219}Rn (action). Without exception products of their decay represent isotopes of heavy metals, which are quickly attached to aerosol particles, forming so-called affiliated ions of radon. The energy of particles emitted by them is within the limits of 4 – 9 MeV, the length of free pass does not exceed 6 cm. For example, an individual α-particle arising during radon ^{222}Rn decay and having average energy E_α=6 MeV can make in atmospheric air ~$2 \cdot 10^5$ electron-ion pairs. Usually the flux of radon makes on the surface of the Earth is less than 1.4 atom/cm^2c (Fleischer 1980), but in some areas, for example, in geo-active areas at breaks of the Earth's crust, an emanation of radon from the ground can be essentially higher (Pulinets et al. 1997). β-emitters generate fast electrons with average energy ~0.249 MeV and free pass in air ~1 m. Basically they arise under decay of ^{85}Kr in the atmosphere, and both ^{90}Sr and ^{90}Y on the surface of the Earth. The contribution of γ-radiators in air ionization at the surface of the Earth under usual conditions, as a rule, creates the soil. Basically this radiation is a result of radium and thorium decay.

The cosmic radiation under the usual activity of the Sun generates ions in an atmosphere at the surface of the Earth with a speed ~2 $cm^{-3}s^{-1}$, at a height of 20–35 km ion production velocity grows up to 5 $cm^{-3}s^{-1}$, however, after intensive solar flares the ion production in an atmosphere at the expense of corpuscular radiation essentially grows (Stozhkov et al. 1997).

All these natural sources of radioactivity contribute to background ion production in the troposphere with a velocity ~10 10 $cm^{-3}s^{-1}$ (or 20 mR/hour) (Smirnov 1992). The main contribution to excess over average velocity of ion production is provided by various natural sources, for example, emanations of radioactive aerosols in geoactive zones, and the anthropogenic sources such as emergency situations at chemical and nuclear enterprises.

In all cases the ionization of air by radiation of radioactive elements occurs differently. It can be radiation of separate high-energy particles or quantums from the removed source, or it may be radiation from the "hot" radioactive aerosols uniformly distributed in the air volume, or these are emanations of radioactive gas (^{222}Rn) from the ground (^{85}Kr), from a pipe of the atomic power station.

1.4.3 The Primary Processes of Atmospheric Air Ionization

The main action of the ionizing radiation on the air is provided by the secondary electrons appearing during the ionization of atoms and molecules. The energy of these electrons is within the limits between the thermal energy up to the energy of the primary particles or γ-quantums. Most of the secondary electrons have energy lower than the energy of ionization but these electrons are able to provide the molecular excitation which facilitates the following molecule dissociation or ionization by other particles and electrons. The ion formation near the ground surface under action of the ultraviolet radiation, as happens in the ionosphere, is unlikely because the probability of the dissociative ionization under action of ultraviolet radiation is by one–two orders of magnitude lower than that of molecule ionization by electron impact (Przerzecki and Dmitriev 1978).

Under the action of ionizing radiation the ionization occurs in the following way: during a time period of the order of 10^{-15} s the molecule excitation takes place; then during a time period of the order of 10^{-13} s the molecule being in an unstable excited state dissociates into the positive ions and electron or into the atom and splinter ions (Pikaev 1986). Under an increase of the ionizing particle energy the dissociative ionization takes place with the splitting of molecules of oxygen and nitrogen as occurs in the upper atmosphere. In the lower layers of the atmosphere the ionization is simpler and consists of electron detachment, not dissociation. To form the ion pair energy of ~34 eV is lost. The potential of ionization for the air molecules does not exceed half of this energy: it makes 17 eV for the nitrogen and 15.5 eV for the oxygen. Therefore, the ionization work corresponds to less than the half of the absorbed energy. The ionization of atoms in the deeper levels cannot happen so often so as to explain such a discrepancy, and it is explained mainly by the high level of energy spent on excitation (Curie 1960).

As a result we have in the primary moments of air ionization the following products (Smirnov 1992):

$$N_2 \rightarrow 15{,}5 \text{ eV} \rightarrow N_2^+ + e \tag{1.15}$$

$$N_2 \rightarrow 24{,}3\ \text{eV} \rightarrow N^+ + N + e \tag{1.16}$$

$$O_2 \rightarrow 12{,}2\ \text{eV} \rightarrow O_2^+ + e \tag{1.17}$$

$$O_2 \rightarrow 19{,}2\ \text{eV} \rightarrow O^+ + O + e \tag{1.18}$$

The output of those or other ions depends on the energy of ionizing electrons, for example, in normal air for electrons with energy 40 eV the probability of reaction (1.15) makes 75%, and reaction (1.16) only 7%. The share of molecular oxygen ions is about 15% from the all formed ions, and the share of atomic oxygen ions is only 2.6%.

Thus, under action of radiation during the initial moment there is plenty of O_2^+ ions – both as a result of direct ionization, and as a result of recharging reaction with initial ion N_2^+:

$$N_2^+ + O_2 \rightarrow O_2^+ + N_2 \tag{1.19}$$

and electrons which quickly attach to atoms of oxygen since oxygen possesses significant affinity to electrons.

The key parameter describing a negative ion is the *electron bond energy* – the energy which is necessary to spend to detach an electron. This value is called *the energy of affinity to an electron* for the given atom or a molecule and is designated as EA (electron affinity). The larger EA of the electron to atom or a molecule, the wider area of parameters of weakly ionized gas where the corresponding negative ion can exist.

If before collision the electron had kinetic energy E_e, and the electron affinity of an atom or a molecule equals E_{af}, then after capture the energy $E_e + E_{af}$ is released. We may assume, that this abundant energy is transformed in radiation or is passed to the third body. Therefore in dense layers of the lower atmosphere the three-particle reaction is most probable:

$$e + O_2 + O_2 \rightarrow O_2^- + O_2 \tag{1.20}$$

where as the third body the molecule of oxygen acts. The effectiveness of the nitrogen molecules in this case is 40 times less. The probability of transfer of energy depends on the nature of the third body and on the probability of a corresponding elementary process and will be larger, if the third body would be the atom or a molecule, instead of a free electron. The molecules having several internuclear degrees of freedom, will be the most effective third bodies since in this case the number of various ways by which abundant energy can be transferred, is larger. If the molecule of oxygen catches an electron directly, without participation of the third body abundant energy results in a molecule in the excited state, and at low pressure the molecular ion has not time to release the abundant oscillatory energy by collisions and in most cases dissociates. The reaction of the dissociative attachment of an electron

$$e + O_2 \rightarrow O^- + O \tag{1.21}$$

is more characteristic for the ionosphere, than for a ground layer of an atmosphere. Free electrons can also be attached to atoms of the metals, which are emanating from faults of the Earth's crust, forming negative ions. The greatest bond energy responds to negative ions with the closed electronic shell, such ions are negative ions of halogens (Smirnov 1992). Thus, in a near ground layer of air the primary free electrons, positive and negative elementary ions appear.

1.4.4 The Secondary Processes and Ion Clusters Formation

As a result of fast ion-molecular reactions during the interval of order 10^{-7} s the main elementary tropospheric ions will be formed: O^-, O_2^-, NO_2^-, NO_3^-, CO_3^- and O_2^+, NO^+, H_3O^+ (Boyarchuk and Svirko 1996; Boyarchuk 1997, 1999). The concentration of electrons is so insignificant, that they can be neglected. The large amount of water vapor molecules contained in the troposphere ($\sim 10^{17}$ cm^{-3}), having a noticeable dipole moment p=1,87 D, leads to hydration of elementary ions and formation of ion complexes of a type $NO_2^- \cdot (H_2O)_n$ and $NO_3^- \cdot (HNO_3)_n (H_2O)_m$ and $O_2^+ \cdot (H_2O)_n$, $NO^+ \cdot (H_2O)_n$, $H^+ \cdot (H_2O)_m$ and $H_3O^+ \cdot (H_2O)_n$ which happens rather fast. The ions $NO_3^- \cdot (H_2O)_n$, $NO_3^- \cdot (HNO_3)_n (H_2O)_m$ and $H_3O^+ \cdot (H_2O)_m$ could be regarded as the main ions of the troposphere with a high level of probability. The average time of life of these ions reaches 30 – 40 minutes and more in normal conditions (Smirnov 1992). In hydrated ions, the cluster shell consisting of 3 up to 5 molecules of water, has the bond energy near 4 eV, it should not be destroyed by impacts at a temperature below 1,000 K. It is natural to assume, that such a shell can hinder a recombination. It was shown recently that recombination of hydrated ions slows down in the lower troposphere (0 – 35 km) and the ion balance equation is linear, not a quadratic one (Stozhkov et al. 1997). The formed heavy ion clusters can give rise to generation of a strong electric field in the troposphere which will be considered in the following chapters.

1.5 The Global Electric Circuit and Atmospheric Electricity

In our ordinary life, the statement that the vertical atmospheric electric field near ground surface is of the order of 100 V/m encounters sincere astonishment. People usually ask, what we can do to connect up the electric bulb to air? Unfortunately not, such large electric field exists in the air due to a very low air conductivity and the mean vertical current is of the order of 2 pA/m^2. The people do not feel this field due to their internal large conductivity, our blood is a highly conducting liquid and the field is short down. But we all live in an electrical world between the plates of the giant electrical capacitor where one plate is ground and the other one – is the ionosphere. The potential difference between the plates is 260–280 kV, and the ground is the negative plate. This means that the electric field is directed down to the ground. The capacitance of our natural capacitor is of the order of 1 farad, and the resistance of the circuit between the Earth and the ionosphere is of the order of 200 Ω. There exists the classical paradigm of the global electric circuit (Roble and Tzur 1986). But recent studies have demonstrated that the picture is not so clear as it seemed at the end of the 1980s (Bering III et al. 1998). Some inconsistence in the parameters measured experimentally and that which is expected from the model forced researchers to look for other elements of the global electric circuit, and we believe that seismo-ionospheric coupling through the atmospheric electric field is one of the missing details. It will be shown later that the processes in the air over seismically active areas are very similar to those in thunderstorm clouds, including the charge separation and the large-scale intensive electric field generation.

1.5.1 Field Generators and Consumers

The theoretical estimations show that there exist three main sources of DC electromotive force: thunderstorms, the dynamo mechanism of the solar wind's interaction with the Earth's magnetosphere, and the tidal movements within the thermosphere involving the charged particles and creating the ionospheric dynamo (Roble 1991). So as not to complicate the discussion and taking into account that the contribution from thunderstorms is a factor of three larger than the other ones, let us consider the classical picture of the global electric circuit (Fig. 1.22)

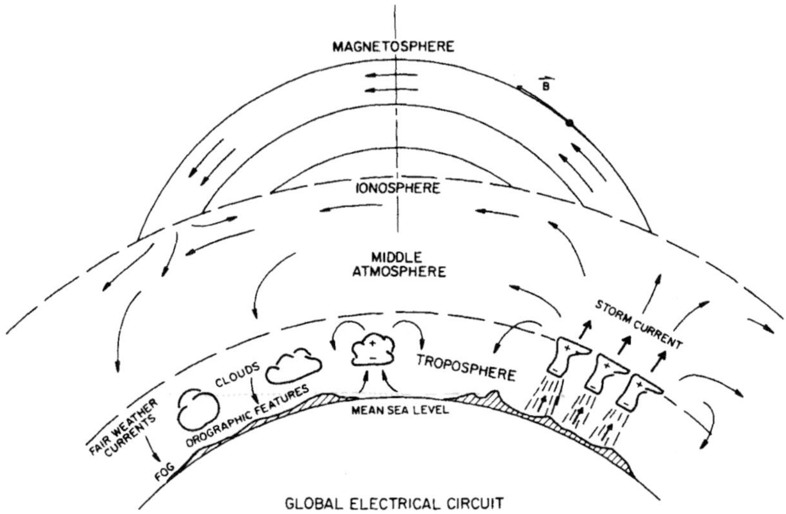

Fig. 1.22 Schematic presentation of the global electric circuit (After Roble and Tzur 1986)

To understand the generation mechanism we also require the height profiles of atmosphere conductivity (Fig. 1.23). We see from the figure that the air conductivity growth with height demonstrates a sharp increase at the altitude near 65 km. Estimations show (Bering III et al. 1998) that the resistance of the air cylinder of 10 km diameter from ground to the lower side of the thunderstorm cloud (near 10 km) is by order of magnitude higher than the resistance from the top of cloud to the ionosphere (400 MΩ and 40 MΩ respectively). This will naturally cause the so-called Wilson storm current from the upper side of the thunderstorm current into the ionosphere. The observed values of such a current for the individual thunderstorm vary from 0.09 up to 3.4 A with the mean value of 1.7 A. The current under cloud has more complex structure. Negative charge is carried to the ground by thunderstorm discharge, precipitation and the atmosphere conductivity current. The global current providing the potential difference between the ground and ionosphere is 750–2,000 A (Roble and Tzur 1986).

The electromotive current of the thunderstorms shown on the right side of the Fig. 1.22 is closed by the return current in the areas of fair weather (left side of Fig. 1.22). The fair weather system is studied less than the thunderstorm field and current system. In the classical models it is supposed that the Earth is flat and an equipotential surface which is very far from the truth.

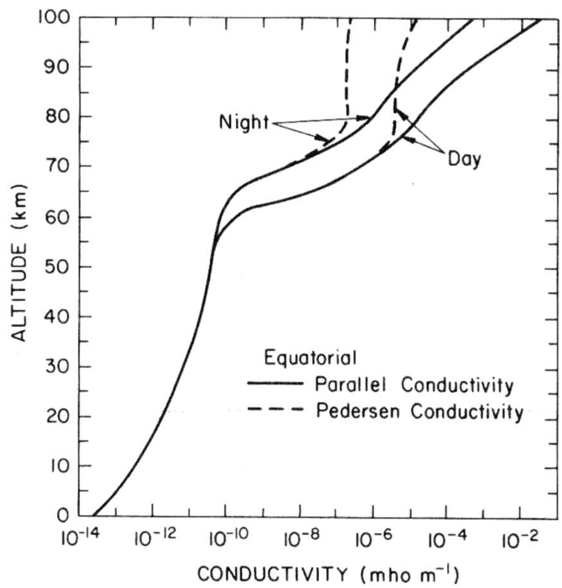

Fig. 1.23 The day and night atmosphere conductivity height profiles in equatorial atmosphere (after Tzur and Roble 1985)

According to daily variation of the thunderstorm activity the atmospheric electric field experiences the oscillations with the minimum near 3 h UT and maximum near 19 UT. For the first time this UT dependence was obtained during electric field measurements onboard the "Carnegie" scientific vessel in 1928–1929 and was named the Carnegie curve (Fig. 1.24). It is natural to expect the correspondent variation of the ionosphere potential. It was measured by balloon soundings (Mühleisen 1977). The seasonal variation of the atmospheric electric field was detected as well by the electric field measurements in Antarctica (Cleary et al. 1997).

1.5.2 Electrode Effect

In fair weather conditions the electric field directed down drives the positive ions downward and negative ions upward. These ion fluxes compose the fair weather current. But near the ground surface the situation changes. If we have no ion production from the ground, the flux of the negative ions from below disappears and the net positive charge accumulates over the ground surface creating the retarding potential and reducing the atmospheric electric field in the near ground layer of a few tens of centimeters thickness (Hoppel 1967; MacGorman and Rust 1998). This positive charge over the ground surface looks like a plate electrode, and the near ground layer is called the electrode layer. Crozier (1965) made experimental vertical profiles of the electric field from the ground surface up to 3 m and demonstrated few tens of centimeters thickness (Hoppel 1967; MacGorman and Rust 1998). This positive charge over the ground surface looks like a plate electrode, and the near ground layer is called the electrode layer. Crozier (1965) made experimental vertical profiles of the electric field from the ground surface up to 3 m and demonstrated the peak charge density at the height of 0.25 m with several hundreds of particles in a cubic centimeter. But as shown by later experimental measurements and theoretical estimations this layer is very variable and depends on many factors: winds, emanation from the ground, surface irregularities, vegetation, etc. Just here starts the process of anomalous electric field generation before the strong earthquakes (Boyarchuk et al. 1997a) which will be discussed in following chapters.

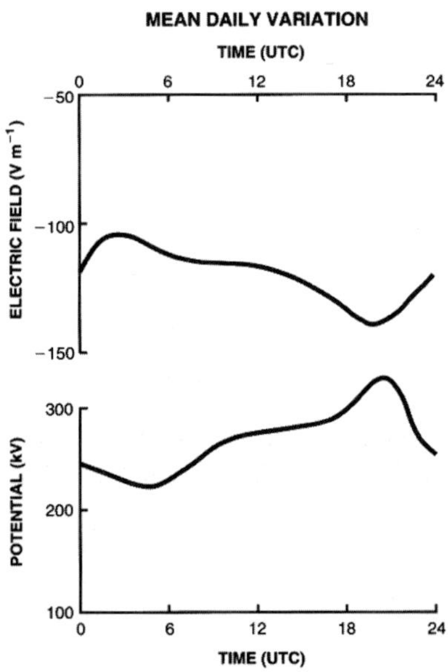

Fig. 1.24 Main diurnal variation of the atmospheric electric field (*top panel*). Negative values reflect the electric field direction down. The ionosphere potential as measured on balloons (*bottom panel*). (After MacGorman and Rust 1998)

1.5.3 Variability in the Boundary Near-Ground Atmospheric Layers

Among the sources of the variability of atmospheric electricity we should mention the non-thunderstorm cloudiness, turbulence in the planetary boundary layer, air pollution, aerosols of different origin (the volcanic eruptions are very strong contributors), radiolytic processes including natural radon and daughter products degassing from the Earth's crust, radioactive pollution (nuclear tests, emergencies like the Three-Mile Island or Chernobyl accidents), dust and snow storms, typhoons, orographic effects, etc. The necessity for accurate estimation of all of these effects was indicated by Roble (1991). Some particular cases were reviewed in Isräel (1973), Reiter (1985), Krider and Roble (1986) and Volland (1995). The attempt to estimate the effects of these irregularities of the atmospheric electric field on the ionosphere variability was made by Pulinets et al. (1998a). The special case of the Three-Mile Island atomic power reactor accident was reviewed in (Boyarchuk et al. 1997b).

Among the results of the cited works we'd like to mention the following. Radon emanation from the crust drastically changes the parameters of the near ground layer (Willet 1985). It can change by several times the value of the electric field within the layer up to the reverse sign of the electrode effect. Ionization effectiveness of the ^{222}Rn and its daughter products could be 10 times larger than that of the cosmic rays (Martel 1985), which can effect the thunderstorm activity (Orville and Spenser 1979; Turman and Edgar 1982). The surface roughness also has an effect on the near ground electric field variability.

Another very important effect is the effect of aerosols. The experimental measurements show that in the rural areas polluted by aerosols the atmospheric electric field is up to 4 times higher than in clear air (400 V/m versus 100 V/m), (Table 1 in Roble and Tzur 1986). One of the major effects of aerosols is the atmosphere conductivity decrease. Such effects were detected in the Northern hemisphere due to increased volcanic activity shown by measurements taken over the period 1907 to 1972. (Meyerott et al. 1985; Makino and Ogawa 1985).

The size of the particles (aerosols, ion clusters) also plays a very important role in the local electricity balance due to the problem of particle mobility. The mobility k is determined as the average velocity acquired by the charged particle under action of the external electric field (MacGorman and Rust 1998):

$$k = v/E\ [\mathrm{m^2V^{-1}s^{-1}}] \qquad (1.22)$$

The mobility of the large ions is three orders of magnitude lower than that of small ions. This means that if under some conditions the small ions will be replaced by the large ions (as happens during the hydration process described above), the conductivity will drop practically to zero, which will sharply increase the electric field.

1.5.4 Concluding Remarks

From the discussion above we can conclude that the near ground atmospheric layer electricity is very variable. The sharp changes in the electric field near the ground may be due to several factors. Among them is radon emanation from the Earth's crust, air contamination by aerosols, and large heavy charged cluster population decreasing air conductivity. All these factors contribute to the physical mechanism of seismo-ionospheric coupling.

Chapter 2 Ionospheric Precursors of Earthquakes as they are seen from the Ground and from Space

Among different precursory phenomena mentioned in the publications on earthquake predictions the ionospheric ones are probably the youngest. The first publications concerning the ionospheric effects connected with earthquakes were published just after the Alaska "Good Friday" earthquake in 1964 (Bolt 1964; Dohn and Posmetier 1964; Moore 1964; Davies and Baker 1965; Leonard and Barnes 1965; Row 1966; Hirshberg et al. 1967). By analogy with seismology, the wave movements were looked at as launched by ground movements before and after the earthquake. That's why the acoustic-gravity wave – ionosphere coupling was the first seismo-ionospheric coupling mechanism that appeared (Yuen et al. 1969; Weaver et al. 1970; Boborykin 1972; Birfeld 1974). The after earthquake effect was demonstrated by direct comparison of the seismograms and records of Doppler shift registered by the ionosonde and reflecting the vertical oscillation of the ionospheric layer caused by the acoustic oscillations from the earthquake (Yuen et al. 1969). But it should be mentioned that the effects *before* the earthquake were marked as well (Moore 1964; Davies and Baker 1965).

As the first publications really dealing with the ionosphere parameter variations as seismic precursors we can mention Antselevich (1971) studying the variations of the f_oE parameter before the Tashkent earthquake in 1966 and Datchenko et al. (1972) where the ionosphere electron variations before the Tashkent earthquakes in 1966 were also studied. The case study papers started to appear regularly. They used mainly the data of ground-based ionosondes but the first papers using satellite data also appeared (Gokhberg et al. 1983b).

The first years of studies of the seismo-ionospheric precursory phenomena were characterized by a mainly phenomenological approach without a solid physical background. Case by case, probe by probe of different processing techniques the information was accumulated. This long history one can find in different reviews (Liperovsky et al. 1990; Gaivoronskaya 1991; Liperovsky et al. 1992; Parrot et al. 1993; Pulinets et al. 1994; Gokhberg et al. 1995) and we will save some space in not repeating these publications. Let us look directly at what the seismo-ionospheric precursors are. We will demonstrate the examples of ionospheric variations registered before several famous earthquakes as they look on the records of the ground based ionosondes and from onboard the artificial satellites. Some data of space plasma parameter variations before earthquakes will be demonstrated too. GPS total electron content variations will be demonstrated and compared with the critical frequency variations.

2.1 Good Friday 1964

Among the largest earthquakes of the 20th century is the Alaska Good Friday earthquake of 1964 and has second place after the Chilean earthquake of 1960 (see Fig. 1.5). It occurred on Friday, March 27 (local time), 1964 and this was the Friday before the Easter holiday. The energy released was the largest to strike North America in the 20th century. It happened at 17:36 local time (03:36 UT 28th of March) at a point with coordinates 61.1 N and 147.6 W, and had a magnitude M_w=9.2. The largest areas (about 200,000 square kilometers) of vertical land movements in records to date were registered. The rupture in the rocks extended horizontally for 800 km almost parallel to the Aleutian Trench. Figure 2.1 shows one example of the earthquake damage.

Fig. 2.1 The street in Valdez town close to the epicenter of the Good Friday Alaskan earthquake of 1964

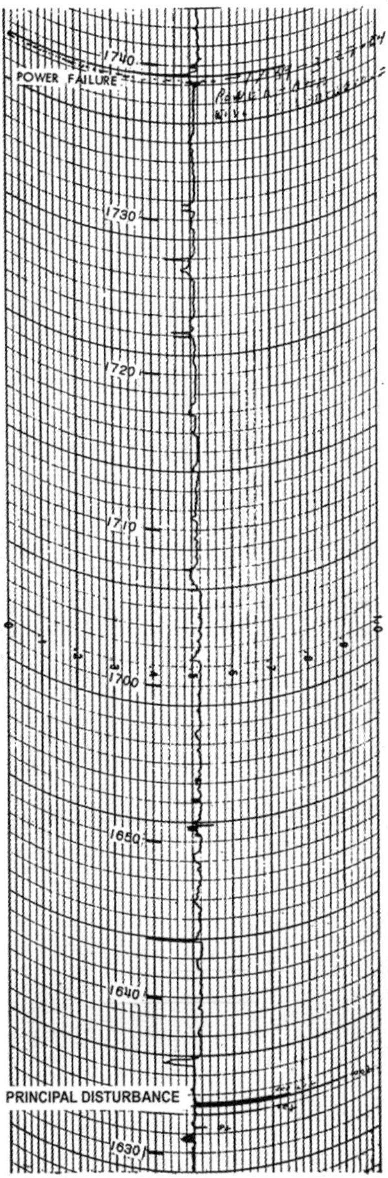

Fig. 2.2 The positive magnetic disturbance (*to the right*) of the order of 250 γ registered at 0132 UT (1632 Alaska standard time) two hours before the Alaska earthquake of 1964 by the magnetometer installed at Kodiak

Let us try to reconstruct the history of studies on the Alaska earthquake. First results on some strange phenomena preceding the earthquake were published by Moore (1964) and Davies and Baker (1965). In the first paper the anomalous variation of the magnetic field registered at Kodiak (30 km from the rupture fault) was reported (Fig. 2.2), and in the second one – the anomalous Doppler shift measured at Boulder from the vertical reflection on three HF frequencies was demonstrated (Fig. 2.3). The first scattering on the 5 MHz record at Boulder appeared at 0130 UT, and pulsation's onset (scallops) – at 0232 UT. It is interesting to note that the pulsation anomalies appeared at the same time as the magnetic anomaly reported by Moore. The ionosonde data were analyzed at that time only for the purpose of the after earthquake effect studies (Leonard and Barnes 1965). The strong F-spreading effect was detected on the Adak ionograms after the seismic shock. Summarizing the first results we can conclude that the anomalies appeared a few hours before the seismic shock as on ground based records (magnetic field), also within the ionosphere (Doppler measurements). The strong acoustic emission excited by the huge movement of the Earth's surface of 200,000 km^2 stimulated the formation of irregularities within the ionosphere registered as spread reflections on the Adak ionograms.

The more detailed analysis of the ionospheric data was carried out much more recently (Khegai et al. 2002; Pulinets and Legen'ka 2003). In the first paper by Khegai et al., the data of 8 ionospheric stations situated at a distance from 124 up to 4,900 km from the epicenter (Table 2.1) were analyzed. With regard to the magnitude-earthquake preparation zone radius relation (1.8), all ionospheric stations are inside the earthquake preparation zone. Pulinets and Legen'ka (2003) made the combined analysis of the ground measurements and topside sounding was from the Alouette satellite. This combination permitted the researchers to obtain the dense measurements grid for ionospheric mapping. They built the maps of the critical frequency distribution a few days/hours before the earthquake and tracked the dynamics of the large irregularity formed in the ionosphere.

Taking into account that the earthquake epicenter is within the auroral zone, one should carefully check the solar-geomagnetic activity to exclude ionospheric variations stimulated by some disturbance in near-earth space.

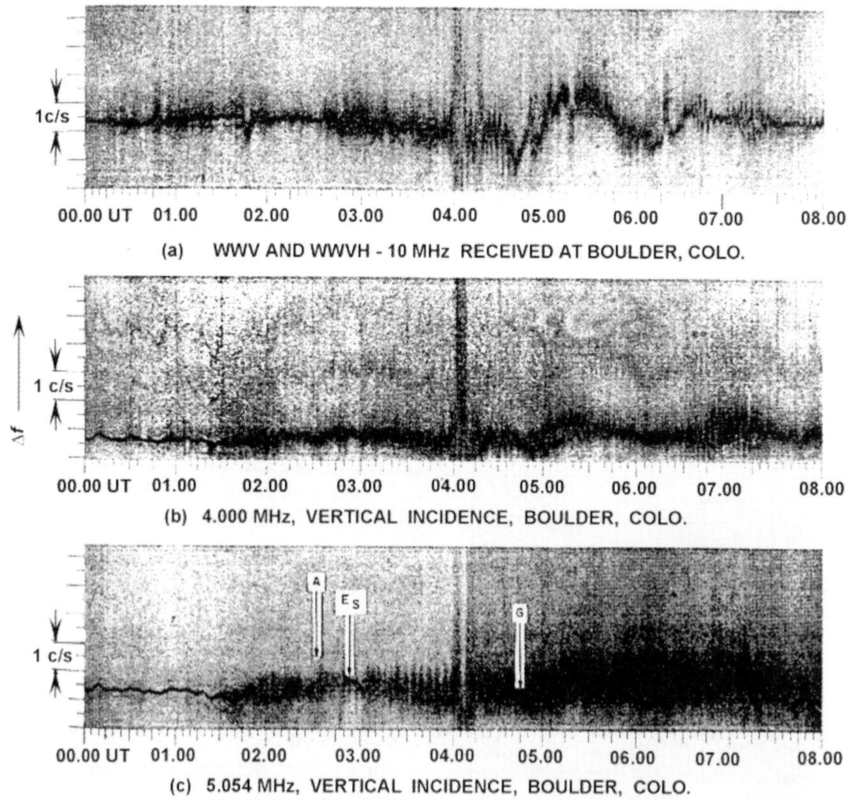

Fig. 2.3 Frequency variations (Doppler shift due to vertical movements of ionospheric layer) registered on 3 frequencies (10 MHz, 4 MHz and 5.054 MHz from *top to bottom*) during the time around the main shock of the Alaska earthquake of 1964

2.1.1 Geophysical Conditions Around the Time of the Alaska Earthquake in 1964

The geomagnetic activity during the period between the 5th and 30th of March was quiet except for the period 22nd–25th of March. The geomagnetic disturbance that can be qualified as a substorm (the D_{st} index did not exceed –50 γ) started on the 22nd of March and all ionospheric stations from the region demonstrated the negative disturbance of the critical frequency. During the disturbance period the *AE* index had an oscillatory character but just before the earthquake dropped to the extremely low value of 33 nT which corresponds to the very quiet magnetic situation. At the same time, the variation in the solar radiation flux $F_{10.7}$ did not exceed 5% of its mean value. So we can conclude that a few days before the earthquake the geomagnetic situation in the region can be characterized as a transition from the slightly disturbed to calm. This means that detected large scale irregularities in the ionosphere cannot be interpreted as a result of geomagnetic disturbance.

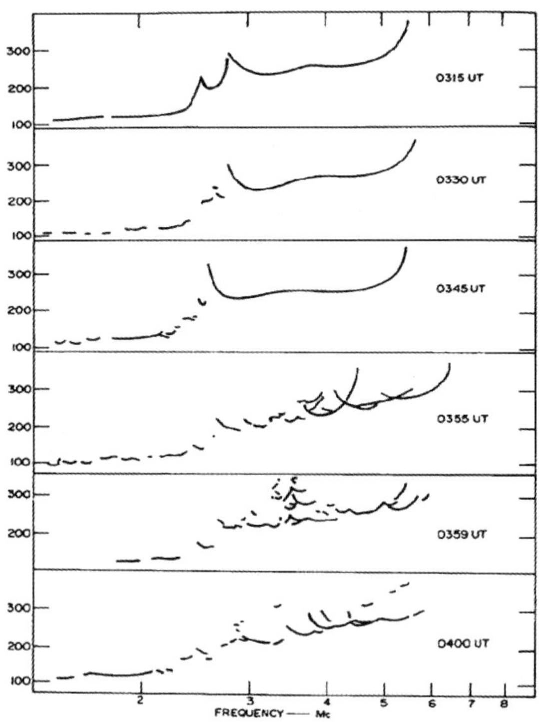

Fig. 2.4 Sequence of the Adak ionograms demonstrating development of irregularities within the ionosphere after the seismic shock of the Alaska earthquake of 1964

2.1.2 Ground and Space Portraits of the Alaska Earthquake Precursor

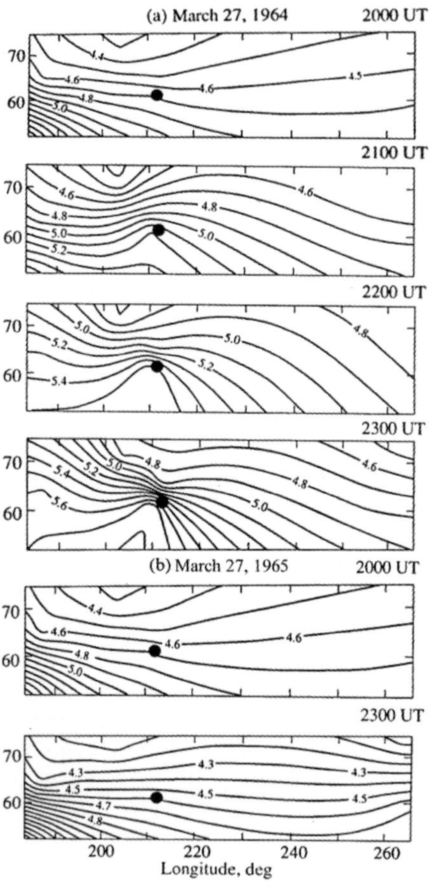

Fig. 2.5a,b Isolines of the critical frequency *foF2* (MHz) for (**a**) 2000, 2100, 2200, and 2300 UT on March 27th, 1964, in the vicinity of the Alaska 1964 earthquake epicenter (*black circle*), (**b**) the same but for 2000 UT and 2300 UT on March 27th,1965

Table 2.1 shows the ionospheric stations whose data were used for analysis of the ionospheric disturbance before the earthquake and their position in relation to the Alaska earthquake epicenter. The data quality was poor which did not permit us to calculate the representative monthly median for all hours of local time, especially for the stations closest to the epicenter. That's why Khegai et al. (2002) used as a reference level the data from March 1965, characterized by the identical level of solar and geomagnetic activity.

To demonstrate the local character of the ionospheric disturbance and its connection with the future epicenter position the spatial distribution of the critical frequency (which is equivalent to the peak electron concentration) using the kriging technique (Oliver and Webster 1990) was built (Fig. 2.5a). One can see the irregularity development on the hourly snapshots with the maximum distortion at 2300 UT which corresponds to 1300 LT. In Fig. 2.5b the same distributions but for March 27th of 1965 are built for 2000 UT and 2300 UT, respectively. If distributions for 2000 UT of the 1964 and 1965 years coincide completely, the difference for 2300 UT is obvious showing the strong distortion of the electron density on March 27th of 1964 in the vicinity of the future earthquake epicenter shown as a filled black circle in the figures.

Table 2.1. Ground based ionosondes the data of which were used for analysis

Station name	Geographic longitude E, deg	Geographic latitude N, deg	Distance to the epicenter, km
Anchorage	210.1	61.2	124
College	212.0	64.9	423
Barrow	203.2	71.2	1192
Providence Bay	186.6	64.4	1365
Adak	183.4	51.9	2090
Resolute Bay	265.1	74.7	2536
Churchill	265.8	58.8	2899
Narsarsuaq	314.6	61.17	4904

The picture is even more convincing looking from space where in that time the Alouette-l satellite was in orbit and was sounding the ionosphere over Alaska. Alouette-1 was the first of a series of satellites with a topside sounder onboard created within the frame of the US-Canadian Alouette-ISIS program (Florida 1969). The sounding data were used from the Alouette Catalogue (1964). The advantage of the satellite data is the regular grid of soundings (Alouette-1 collected ionograms every 18 s, i.e. every 140 km along its orbit). The specificity of the high inclination satellites is that they fly in quasi-solar synchronized orbit which means that all the time they are fixed in the specific sector of local time. For the period of the Good Friday earthquake the Alouette-1 satellite passed over Alaska at 12–13 LT when the maximum deviation was marked by the ground based ionosondes. The interpolation between the neighbor orbits gave the possibility to obtain the so-called local time maps. To separate the irregularity associated with the earthquake preparation the critical frequency deviation was calculated $\Delta foF2$:

$$\Delta f_oF2 = (f_oF2_{cur.} - f_oF2_{ref}) \times 100 / f_oF2_{ref}, \; [\%] \qquad (2.1)$$

where f_oF2_{cur} – the current value of the critical frequency, and f_oF2_{ref} – the reference value. As was mentioned in Chapter 1 the most appropriate way to present the ionospheric disturbances in the *F2* region is to calculate the deviation (in %) from the undisturbed, or reference level. The relative variations calculation is a more physical way of presentation than the absolute ones (in MHz) because of the large difference in daytime and nighttime ionospheric parameters. One MHz difference for daytime and nighttime conditions can in relative values be 10% and 100% respectively. As a reference value for ground based measurements the monthly median is used generally. There exists a discussion in ionospheric circles as to what to select for the reference because the monthly median contains the values of the disturbed days too. Some scientists propose we use the quiet days variations as a reference level (Gulyaeva 1995). Sometimes the running monthly median is used. Our opinion is the following. To separate the ionospheric precursor is a kind of pattern recognition art, including the procedure of the signal selection on the noise background, so the different approaches could be used in different specific geophysical conditions for different levels of the ionospheric variability. In the following examples of experimental records we will indicate which parameter was used as the reference level. As concerns the satellite data, it is impossible to calculate the monthly median. Usually the quiet day distribution is used as a reference level. In the case of the Alaska earthquake the daytime data of the Alouette-1 satellite were available for the 17th, 19th, 23rd and the 27th–29th of March. This permitted us to study the ionosphere state in quiet conditions (17th, 19th of March, the reference level), during the time of the maximum of the magnetic disturbance (23rd of March), and the days before and after the earthquake $\Delta t \approx -30.5$ h (26.03), $\Delta t \approx -6$ h (27.03), $\Delta t \approx +18$ h (28.03). Here signs "–" and "+" indicate the time before and after the seismic shock respectively.

Even a simple comparison of the along orbit measurements of the critical frequency for neighbor orbits (Fig. 2.6) on the 27th of March shows the significant difference. The electron concentration along the orbit whose longitude is closer to the epicenter is much higher demonstrating the large latitudinal gradient. It should be noted also that the irregularity also extends to the South from the epicenter. This issue will be considered during the physical mechanism of seismo-ionospheric coupling discussion in Chapter 4

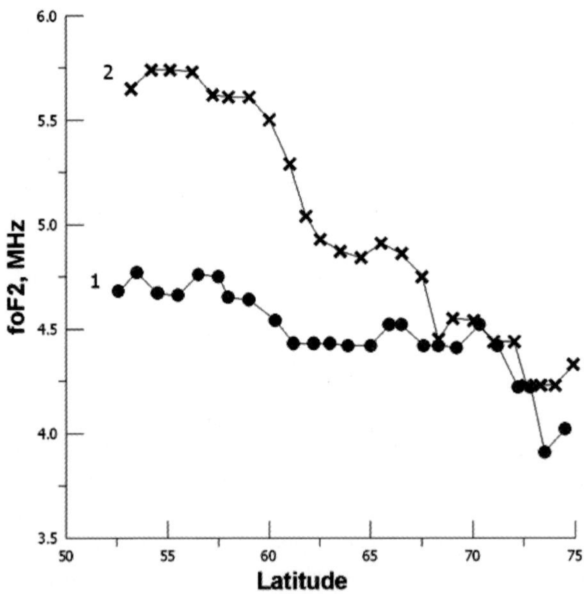

Fig. 2.6 Critical frequency distributions along the Alouette-1 satellite orbits on the 27th of March 1964. 1 –k_p=2, UT=20.46, LT=12.17, λ=–127.4 °, Δλ=20.2 °, Δt=–30.5 h, 6–k_p=1, UT=22.31, LT=12.15, λ=–154.1 °, Δλ=–6.5 °, Δt=–29 h

Taking into account the identity of the critical frequency measurements from the ground and from space the whole available volume of information from the Alouette-1 topside sounder and from the ground based ionosondes was used to build the regional distribution of the critical frequency deviation for three consecutive days (26th, 27th and 28th of March). The results of data processing are presented in Fig. 2.7. One can see that the area of the positive deviation appears in the vicinity of the future epicenter on the 26th of March, approximately 29 hours before the seismic shock (top panel). Then the intensity and size of the irregularity increases significantly (reaching in peaks 35%) at the same local time but on the next day, the 27th of March, 5 hours before the seismic shock (middle panel). The day after the earthquake, the 28th of March, the irregularity still exists but its intensity drops up to the value of ionosphere variability near 20%.

What can we conclude at first glance? The most important thing is that the ionospheric precursor can be detected at least 29 hours before the seismic shock. The maximum of the disturbed area does not coincide exactly with the vertical projection of the epicenter, but the ionospheric irregularity surely has the regional character, and it is "tied" to the area of the seismic activity. At the phase of the maximum development of the irregularity its longitudinal size is of the order of 40° (nearly 4,000 km) which is consistent with estimations of the earthquake preparation zone (Table 1.3). One question appears: is the positive deviation the common feature of the ionospheric precursors of earthquakes or not? The data analysis of many cases of ionospheric precursors of earthquakes shows that it is not. The precursors could have both a positive and negative sign of deviation from the undisturbed state. The common feature is only the precursors dynamics. The appearance a few days before the earthquake, reaching the maximum phase (usually one day to a few hours before the seismic shock) and disruption after the shock.

To demonstrate this we present one more example of the ionospheric precursor dynamics before the earthquake in the New Guinea Islands region which happened on the 16th of July 1980 at 1956 UT (Lat=3.2 S, Long 143.3 E, M=7.3, H=54 km). A similar procedure of data processing was used for the construction of the critical frequency deviation maps. The data of the Intercosmos-19 satellite topside sounder were used (Pulinets 1998a). Satellite data were available for the passes of the satellite on the 14th of July 1980 (48 hours before the shock), 15th of July (24 hours before the shock), and 17th of July (24 hours after the shock). The local time of the satellite pass over the seismically active area was 5–6 LT. Results are presented in Fig. 2.8.

From the presented examples we can conclude that the sign of the precursor is a result of some yet unknown factor which is connected with the physical mechanism of the ionospheric precursor generation. This issue will be discussed in Chapter 4.

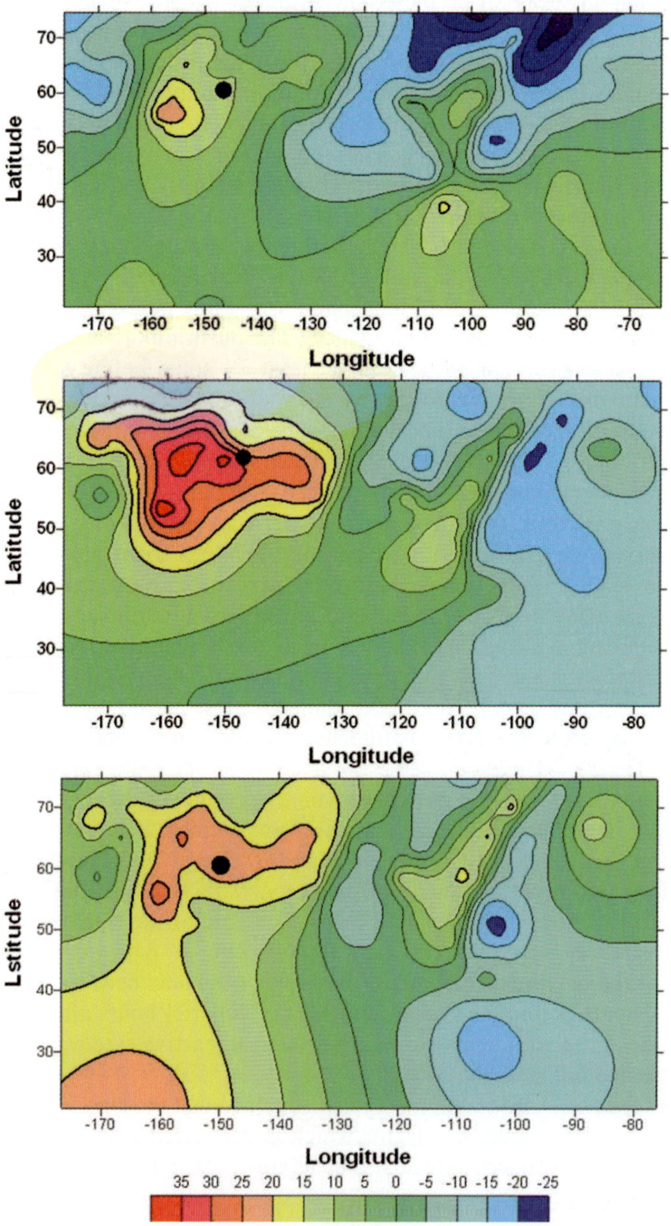

Fig 2.7 Consecutive distributions of the deviation of critical frequency *ΔfoF2* by the data of the Alouette-1 satellite and ground based ionosondes. *Top* – 26.08.1964 Δt≈–30.5 hours), *middle* – 27.03.1964 Δt≈–6 hours, *bottom* – 27.03.1964, Δt≈+18 hours. • epicenter position

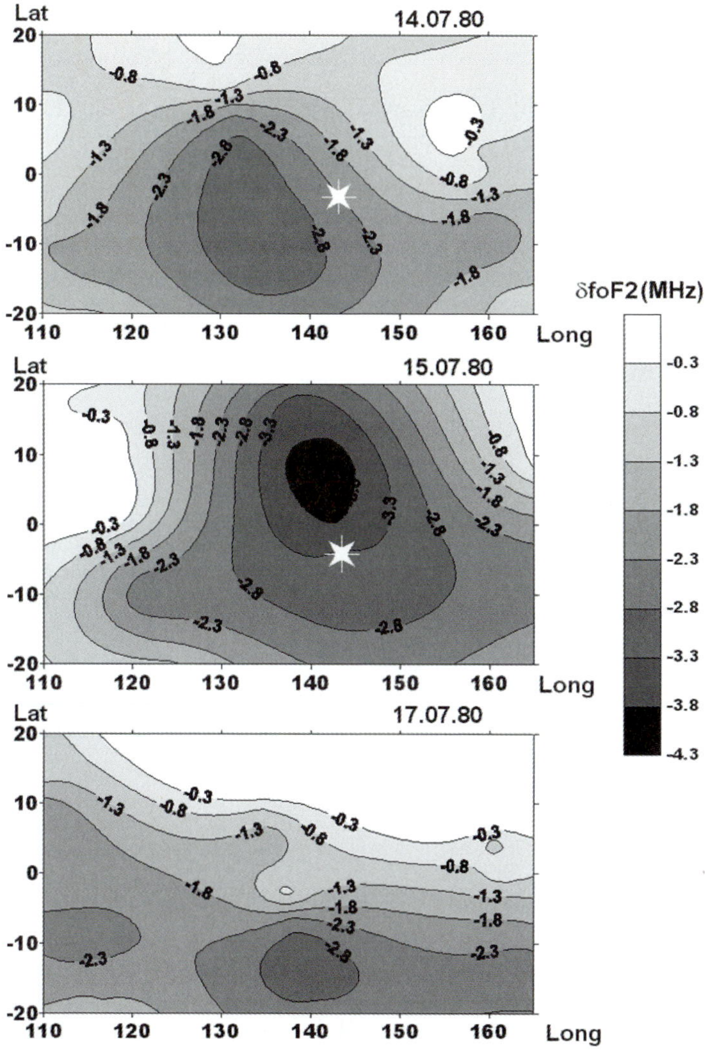

Fig. 2.8 Critical frequency deviation (in MHz) registered by the Intercosmos-19 satellite before the earthquake M=7.3 at New Guinea Islands (3.2 °S, 143.3 °E) on 16.07.1980 at 1956 UT. *Top* – 48 h before the shock, *middle* – 24 h before the shock, *bottom* – 24 h after shock

Sometimes the critics indicate the high variability of the ionosphere, especially during the magnetic storms, and use this as an argument against the possibility of ionospheric precursor detection. Of course, probably very complex cases could be found, but statistics show that the magnetic storms happen usually before or after the earthquake (this issue will be discussed in Chapter 6), but practically never coincide. However, the geomagnetic storm has a global character (Szuszczewicz et al. 1998) and the disturbances in the ionosphere are registered practically everywhere over the globe, contrary to the earthquake precursors which are local as was demonstrated in Figs. 2.6–2.8. The local character of the ionospheric precursor was demonstrated also by Pulinets et al. (1991) and Hegai et al. (2002) for the Alaska 1964 earthquake. To demonstrate how different is the result of the magnetic storm in the ionosphere we present the distribution for the same local time, same area, but for March 23rd, 1964, during the main phase of the magnetic disturbance.

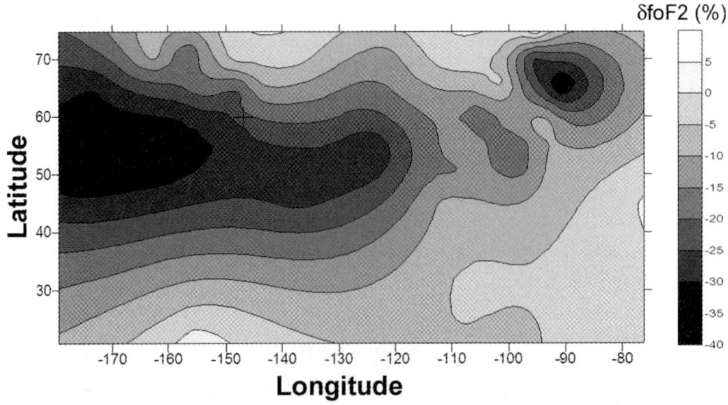

Fig. 2.9 Critical frequency deviation during the main phase of the magnetic disturbance on 23.03.1964 registered by the Alouette-1 satellite. The intensive trough is formed during local noon time with the maximum deviation reaching – 40%

2.2 Ionospheric Precursors Registered by Ground Based Ionosondes

Starting from 30th the ground-based ionosondes during several decades were the main instrument for ionosphere studies (Bibl 1998). Regardless of the new more advanced techniques that have appeared they still remain irreplaceable for many fundamental studies and applications. And one of them is the registration of the ionospheric precursors in seismically active areas (Gokhberg et al 1989; Zelenova and Legen'ka 1989; Liperovsky et al. 1990; Pulinets et al. 1994; Pulinets 1998b; Ondoh 1999; Liperovsky et al. 2000; Liu et al. 2000; Chuo et al. 2001; Chuo et al. 2002; Pulinets et al. 2002). The most difficult task is to identify variations registered by the ionosonde as ionospheric precursors. Let us try one exercise to make sure that they are seen by ground based ionosondes at all.

2.2.1 Ionospheric Precursors Visualization

To detect the ionospheric precursors the data of the European network of ionosondes was used (Pulinets 1998b) for the period of seismic activity in Central Italy in May 1984 (Table 2.2).

Table 2.2. Parameters of the seismic events under study

Date	Time UT	Lat	Long	Depth	M	Area
07.05.84	1750	41.8N	13.9E	16	5.8	Italy: S-Central: Abruzzo
11.05.84	1042	41.8N	13.9E	15	5.4	Italy: S-Central: Abruzzo
13.05.84	1245	43.0N	17.8E	34	5.1	Balkans NW

Taking into account that the magnitude of the earthquakes presented in Table 2.2 was not very large we can also expect that the modified area within the ionosphere will not be large, and only the ionosondes closest to the epicenters will "feel" the earthquake preparation. To check this hypothesis the data of several European ionosondes within the latitudinal band 5° and longitudes from 0° to 25° E were used (see the map in Fig. 2.10)

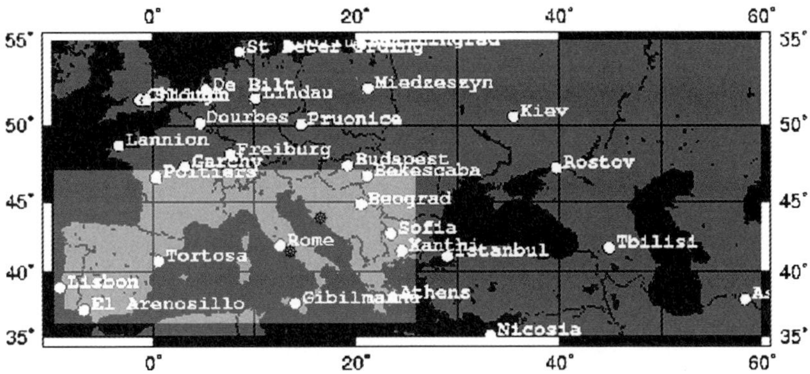

Fig. 2.10 The European ionosondes network. *Gray tone* indicates the area and positions of ionosondes used in the precursors visualization. ✽ – epicenters position

The Δf_oF2 was calculated for every station falling inside the selected longitudinal bar for the period from the 2nd to 17th of May 1984. The latitudinal interval 5 ° was presented as a horizontal line and than by interpolation between the stations on different longitudes the Δf_oF2 dynamics for the given period was presented in the form of a color coded picture (Fig. 2.11). The vertical axis presents the time (right axis – calendar dates, left axis – hours). Horizontal axis – geodetic longitude. The idea of processing is the following. If the source of the variations in the ionosphere has a global origin, all the stations will react almost identically which will give as a result the horizontal bar of the same color for all stations, if the ionosphere is disturbed locally, these variations will be characteristic only for the longitudinal interval where the source is situated. Arrows on the horizontal axis mark the longitudinal interval where the epicenters of earthquakes are situated. As we see from the picture, just this interval differs by its activity from the other longitudes, which means that we observe the precursors because they appear five days before the first seismic shock. For the third earthquake we see how the bar of precursors activity turns to the right according to the increasing longitude of the third earthquake.

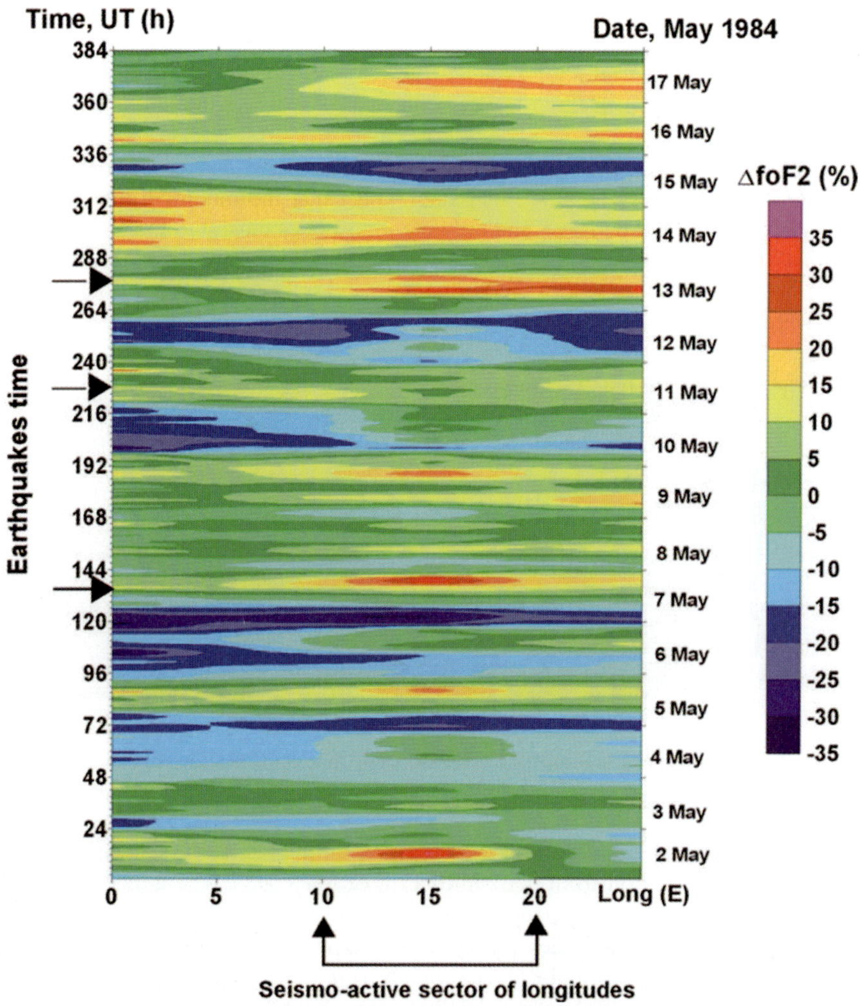

Fig. 2.11 Deviation of the critical frequency *foF2* (*color coded*) in the longitude sector 0 °–25 °E at latitudes 35 °–45 °N as a function of time for a period of strong seismic activity. *Black arrows* at the *left axis* indicate the moments of earthquakes (see also Table 2.2)

2.2.2 The Diversity of the Ionospheric Precursors Shape

As we have seen from the satellite data, the sign of the critical frequency deviation can be different. One of the possible reasons of these precursors sign deviations is their dependence on local time (Pulinets et al. 1998b). But other reasons are possible which will be discussed in Chapter 4.

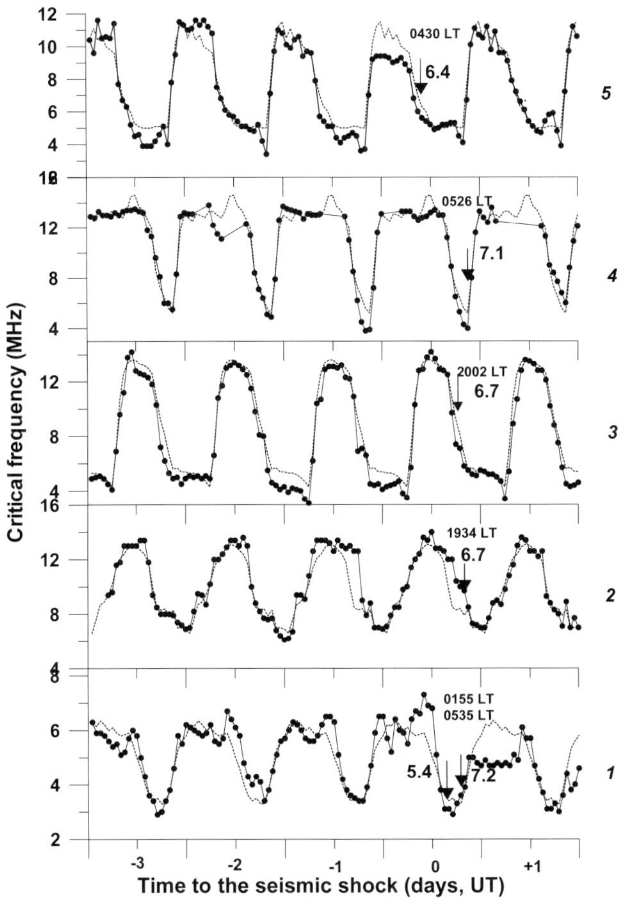

Fig. 2.12 Examples of comparison of daily variations of *foF2* (*points*) with their monthly median values (*stroke*), three days before and one day after the earthquake. By the data of ground-based stations of vertical sounding for various earthquakes: 1-Milkovo (5.09.1971, 18.35 UT, *M*=7.2,); 2-Rome (23.11.1980, 18.34 UT, *M*=6.7); 3-Norfolk (19.06.1980, 08.34 UT, *M*=6.4); 4-Vanimo (16.07.1980, 05.26 UT, *M*=7.3); 5-Norfolk (14.07.1980, 16.15 UT, *M*=6.6). *Arrows* specify the moments of earthquakes

But the precursors registered on the ground differ not only by sign but also by many other parameters, including their shape, time of advance of their appearance before the seismic shock, duration time, amplitude. We will depict several examples to give an idea of their diversity and the complexity of the task of their identification. The precursors identification technique will be described in Chapter 5.

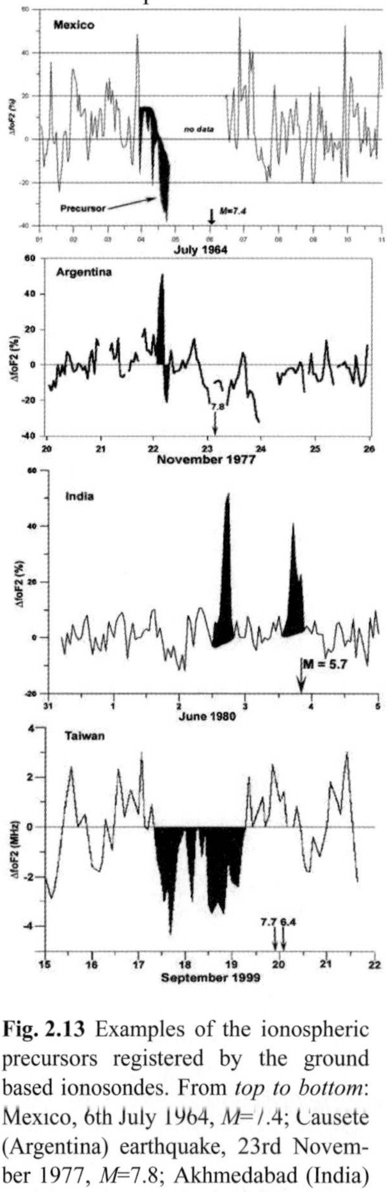

Fig. 2.13 Examples of the ionospheric precursors registered by the ground based ionosondes. From *top to bottom*: Mexico, 6th July 1964, *M*=7.4; Causete (Argentina) earthquake, 23rd November 1977, *M*=7.8; Akhmedabad (India) earthquake 23rd July 1999, *M*=6.6; Chi-Chi (Taiwan) earthquake. 20th September 1999, *M*=7.7

Figure 2.12 demonstrates in the raw data of the ionospheric stations the look of the precursors of strong earthquakes with a magnitude higher than 6 (Pulinets et al. 1998b). To pick out some specific properties of the precursors different types of presentation are possible, for example, in the form of critical frequency deviation. Such examples for some strong earthquakes are presented in Fig. 2.13. The earthquake parameters are given in the figure caption. One can see how different ionospheric precursors can be. Sometimes they are only negative, sometimes – only positive, sometimes – of both signs. One can find more complex cases shadowed by ordinary daily variations, and people without experience cannot find them (Kouris et al. 2001). But the detailed and robust data analysis demonstrates that the information on the preparation of strong earthquakes is always present in the ionospheric data. Chapter 5 will be completely devoted to the different techniques of ionospheric precursor identification, including the case of the Irpinia earthquake questioned in the paper by Kouris et al. (2001).

Among the likely reasons of the ionospheric precursors diversity we can name the following: different focal mechanisms of earthquakes, different earthquake zones (subduction, intraplate, zone of the volcanic activity, etc.), different structure of the crust in the seismic zone (granites, sand, clay, etc.), temporal variations of the radon emanation and anomalous electric field, the weather (wind, rain, snow), the season (ionosphere parameters seasonal dependence), solar and geomagnetic activity, latitude and longitude of the epicenter, relative position of the ionosonde and epicenter. Some of them have regular character, for example, seasonal and solar cycle dependencies, the other ones, such as electric field variations are sporadic. The role of these factors will be discussed in Chapters 4 and 5.

After the reader obtained the first impression on the ionospheric precursors let us consider existing experimental data on the pre-earthquake effects in different ionospheric layers.

2.3 Precursory Effects in the *D*-Layer

Taking into account that the *D*-layer is very difficult for studies due to its extreme variability and very low electron concentration, we have only indirect manifestations of the *D*-layer variations before earthquakes. One of the first experimental results concerning the *D*-layer reaction on the earthquake preparation was the change of the parameters of VLF signals received by ground-based receivers from remote transmitters, mainly from the Omega navigational transmitters (10.2–13.6 kHz) (Gokhberg et al. 1987, 1989b; Gufeld et al. 1992). The amplitude and phase distortions of the received signal were interpreted as influence of the waveguide earth-ionosphere where the VLF signal propagates. It was proposed that the distortions are due to ionosphere lowering over the region of the earthquake preparation (Fig. 2.14). It was believed that by triangulation it would be possible to determine the position of the future epicenter, and from the distortions amplitude – to determine the earthquake magnitude. The time of the earthquake had a relation with the time of the VLF signal anomalies appearance. As a result, the registration of variations of the VLF signal parameters during its subionospheric propagation over the earthquake preparation zone was proposed for short-term earthquake prediction (Gufeld et al. 1994).

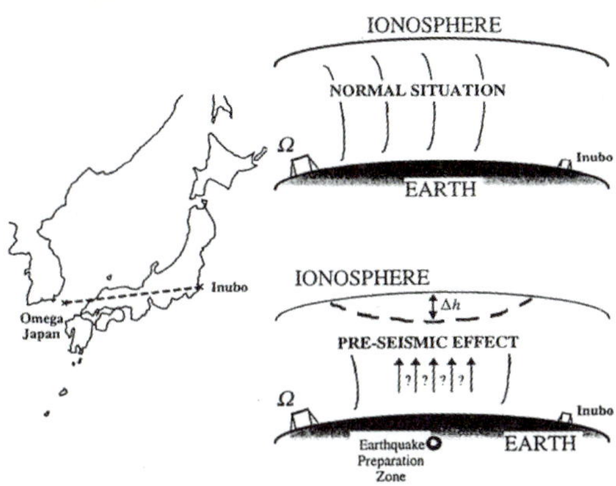

Fig. 2.14 Schematic presentation of the ionosphere lowering over the earthquake preparation zone as a source of the VLF Omega signal distortions

This technique was modified later to apply it in Japan (Hayakawa 2001). It uses not the arbitrary fluctuations in the received VLF signal but the variations of the terminator time (sunrise and sunset) observed on the VLF phase records a few days before the earthquake (Molchanov et al. 1998a, 1998b). Examples of the records are shown in Fig. 2.15. This method was criticized by Rodger et al. (1999) and Cliverd et al. (1999) using the same technique at long passes from North America to Antarctica permitting to monitor the effects of the earthquakes in Southern America, but it should be noted that they used the other geometry of the signal passes in relation to the terminator orientation in comparison with the Japanese experiments which may lead to the difference in the results obtained.

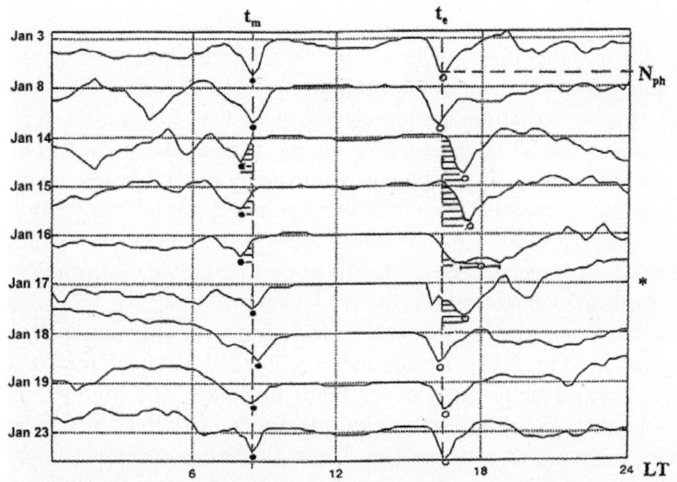

Fig. 2.15 Sequential plots of the terminator time variations of the Inubo (10.2 kHz) receiving station signal phase before the Kobe earthquake of 17th January, 1995 (*asterisk*). The time of sunrise and sunset is designated by t_m, and t_e. The shift of the terminator time is marked by *shadowing*. Radio pass configuration is shown, in Fig. 2.14

Nevertheless, the investigations with VLF propagation were continued in Japan and received new interpretation connected with the AGW influence on the ionosphere (Miyaki et al. 2002) and VLF monitoring is proposed for the planetary studies of the atmosphere-ionosphere boundary including the pre-earthquake effects (Molchanov and Hayakawa 2001).

Regardless of the high sensitivity of the TT (terminator time) technique claimed by Molchanov and Hayakawa (1998b) it seems that the VLF monitoring shouldn't limit itself only by the time around the terminator passing. Looking at Fig. 2.16 nobody can neglect the existence of the strong phase variations of the nighttime registrations of the Omega signal before earthquakes.

So, it seems that VLF sounding should be developed using round the clock registrations because not only the terminator time effect could be responsible for the *D*-layer lowering before the earthquakes. This statement will be supported by the theoretical consideration in Chapter 4.

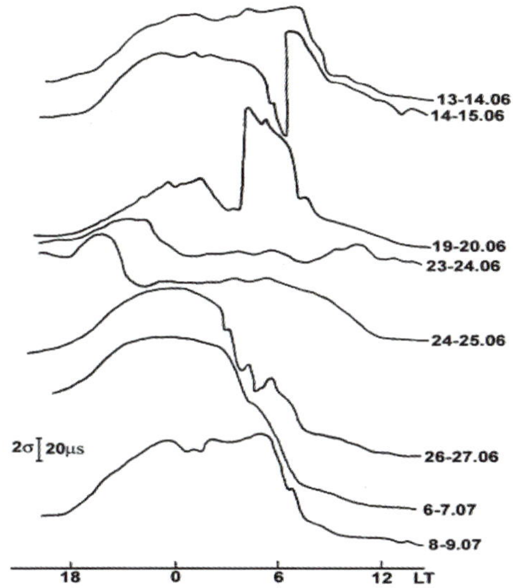

Fig. 2.16 Omega signal phase anomalies registered before the Rudbar (Iran) earthquake on June 20th, 1990, M=7.5 on the Moscow-Reunion pass. (After Gufeld et al. 1992)

2.4 *E*-Layer Reaction to the Earthquake Preparation

According to regular *E*-layer variations before the strong earthquakes there is too low information. Probably the only case described in the literature is the *E*-layer critical frequency increase before the Tashkent earthquake of 1966 (Gokhberg et al. 1984). This example is presented in Fig. 2.17.

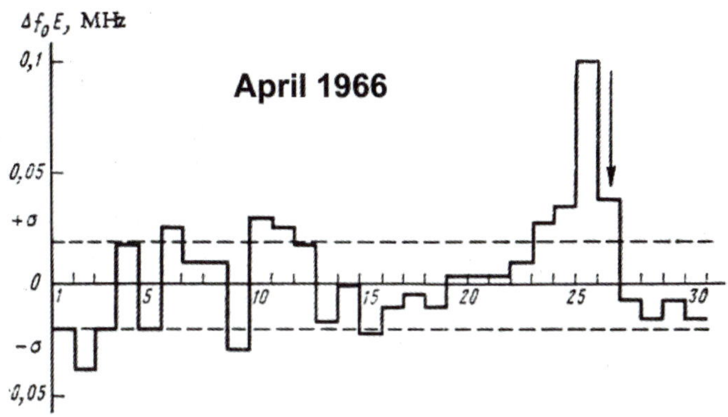

Fig. 2.17 Deviation of the *E*-layer critical frequency before the Tashkent earthquake on April 26th, 1966, M=5.3. The earthquake moment is indicated by the *arrow*

But the main feature marked in the E-layer is the appearance of the strong sporadic E-layers over the impending earthquake area several days before the earthquake. This effect was observed practically in all zones of the seismic activity and was reported by many authors (Ondoh and Hayakawa 1999; Liperovsky et al. 2000; Pulinets et al. 2000; Chuo et al. 2002). The most systematic approach to the study of seismically induced sporadic E-layers could be associated with the Liperovsky group who published recently their extended review (Liperovsky et al. 2000). But their publications, bringing them all together, leave a controversial impression. Along with statistically grounded results on the ionospheric precursors in the E_s-layer (Liperovsky et al. 2000; Silina et al. 2001) the same authors claim doubts over the possibility of using the E_s -layer parameter variations before the earthquake for short-term earthquake prediction (Liperovskaya et al. 2000).

To avoid solar effect interference, the authors use only the nighttime measurements. Their results are limited to the area of Middle Asia (Tashkent, Dushanbe and Alma-Ata ionospheric stations). So, these statistical results have limited meaning keeping in mind different ionosphere behavior in different regions of the globe. Another reason that Middle Asia is not a typical area for seismically active areas, it is an intercontinental area while the majority of earthquakes happen in the near shore subduction zones. As precursory effects the authors use the short-time (15–30 mm) drops of sporadic E-layer blanketing frequency f_bE_s, which characterizes the screening properties of the E-layer (the reflections from higher altitudes are not obtained in the frequency band lower than f_bE_s), and introduced by the authors semitransparent of the E_s, which is expressed in the form:

$$\Delta f_bE_s = (f_oE_s - f_bE_s)/f_bE_s \tag{2.2}$$

where f_oE_s – the critical frequency of the sporadic E-layer.

Liperovsky et al. (2000) propose as the main precursory features in E_s, the short time variations of the blanketing frequency observed during a few nights before the seismic shock, a steady decrease of f_bE_s 1–2 nights before the earthquake, and the corresponding increase (looking at formula 2.2) of the f_bE_s.

We regard the more physical approach is connected with the tracking of the E_s -layer critical frequency f_oE_s variations before earthquakes as a real plasma characteristic of the sporadic E-layer. As an attendant measurement to the VLF emission studies before the earthquake, the precursory f_oE_s variations were demonstrated by Parrot et al. (1985) and Parrot and Mogilevsky (1989). Passing to more recent results, the sharp increase of f_oE_s two days before the Kobe earthquake registered at the Kokobunji ionospheric station is demonstrated in Fig. 2.18 (Ondoh 2000).

Chuo et al. (2002) made statistical studies of the variations of f_oE_s before Taiwan earthquakes with a magnitude higher than 6, and discovered that the most often sharp variations of f_oE_s are observed during terminator time within the interval 5 days before the seismic shock. This is illustrated in Fig. 2.19, where a comparison of daily variations in pre-seismic periods (left panel) is made with the corresponding medians for the same interval of time (right panel).

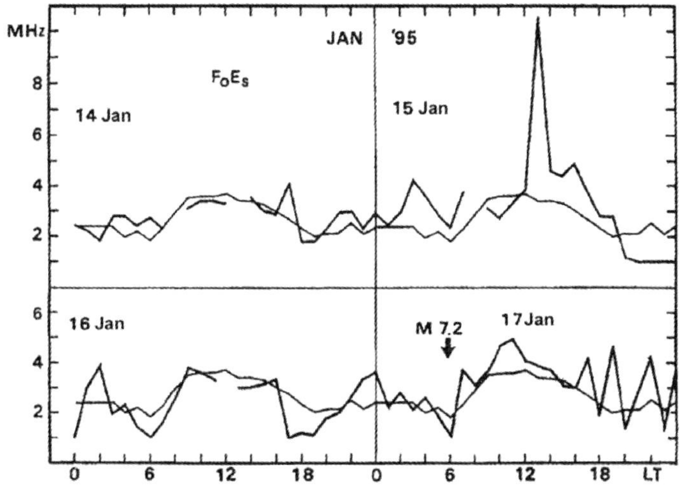

Fig. 2.18 The variations of f_oE_s, from January 14th to January 17th 1995 registered at Kokobunji ionospheric station. *Thin line* – monthly median, *thick line* – raw data of f_oE_s, The Kobe earthquake (M=7.8) moment is indicated by the *row*

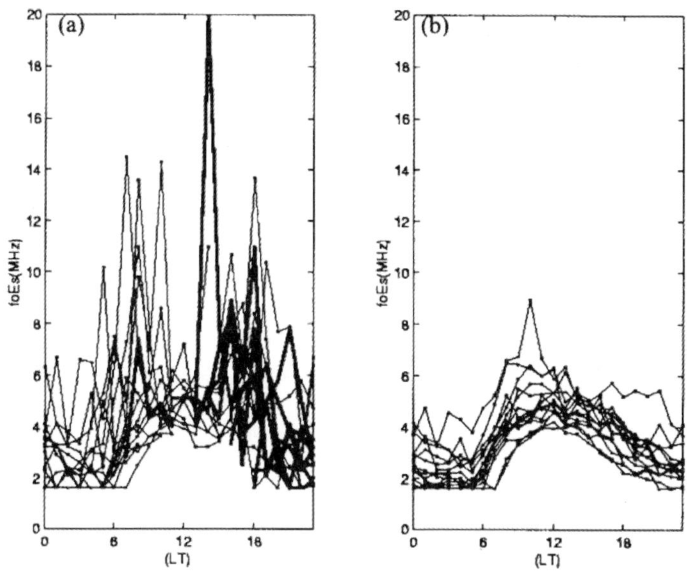

Fig. 2.19 *Left panel* – superimposed daily variations of f_oE_s for disturbed intervals before strong earthquakes in Taiwan (List of earthquakes given in Table 1, Liu et al. 2002). *Right panel* – superimposed medians corresponding to the daily variations presented in the *left* panel

The most interesting conclusion from the Chuo et al. (2002) paper is that the terminator time increase of the sporadic *E*-layer critical frequency automatically leads to the increase of the electron concentration in the *D*-layer leading to the effect of terminator time variations in VLF propagation signals observed by Molchanov and Hayakawa (1998b).

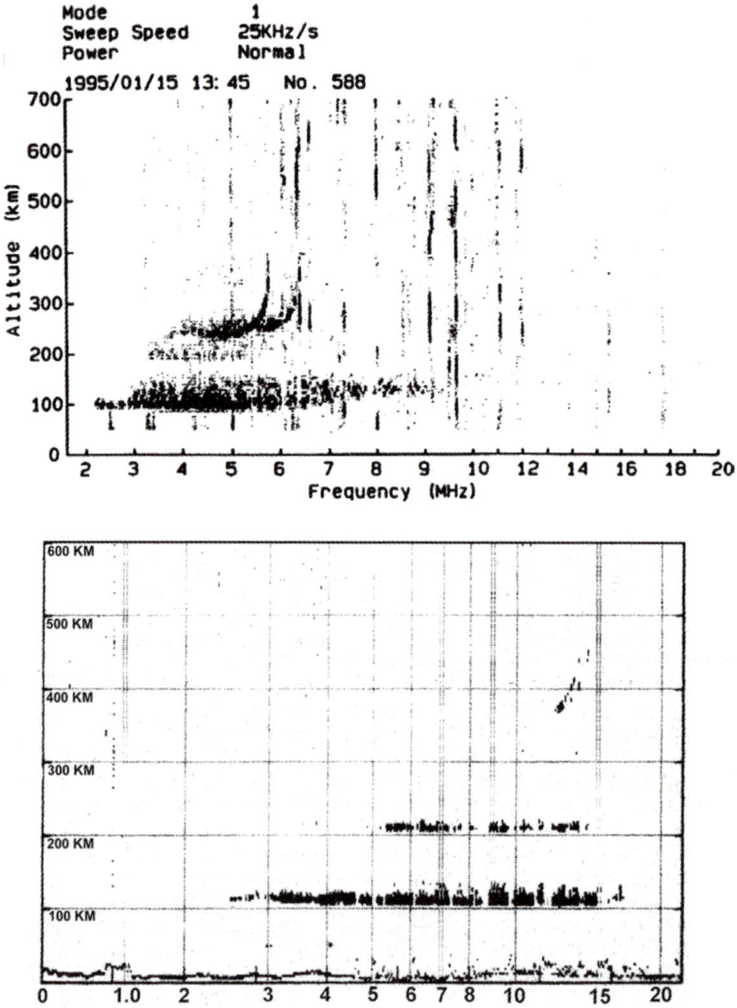

Fig. 2.20 *Top panel,* the example of a sporadic *E*-layer registered at Shigaraki ionospheric station 2 days before the Kobe earthquake 17 Jan 1995 (*M*=7.8) (after Ondoh 2000). *Lower panel,* the example of the sporadic *E*-layer registered at the Chung-Li ionospheric station 3 days before the Rei-Li earthquake on 17 Jul 1998 (*M*=6.2), Taiwan (after Pulinets et al. 2000)

Ionogram examples of the strong sporadic *E*-layer observed before the earthquake are presented in Fig. 2.20. The upper panel shows the ionogram of the Shigaraki ionospheric station at 1345 JST when the maximum development of sporadic *E*-layer was observed (Fig. 2.18). An even more pronounced effect was registered before the earthquake in Taiwan on the 17th July 1998, *M*=6.2 (lower panel of Fig. 2.20). The critical frequency of the sporadic *E*-layer exceeded the *foF2* and reached 17 MHz. In addition, the *E*-layer was not semitransparent as in the upper example and screened practically the whole *F*-layer up to 12.5 MHz. Only the very last portion of the *F*-layer traces can be observed on the ionogram.

Summarizing the effects in the ionospheric *E*-layer we can state that regardless of the large number of publications it is difficult to separate the definite regularities of the observed variations, except that before some strong earthquakes the intensive sporadic layers are observed within the few days before the earthquake. Most promising is the sporadic *E*-layer critical frequency increase during terminator time within the interval of 5 days before the earthquake (Chuo et al. 2002) as a precursory event having a stable time scale (5 days) and possibly related with the other precursors – terminator time effect in the *D*-layer.

2.5 The *F*-Layer, Upper Ionosphere and Magnetosphere

The ionospheric precursors in the peak of the *F*-layer were demonstrated in Figs. 2.3–2.13. We can conclude from the figured examples that they appear as unusual deviations from the normal daily variations of the electron concentration and for the strong earthquakes can occupy the significant region of the ionosphere over the imminent earthquake epicenter projection onto the ionosphere. The modified region can extend up to several thousand kilometers in diameter. In high and middle latitudes, when the geomagnetic field line has a significant vertical component which facilitates the electric field penetration into the ionosphere, the modified region looks like isolated irregularity but in the equatorial region where the geomagnetic field lines are horizontal, the situation is more complex.

2.5.1 Ionospheric Precursors in Equatorial Latitudes

The equatorial anomaly described in Chapter 1 is a specific belt of electron concentration irregularity in the shape of a crest appearing on both sides of the geomagnetic equator electrodynamically driven by the crossed geomagnetic and electric fields. The east directed electric field responsible for its formation appears in the ionosphere during the afternoon hours, which causes the equatorial anomaly development. If the earthquake preparation area lies between the crests of the equatorial anomaly close to the geomagnetic equator the anomalous electric field will interact with the eastward electric field, which will result in the equatorial anomaly shape distortions described in Pulinets and Legen'ka (2002). The most often observed modification is expressed in the form of the crests disappearance before an earthquake. What is interesting is that the anomaly reacts as a whole structure regardless of the epicenter position, what this means is that both crests are subjected to modification.

This is demonstrated in Fig. 2.21 where the variations of the critical frequency over equatorial anomaly are shown using the topside sounding data from the Intercosmos-19 satellite. Every curve is the satellite pass close to the longitude of the epicenter at the same local time within the interval from 4 to 1 days before the shock *M*=5.2, which occurred on 24.05 1980 at 0824 UT at the point with coordinates 3.6 N and 126.5 E. The shape of the equatorial anomaly drawn by the satellite is the similar for all passes except one (bold black triangles) demonstrating the absence of the equatorial anomaly crests.

The dynamics of the crests in time was tracked by continuous observations of TEC by the network of GPS receivers in Taiwan at the same longitudinal sector as presented in Fig. 2.21 satellite data (Liu et al. 2001, 2002). The TEC data show that during 1–4 days prior to the three strong Taiwan earthquake onsets, the TEC values decrease and the anomaly crests move toward the equator.

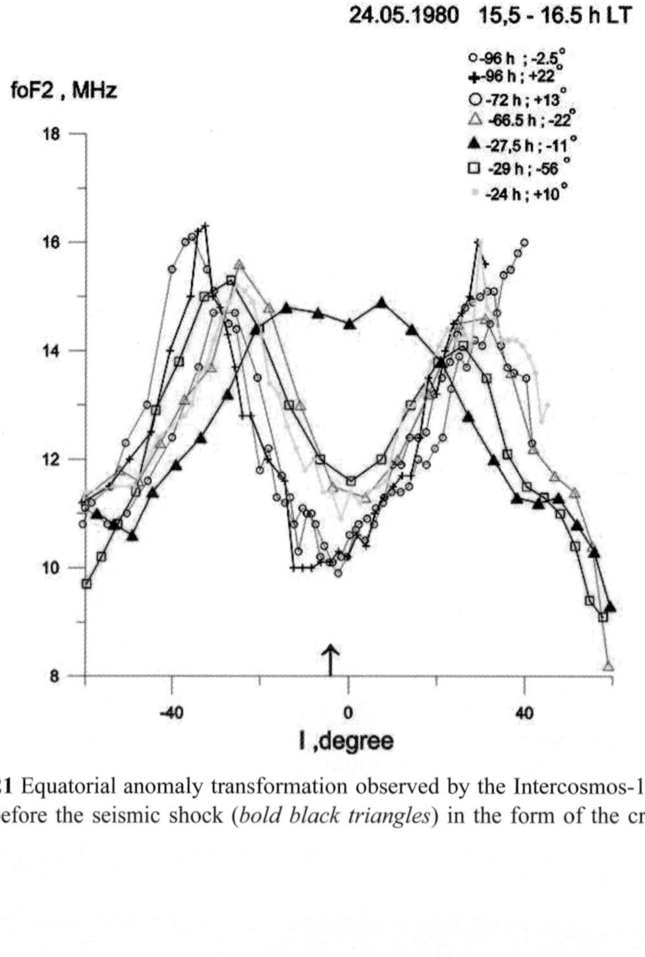

Fig. 2.21 Equatorial anomaly transformation observed by the Intercosmos-19 satellite 27.5 hours before the seismic shock (*bold black triangles*) in the form of the crests disappearance

2.5.2 Vertical Re-Distribution of Electron Concentration Before Earthquakes

Up to now we looked at the precursors on different fixed levels of the ionosphere: *D*-layer, *E*-layer, peak of the *F*-layer. It is interesting to monitor the changes in electron concentration with altitude. The vertical ionospheric sounding gives us such opportunity. Depuev and Zelenova (1996) analyzed the vertical electron profile dynamics using the data of ground based ionospheric stations. They found the very strong difference between the IRI-CCIR prediction (even using the real peak parameters of the profiles) and experimental vertical profiles registered one day before the two shocks of M=5.4 and M=7.2 respectively following with the interval of 4 hours one from another on the 5th of September in 1971 (6th September in LT). The second earthquake is known as the Moneron earthquake (Zelenova and Legen'ka 1989). The main difference in the profile shape is an increase of the scale height (in the case of parabolic approximation of the *F*-layer shape it is the

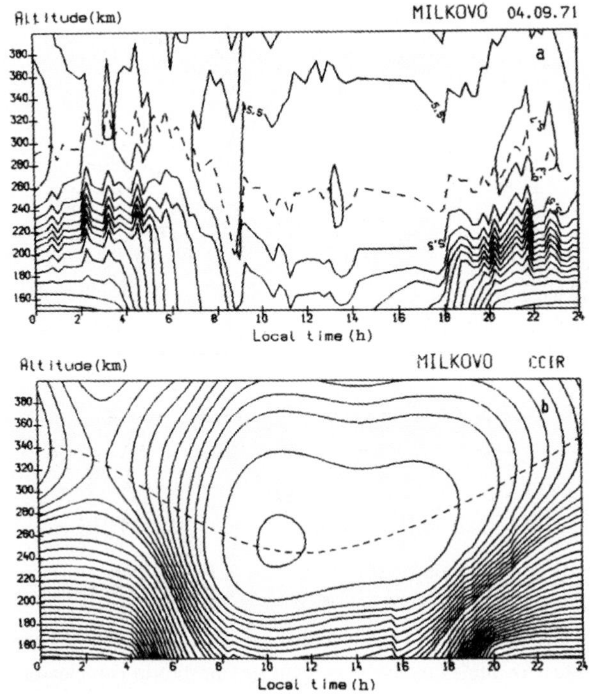

semithickness of the parabola).

Fig. 2.22 Comparison of daily dynamics of the vertical distribution of electron concentration one day before two earthquakes M=5.4 and M=7.2 respectively, following with an interval of 4 hours one from another. *Top panel* – experimental profiles, *bottom panel* – IRI model

It is well seen in the Fig. 2.22 where the vertical profiles are presented by the isolines of the reflection frequency during the 4th of September 1971. The top panel is the experimental data, and the bottom one – the model profiles using the real peak parameters. The second result is the appearance of periodic oscillations during nighttime. They will be discussed in the next paragraph.

The results of ground based sounding well comply with the results of topside sounding which also shows the increase of the topside profile scale height while the satellite was passing over the area of the strong earthquake preparation one to a few days before the seismic shock (Pulinets and Legen'ka 2002; Pulinets et al. 2003b). Figure 2.23 demonstrates the undisturbed (squares) and pre-earthquake (triangles) topside profiles for the case of peak density distribution shown in Fig. 2.8. One can observe also the peak height changes on the profile, which is the case also for the ground-based results of Depuev and Zelenova (1996).

Due to high field aligned conductivity of the ionosphere the anomalous electric field creating irregularities in the *F*-layer of the ionosphere easily penetrates on the high levels and is able to modify the electron concentration also in the magnetosphere. The correspondent calculations were made by Kim and Hegai (1997) and will be discussed in Chapter 4.

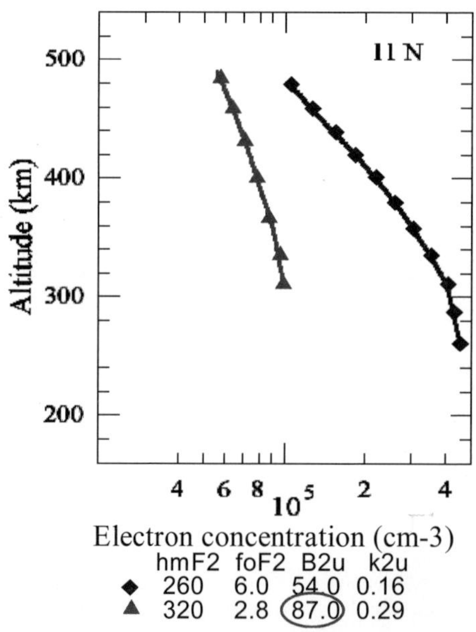

Fig. 2.23 Examples of the vertical profiles of electron concentration modification before the strong earthquakes. *Triangles* indicate the modified profile in comparison with the undisturbed one (*squared*). The modified profiles are characterized by increased value of the $B2_u$ parameter which is the profile semithickness according to Depuev and Pulinets (2001) model

2.5.3 Short-Period Variations in the Ionosphere Before Earthquakes

So far we have talked about the large-scale irregularities appearing in the ionosphere within the interval 1–5 days before the earthquakes. Figure 2.22 clearly indicates the existence of the short-period variations within the ionosphere in the time interval preceding the earthquake. Zelenova and Legen'ka (1989) made the spectral analysis of the critical frequency deviation Δf_oF2 for the period of several days preceding the Moneron earthquake (Fig. 2.24)

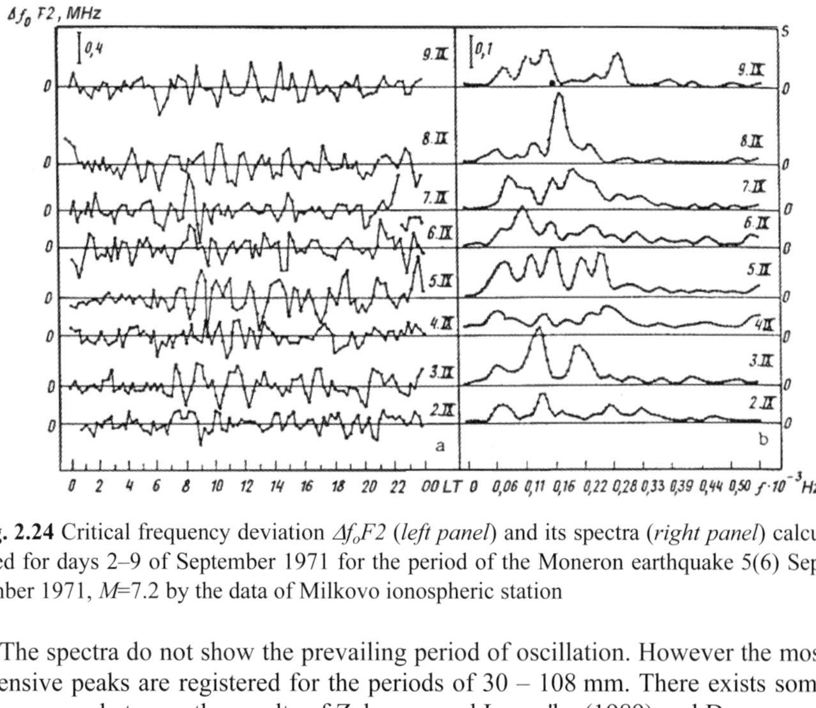

Fig. 2.24 Critical frequency deviation Δf_oF2 (*left panel*) and its spectra (*right panel*) calculated for days 2–9 of September 1971 for the period of the Moneron earthquake 5(6) September 1971, M=7.2 by the data of Milkovo ionospheric station

The spectra do not show the prevailing period of oscillation. However the most intensive peaks are registered for the periods of 30 – 108 mm. There exists some discrepancy between the results of Zelenova and Legen'ka (1989) and Depuev and Zelenova (1996). If the first authors claim more intensive variations during the daytime, the second authors observe more intensive variations during the nighttime.

2.5.4 *In-situ* Satellite Observations of the Ionospheric Precursors

Local probes installed onboard artificial satellites are able to register the variations of near-earth plasma parameters associated with the preparing earthquakes. Probably the first reports on the satellite measurements of ionospheric precursors of earthquakes could be attributed to the beginning of the 80th (Migulin et al. 1982; Larkina et al. 1983; Gokhberg et al. 1983a, 1983b). Following we attribute the list of the satellites where ionospheric precursors were registered: Alouette-1 – 1964,

Cosmos-274 – 1969, OGO-6 – 1969, OVI-17 – 1969,ISIS-2– 1971, AE-C – 1973, GEOS-1 and 2 – 1978, Intercosmos-19 – 1979, Aureol-3 – 1981, Intercosmos-Bulgaria 1300 – 1982, Salyut-7 – 1985, Meteor-3 – 1986, Cosmos 1809–1987, Activny -(Intercosmos-24) -1989, TOPEX/Poseidon – 1996, MIR Space Station, GPS – 1994–2003. Conditionally the precursors observed by satellites can be divided into three groups: electromagnetic fields and noises, plasma parameter variations and particle fluxes. To have at least a general idea of the effects observed by satellites, let us consider them in turn.

2.5.4.1 Electromagnetic Fields and Noises

Probably the first essential break-through in the area of short-term earthquake precursors was the discovery of abnormal splashes of VLF-ELF-oscillations at passes in middle and low latitudes in wave experiment (0.1–16 kHz) on the artificial satellite "Intercosmos-19" (Migulin et al. 1982; Larkina et al. 1983). Their possible connection with superficial earthquakes (h<60 km) with magnitude M>5.0, taking place basically in a magnetically quiet time was analyzed. The statistical data processing permitted the researchers to reveal the main regularities of the observed emission, which can be formulated briefly as follows (Larkina et al. 1988, 1989):

- with reliability of the order 0.9 it is shown, that above an epicenter of a preparing earthquake at heights of the top ionosphere the increase in intensity (up to 20 dB) of the VLF radiation is observed at crossing by the satellite of the geomagnetic force lines projected on an ionosphere near to an epicenter of a strong earthquake;
- abnormal splashes were registered for tens of minutes – tens of hours before the earthquake;
- the maximal amplitudes of splashes were observed on frequencies f-15 kHz more often, on frequencies up to 140 Hz signals are less intensive. Plasma fluctuations were basically electrostatic (since the relative amplitude of electric components much more exceeded the relative amplitude of the magnetic one);
- the statistically determined sizes of a zone of registration of VLF emissions in relation to an epicenter of an earthquake make ±2 ° in latitude and ±60 ° in longitude (Fig. 2.25).

The retrospective data of OGO-6, satellite magnetometer were analyzed by Gokhberg et al. (1983a, 1983b), and anomalous splashes of ELF emission were detected a few hours before the shock while the satellite passed over the earthquake epicenter in three of the six cases studied.

Parrot and Lefeuvre (1985), Parrot et al. (1985), Parrot and Mogilevsky (1989), Parrot (1994a, 1994b) performed a statistical analysis on the data from the GEOS-2 satellite which is a geostationary satellite as opposed to the low Earth orbit of other satellites investigating the ionosphere. GEOS-2 data as well as the data from the AUREOL-3 low orbit satellite show positive correlations between ELF/VLF electric and magnetic field recordings and the proximity to earthquake epicenters. It was also observed that anomalies could be observed for a long distance along the magnetic meridian passing over the epicenter of the earthquake corresponding roughly to a North-South direction. Only the data recorded between the +45° and –45° latitude were considered, to avoid high levels of natural noise closer to the auroral latitudes.

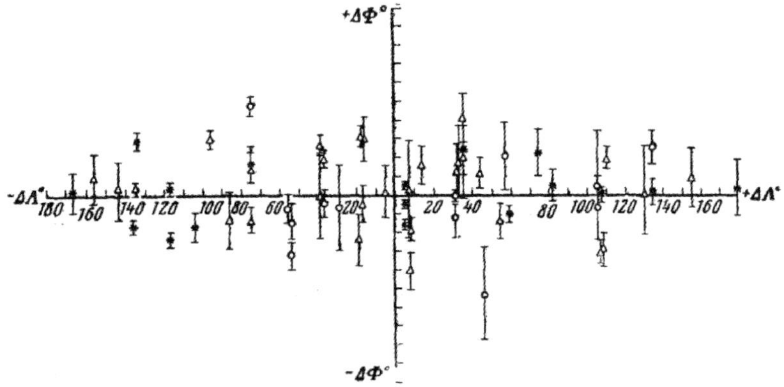

Fig. 2.25 Statistical spatial (latitude-longitude) distribution of the VLF emission (electric component on 4,650 Hz) associated with earthquakes by Intercosmos-19 data

The Russian Intercosmos-24 satellite measured ELF-VLF emissions during its 3,000 orbits in 1989–1991. Molchanov et al. (1993) analyzed data from 180 orbits during November and December of 1989, in which twenty-eight rather strong earthquakes took place (5.2<*Ms*<6.1). Among their main conclusions are:

- ULF-VLF (f<1,000 Hz) and VLF (10 kHz<f<15 kHz) emissions are typically observed as bursts above the earthquake epicenters,
- Only VLF emissions are observed far from the epicenter,
- These emissions may start several days before but emission occurrence probability is at a maximum at 12–24 hours before the earthquake.

The detailed studies of quasi-DC electric fields and ELF emissions were carried out on low orbiting satellites Intercosmos-Bulgaria-1300 and Cosmos 1809 (Chmyrev et al. 1986, 1989; Bilichenko et al. 1990; Serebryakova et al. 1992). At equatorial latitudes the vertical electric field of 3–7 mV/m and magnetic pulsations with a frequency of 1 Hz and amplitude 3 nT were observed at an altitude near 800 km ~15 min before the earthquake of 21 January 1982. The size of a zone of registration makes 40–100 km along a trajectory of the satellite, and the maximum of intensity lies on the *L*-shell corresponding to a projection of an epicenter of earthquake on a height of 100 km of the ionosphere. It is interesting to note that the Aureol-3 satellite flew by above the area of the same earthquake of 21.01.1982 and registered the ULF emissions within the bands 40–100 Hz and

100–200 Hz 4 and 48 hours before the seismic shock (Galperin et al. 1992). We shall note a curious feature in the behavior of fields. While the satellite is inside the geomagnetic field tube of an epicenter of a preparing earthquake or near to it, the relation $E_{\perp} >> E_{\parallel}$, is fulfilled where $E_{\perp}$ and $E_{\parallel}$ accordingly are perpendicular and field aligned (in relation to the geomagnetic field line direction) components quasi-DC electric field (1–10 Hz). When the satellite is out of the tube or far in latitude from the epicenter, $E_{\perp} < E_{\parallel}$. On frequencies at least up to 1 kHz before the earthquake, as a rule, $E_{\perp} >> E_{\parallel}$, and after the earthquake $E_{\perp} << E_{\parallel}$ (Liperovsky et al. 1992).

Similar measurements have been carried out onboard the Cosmos – 1809 satellite, for a series of aftershocks (about 40 pulses of different intensity were registered), of the Spitak earthquake (07.12.1988, *M*=6.7), for the period from 13.01 to 17.02.1989. Electromagnetic radiation of frequencies of 100–400 Hz was observed in a zone ±6° in longitude (please note, as against ±60° for the data of Intercosmos-19), and the border of the area on latitudes is symmetric concerning the L-shell corresponding to an epicenter (Serebryakova et al. 1992).

The absence of new experimental results slightly decreased the number of publications on the electromagnetic emissions registered by satellites over seismically active areas. Nevertheless, one can find the results of retrospective analysis of VLF emissions data (Larkina et al. 2001).

2.5.4.2 Variations of Local Plasma Parameters Before Earthquakes

Much more interesting for our discussion is to review the results of the satellite measurements of variations of the ionospheric plasma parameters before earthquakes. Probably one of the first reports on the local plasma density and temperature variations measured onboard AE-C and ISIS-2 satellites was published by Gokhberg et al. (1983b), and for a long period of time this was the only data published. This direction of research was in the shadow of the boom in the use of satellite VLF emissions. The situation changed after the first results of topside sounding were published (Pulinets et al. 1991) demonstrating the existence of large-scale plasma irregularities over the areas of earthquake preparation. The statistical studies of plasma measurements started to be performed looking at any correlation with seismic activity. Afonin et al. (1999) and Hayakawa et al. (2000) analyzed a larger database of plasma densities recorded in the 3,000 orbits of Intercosmos-24. They report a reliable correlation between the global distribution of seismic activity and ion density variations in the ionosphere, as measured by the normalized standard deviation (NSD) and the relative normalized standard deviation (RNSD). However, a clear correlation was found only during daytime measurements (10–16 hours of local time), quiet magnetic conditions and an altitude range of 500 to 700 km. The correlation disappeared during magnetic storms. The later statistical studies of Afonin et al. (2000) based on the Cosmos-900 data contained the brilliant confirmation of the Pulinets et al. (1991), Pulinets (1998a, 1998b), and Pulinets and Legen'ka (2003) results on the existence of the large scale irregularities in the ionosphere several days or hours before strong earthquakes. Afonin et al. (2000) conclude that dramatic changes in the ion density distribution pattern occur over earthquake epicenters at night (quieter environment) and at 450 to 500 km satellite altitude. More importantly these changes are of precursor type. They start to be observed about 7 days before a strong earthquake and last for about 7 days afterwards. The effects are more pronounced within + or –40 ° of geomagnetic latitude.

Except large-scale irregularities (from hundreds to thousands km) the small-scale irregularities were observed before the strong earthquake onboard the Cosmos 1809 satellite (Chmyrev et al. 1997, 1999). The special scale of the observed irregularities was ≤10 km. They are localized in the geomagnetic force tube loaned on the epicentral zone at the altitude of the E-layer. The characteristic horizontal size of the disturbed region occupied by these irregularities is 300–450 km at the height of the satellite orbit, which was 960 km. The observed irregularities with magnitude $\Delta N_e/N_e \geq 5\%$ are not typical for the midlatitude ionosphere under normal conditions. It should be noted that the irregularities over seismically active areas with the given parameters were accompanied by intense bursts of electromagnetic emission at the frequencies f~140–500 Hz with the amplitude up to 10 pT. An example of the Cosmos 1809 record of the small-scale irregularities over the Spitak aftershock area is presented in Fig. 2.26.

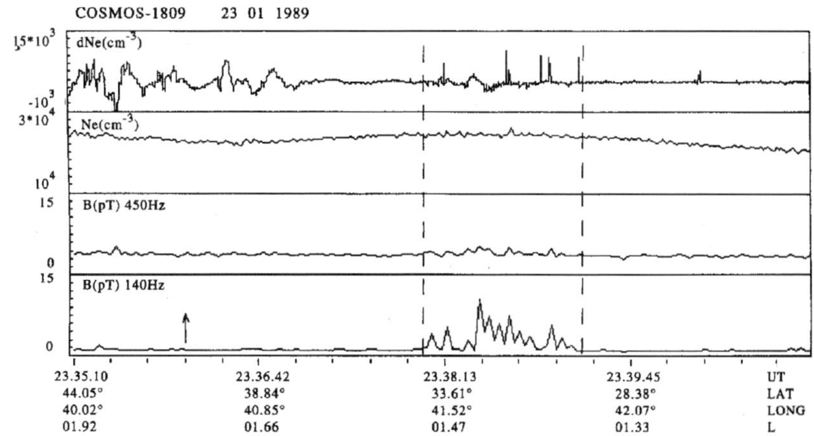

Fig. 2.26 The small scale irregularities (*upper curve*) measured before the Spitak earthquake aftershock onboard the Cosmos 1809 satellite

One of the most intriguing and promising results of local plasma parameter measurements is the changes in the ion composition before earthquakes over the earthquake preparation zone (Boškova et al. 1993, 1994). These changes were observed first onboard the Intercosmos-24 satellite at altitudes between 2,300–2,500 km before the Iranian earthquake of 20 June 1990. They are expressed in the form of increased concentration of the light ions H^+ and He^+ on satellite passes close to the epicenter. Three consecutive passes together with a map demonstrating their location in relation to the epicenter position are demonstrated in

Figs. 2.27 and 2.28 respectively. One can unambiguously see the increase of the H^+ and He^+ concentration along two passes embracing the epicenter position where the most intensive increase is observed on the pass closest to the epicenter.

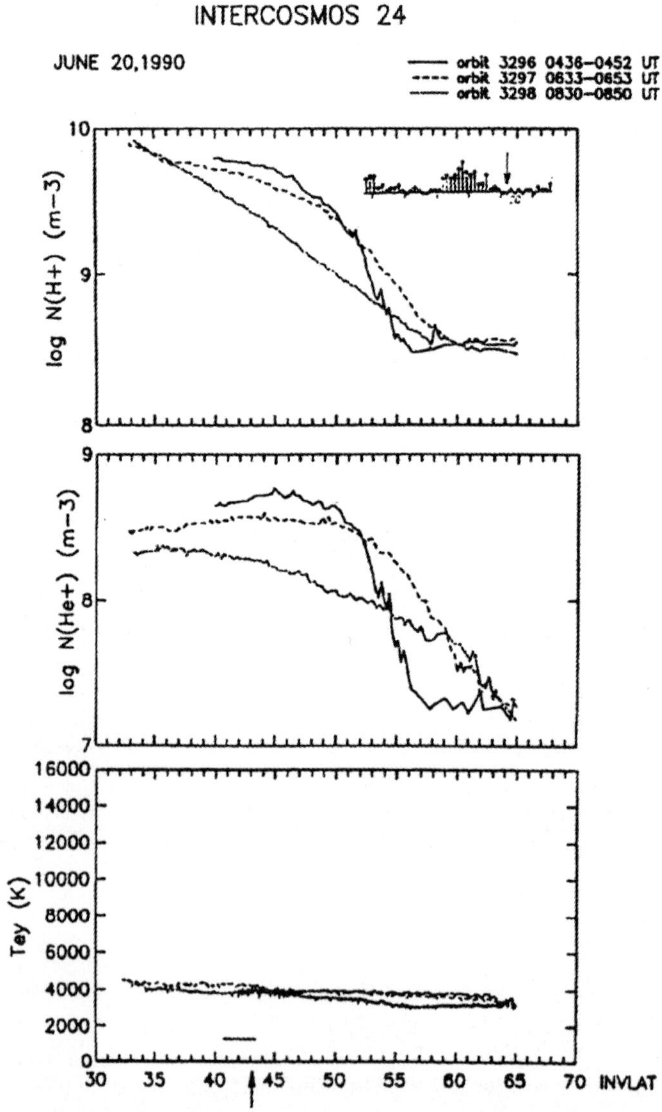

Fig. 2.27 Latitudinal variations of the H^+ (*upper panel*), He^+ (*middle panel*) concentration and electron temperature perpendicular component (*bottom panel*) measured onboard the Intercosmos-24 satellite on 20 June 1990 before the Iranian earthquake M=6.4 which happened on 20.06.1990 at 2100.09 UT

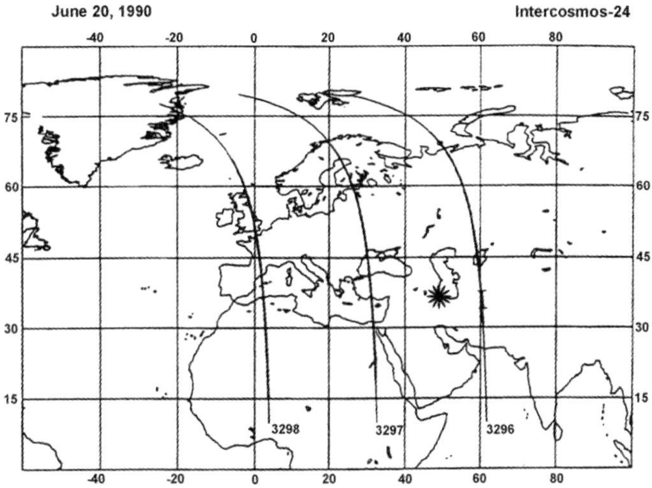

Fig. 2.28 The Intercosmo-24 satellite orbit projections in relation to the Iranian earthquake of 20 June 1990 position marked by *asterisk*

Similar studies of the ion composition but on the lower altitudes were made by Pulinets et al. (2003b) using the AE-C satellite mass spectrometer data. These studies have shown the decrease of the mean ion mass while the satellite passed over the earthquake preparation area, which is equivalent to an increase of light ion concentration. These data will be discussed more carefully in Chapters 4 and 5.

The electron and ion temperature variations over the seismically active zones before earthquakes reported by the data of the AE-C satellite were insignificant (Pulinets et al. 2003b) in comparison with temperature variations observed during magnetic storms. The lower panel of Fig. 2.27 demonstrates the absence of electron temperature variations for the passes where the significant changes of the light ions concentration was observed. These results will be discussed in more detail in Chapter 5.

2.5.4.3 Particle Precipitation's Associated with Earthquakes Registered by the Satellites

As shown above, the VLF and ELF emission bursts of a seismic origin are often registered in the ionosphere and magnetosphere. Besides direct registration of these emissions onboard an artificial satellite additional effects can be registered caused by this emission, and the result of such interactions can be seen not necessarily in the area where it has taken place. This is the well-known particle precipitation from radiation belts stimulated by VLF emission in the magnetosphere (Rycroft 1991).

For the first time indications on the probable interrelation of high energy particle fluxes originating from radiation belts with seismic activity have been received at orbital station Salute–7 (Voronov et at. 1989).

Experiments have been continued onboard the MIR space station, (orbit attitude ~400 km) where 11 more than tenfold increases of the intensity of particles with an energy more than several tens of MeV have been registered, and the major part of the flux (about 80%) consisted of protons with energy E_p >30 Mev. In 8 cases (when there were no geomagnetic disturbances) 2.5–3 hours after each splash in particle intensity an earthquake with M>4 was registered (Voronov et al. 1990; Aleshina et al. 1992). Such protons trapped by a geomagnetic field can undergo the resonant interaction with ELF-waves (in the frequency band of 1–10 Hz).

On an artificial satellite Meteor–3 (height of 1,250 km) electrons with energy E_e>30 MeV and protons with E_p>400 MeV were registered. Thirty-six sharp fluctuations of count rates have been fixed (over five standard deviations) with durations of 1–8 minutes, 34 have taken place in the absence of appreciable geomagnetic disturbances. They were grouped on L=1.1–1.25. About 70% of increases in count rate were registered in conditions of increased global seismicity (above the daily particle precipitation's associated with earthquakes registered by the satellites average number of earthquakes calculated for the period of 2 years). In 70% of records in which the count rate remained on an average level, global seismicity was lower than average. Data processing shows, that the occurrence of a deviation of count rate precedes earthquakes with M>4 approximately by 2–3 hours.

On the Intercosmos-Bulgaria-1300 satellite in the same range of energy 11 cases of deviations of the count rate of high energy trapped particles were also revealed, and in 5 cases within the interval 2.5–3 hours after the count rate increase was fixed earthquakes with M>4 occurred (Galper et al. 1989, 1995). Aleshina et al. (1993) and Pustovetov and Matyshev (1993) specified the presence of a spatial interrelation between coordinates of the earthquakes epicenters and areas of the charged particles precipitation records.

On the Aureol-3 satellite from 20 orbits, for which L=1.25±0.1 increases of fluxes of electrons with energy E_e>100 keV have been registered, in 18 cases on L<1.3 in a longitudinal interval ±600 ° earthquakes with M>4 took place (Galperin et al. 1992).

Asatryan et al. (1991) reported the registration just before the Spitak earthquake of 07.12.1988 the flux of relativistic electrons with energy not less than several MeV. This was registered by a vertical telescope of cosmic rays and the single counter onboard a balloon launched near to Yerevan City. Forty-one minutes prior to the main shock, during the rise of the balloon, starting from the depth of 500 g/sm^2, an increase in the flux of the penetrating particles, much exceeding statistical errors was registered.

Simultaneously with the light ion concentration increase onboard the Intercosmos-24 satellite described above fluxes of electrons with energy E_e=1.7–277 keV were registered (Boškova et al. 1994). The fluxes of energetic particle precipitation associated with an earthquake were registered simultaneously with the seismically induced ELF emissions onboard the Intercosmos-19 satellite (Ruzhin and

Larkina 1996). An example of such registration is presented in Fig. 2.29. The particle burst duration was of the order of 30 seconds.

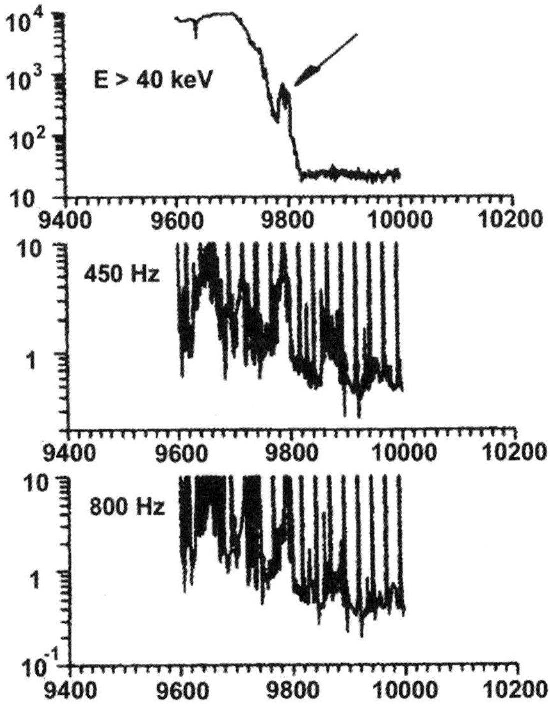

Fig. 2.29 An example of the seismically induced particle precipitation (E>40 keV) indicated by row (*top panel*), registered onboard the Intercosmos-19 satellite on 29th of March 1979 together with ELF emission bursts at 450 Hz (*middle panel*) and 800 Hz (*bottom panel*). Horizontal axis time in seconds

2.6 Optical Emissions Associated with Earthquakes

Optical emissions registered before earthquakes at different heights of the atmosphere are clearly different from the other kinds of precursors described above, and at the same time have a close and direct relation with them. Conditionally, they can be divided in two groups: those in the troposphere, named "earthquake lights" and probably associated with the anomalous electric field, and the other group observed at altitudes from 100 to 250 km and associated with the ionospheric plasma modification before earthquakes.

2.6.1 Earthquake Lights

This mysterious phenomenon has a centuries-old history but up to now having no clear and unequivocal explanation. The reason is that the earthquake lights are a transient phenomenon happening usually a few minutes before the seismic shock, they are described mainly by spectators without any registration permitting us to make an instrumental estimation of its parameters. Well-documented researches of the earthquake lights were made in Japan in the 30th (Terada 1931; Musya 1931, 1932, 1934) and in the 60–70th (Yasui 1968, 1971, 1972). Taking into account that it is difficult to have access to the cited publications, we provide the illustration from the work by Yasiu (1968) in Fig. 2.30. These investigations were reviewed by Derr (1973) and documented by photos from the cited papers. Papadopoulus (1999) tried to summarize information on the luminous and fairy earthquake related phenomena in the Eastern Mediterranean from ancient times up to now. From 30 carefully selected cases he was able make some conclusions on the morphological features of the earthquake lights. Summarizing the Derr (1973) and Papadopoulus (1999) results we can conclude:

- The luminous phenomena are observed in the epicentral area of the impending earthquake with the maximum registered distance of observation of 3 °–4 ° from Derr (1973) and 140 km from Papadopoulus (1999)
- The duration of appearance of luminous phenomena is very short, of the order 10 s–2 min from Derr (1973) and 2–3 s from Papadopoulus (1999)
- In most cases the lights are observed a few seconds (minutes) before the earthquake or during the earthquake. A few cases were reported with the observation 5–6 days prior to the earthquake
- In most cases the lights were observed during nighttime
- The shapes of lights reported are very different: flush of lighting, fire, flames, burning atmosphere, light columns. Where observed the lights are initiated both from the ground and the sky. Sometimes the lights were moving along the active fault
- The earthquakes accompanied by the luminous phenomena were the large destructive events with shallow focus with the epicenter located either on land or at sea. The majority of earthquake lights are seen around mountain summits with a quartz-diorite faulted rock
- The color of the luminous phenomena varies from white to blue and orange

The possible physical explanation of the earthquake lights will be given in Chapter 4.

Fig. 2.30 Examples of earthquake lights registered before the different earthquakes in Japan (after Yasui 1968)

2.6.2 Optical Emissions Registered in the Ionosphere

The appearance of the electric field within the ionosphere leads to initiation of electric current and consequent excitation of the atomic and molecular species of the ionospheric plasma. Most often it was reported on the registration of variations of intensity of two oxygen lines 5,577 Å and 6,300 Å (Fishkova et al. 1985; Liperovsky et al. 1991, 1992; Akhmamedov 1993) before earthquakes. The appreciable intensity changes of the nighttime emissions were found precisely in these lines during seismically active periods. This result is expectable. As we know from the previous discussion, the ionosphere is affected by the seismogenic electric field at the heights of the *E*-region and the heights of the *F*-region. At the same time it is known that the luminescent layer of atomic oxygen OI 5,577 Å (green line) is situated at an altitude 85–110 km, that is, in the ionospheric *E*-region, whereas the layer of emission in the line OI 6,300 Å (red line) is at an altitude 200–300 km, that is, in the ionospheric *F*-region and excitation of both lines is quite natural (Chamberlain 1995; Wayne 2000). It can therefore be expected that photometric observations and the results of vertical sounding of the ionosphere may definitely be related.

Multiyear observations of luminescence were carried out in the former USSR mainly at two stations: at the Abastumani observatory (Georgia) and at Vannovskiy station (Turkmenia). During the period 1974–1982 about 400 earthquakes with a magnitude $M<4$ occurred in a radius ~ 200 km relative to Abastumani. Among these for 181 events there are observations of night airglow in the interval 4 days before and after an earthquake; in addition, for approximately 80 events there are continuous observations in the interval 1 day before and after the earthquake. Excluding days with an increased geomagnetic activity (when the K_p index was >4) and using the superposition of epochs method it was possible to obtain the following results (Fishkova et al. 1985): 3 days prior to an earthquake the intensity of the 5,577 Å and 6,300 Å emissions exceeds the mean seasonal value by approximately 16%; 2 days prior to an event the intensity of 6,300 Å is reduced, remaining high, but the 5,577 Å intensity within the limits of probable error changes relative to the mean seasonal value. A day prior to an earthquake the intensity of the 5,577 Å emission begins to increase, whereas the intensity of the 6,300 Å emission decreases. With approach to the time of an earthquake the intensity of the 5,577 Å emission approximately 6 hours before an event increases on the average up to 7% relative to the mean seasonal value. For the 6,300 Å emission the intensity begins to decrease after approximately 24 hours before and attains a minimum 4–6 hours prior to an earthquake; on the average the decrease is 15% relative to the mean seasonal value. During the first hours after an earthquake it is possible to note an approximately 7% increase in the intensities of the 6,300 Å and 5,577 Å emissions. During the several days which follow the earthquake for the 6,300 Å emission there is a stable decrease (~20%) in intensity, but for 5,577 Å there is a relaxation to the mean seasonal value.

A somewhat different picture is observed for earthquakes with a magnitude $M>4$. Photometric measurements with a duration not less than 3 consequent nights were available for six such events at Abastumani station (Fishkova et al. 1985).

The distance to the epicenters from Abastumani varied in the range from 55 to 1,250 km. It was found that for near strong earthquakes the effects were considerably more impressive. For example, for the earthquake of 24.01.1982 (M=4.1, epicentral distance R=55 km, hypocentral depth h=10 km) the intensity of the 5,577 Å emission at the zenith 5–6 hours prior to the event was greater by a factor 2.5 than the mean seasonal value, whereas the intensity of the 6,300 Å emission 2 hours prior to the event was lower by a factor of 4. In addition, as a rule, for strong earthquakes it is possible to note an increased intensity of the 6,300 Å emission on the night following the earthquake.

A comparison of glow observational data for Ashkhabad (Vannovskiy station) and Abastumani for earthquakes of the Central Asian zone with M>5 (16 events) during the period 1961–1977 led to the conclusion that there is an increase (up to 40%) relative to the mean annual intensity of the 6,300 Å emission the day before an earthquake (Nasyrov 1978). According to Nasyrov (1978), the region of the atmosphere where the influence of impending earthquakes was expressed on the behavior of the 6,300 Å line extends from Turkmenia to Georgia and is ~1,000 km in radius. This value exceeds estimations of the earthquake preparation zone (Table 1.3) but nevertheless supports the idea that the space effects associated with the processes of the earthquake preparation in the final stage occupy the large area testifying to activation of large areas, not only the closest vicinity to the earthquake source. The intensification of the 5,577 Å emission several days prior to the seismic events was noted as well.

The increase and decrease of the emission intensity may correspond to the increase and decrease of the electron density (positive and negative signs of the deviations discussed above). So the difference of the results for the periods 1961–1977 and 1974–1982 should not be surprising. The relative position of the observation points and the earthquake epicenters also should be taken into account, as well as the absolute distance.

It is necessary to mention the results of a spectral analysis of variations in the intensity of emissions of nighttime radiation carried out for several selected earthquakes: Akhalkalak (January 24, 1982, M=4.1, R=55 km), Yerevan (August 2, 1984, M=3.8, R=220 km), Rumyn (August 31, 1986, M=6.7, R=1,250 km) and Dzhavakhet (October 21, 1987, M=3.5, R=106 km) (Toroshelidze et al. 1988). It was found that spectral energy increase in the entire range of periods (from several hours to 12 minutes) and an increase in the number of spectral density maxima in the range of short periods begins several days before an earthquake; these phenomena are especially active several hours before an event.

The aspect measurements for the variations in the line 6,300 Å were carried out near Ashkhabad for the period of the Iranian earthquake of 20 June 1990 (M=7) (Akhmamedov 1993). The asymmetry was determined: the emission intensity increase interpreted by the author as the temperature increase by 350 K was observed in the North and South directions from the observation point, while in the East and West directions the variations were not so pronounced (70 K and 100–150 K respectively). The oscillations with the period of 30–60 minutes were observed with the magnitude of 50–100 K, and the major increase started one hour before the earthquake with the maximum around the time of the seismic shock.

Thus, the results of observations of the night airglow of seismically active regions makes it possible to draw the following conclusions: (1) the glow in the line 5,577 Å several days before an earthquake is characterized by increased variability; (2) at the same time there is a stable (on the average) increase in 5,577 Å intensity, attaining a maximum several hours prior to an earthquake; (3) several days before an earthquake there is an increased intensity of the 6,300 Å emission, which a day prior to the event is replaced by a reduced intensity and attains a maximum several hours before an earthquake.

Examining the main features of the optical emissions mentioned above, it is possible to define two characteristic time scales for active earthquake preparation: several hours and several days which probably reflects the different stages of the earthquake preparatory process.

The spatial picture is more complex because depending on the place where the observation station is located the very same method at first glance may give contradictory results. But this result completely coincides with the ionosonde results: the combination of the relative positions of the observation point to the earthquake epicenter, as well as the changing of the anomalous electric field direction gives the changes of the parameter variation sign.

2.7 Conclusion

The aim of the present chapter is to give the reader the chance to feel the flavor of the ionospheric precursors, the different techniques of their visualization and possible ways to explain their interrelations. Generally speaking, we deal with the local modification of the near-earth plasma by the anomalous electric field penetrating into plasma from the ground level. The locality of the modification is determined by the anomalous electric field source (earthquake epicenter) position, the spatial size is connected with the impending earthquake intensity, and the time scale is determined by the development of the final stage of the earthquake preparation. Taking into account the possibility of the electric field penetration into the magnetosphere we can speak of the plasmaspheric tube modification, when the tube is loaned on the area of the earthquake preparation. The modified tube concept easily permits us to explain the trapping of the VLF emission, stimulated particle precipitation, large scale and small scale plasma density perturbations on the lower levels (ionospheric heights). Now is the time to place the demonstrated experimental results into the physical background. The next two chapters will be devoted to this purpose. We will review the possible mechanisms of the anomalous electric field generation, the process of its penetration into the ionosphere and magnetosphere and the effects of its action on the space plasma. The alternative mechanisms, not regarding the electric field as a source of the observed variations will be considered as well.

Chapter 3 Near Ground and Troposphere Plasmachemistry and Electric Field

3.1 Introduction: Electric Field and Radioactivity

The anomalous electric field as an earthquake precursor and radon emanation always existed together in the list of short-term physical precursors (Scholz et al. 1973) but they were never tied together to understand the ionospheric effects. Of course, one can find a lot of publications claiming that radon emanation changes the air conductivity, and consequently, it should influence the atmosphere electric field, but the chain radon-electric field-ionosphere was never regarded directly. The first ideas appeared when we found the correlation between the radon concentration and variations of the critical frequency around the time of the Nazarbek earthquake on the 11th December 1980 (M=5.1, 41.3 N, 69.2 S), near Tashkent (Pulinets et al. 1997). This correlation was obtained only for the values of the critical frequency taken for specific intervals of local time, which implied the local time dependence of the observed effect.

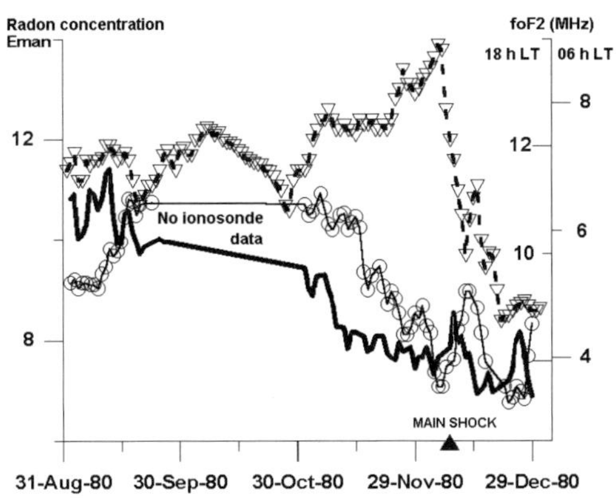

Fig. 3.1 Correlation between radon emanation in the well near future epicenter (*triangles*) and 5-day running mean value of ionosphere critical frequency f_oF2 scaled for 06 LT (*bold line*) and 18 LT (*circles*) for Tashkent ionospheric station. The earthquake time is shown on the figure by the *bold triangle*

One can see in Fig. 3.1 variations of radon concentration (triangles) measured in the well not far from the earthquake epicenter. Two other curves demonstrate smoothed (running average) variations of critical frequency f_oF2 scaled during several months for two selected local time moments (0600 LT – thick line, and 1800 LT – thin line with circles). Anticorrelation of two processes is evident: radon concentration growth (with a sharp maximum a few days before the shock) while concentration in the morning and afternoon hours gradually falls with a sharp oscillation (minimum before the shock and maximum after) in the vicinity of the seismic shock.

This case suggested we start looking at the ionospheric effects of the radioactive pollution because of the similarity of processes. And such a case was found for the Chernobyl atomic plant catastrophe. Fux and Shubova (1995) observed the anomaly of VLF signal propagation (16 kHz) registered on a radio pass Rugby (Great Britain) – Kharkov (Ukraine) during the Chernobyl atomic plant catastrophe. Their record is shown in Fig. 3.2. In the upper panel the regular amplitude and phase daily variations are shown for quiet geomagnetic conditions. In the middle panel one can observe anomalous variations after the sharp increase of reactor temperature was reported (even before the explosion). These variations were of the same kind up to the reactor plying. And the lowest panel shows recovery to the normal variations when emanation of radioactive dust into the atmosphere was stopped. So, one can note that the radioactive dust strongly changes the electromagnetic properties of the waveguide "earth-ionosphere". Very similar variations of VLF signal phase can be observed in registrations of signals passing over the regions of preparing earthquakes (Gufeld et al. 1992). These two experimental results forced us to look more carefully at the role of natural radioactivity in atmospheric electricity.

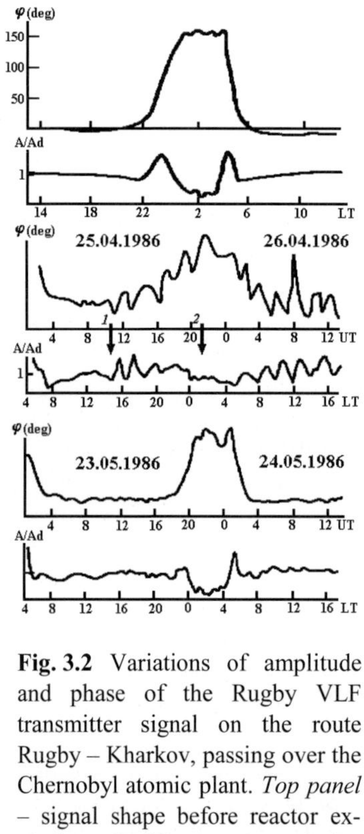

Fig. 3.2 Variations of amplitude and phase of the Rugby VLF transmitter signal on the route Rugby – Kharkov, passing over the Chernobyl atomic plant. *Top panel* – signal shape before reactor explosion. *Middle panel* – during emanation of radioactive substances. *Bottom panel* – after the reactor plying

Regardless of the existing publications on the electric anomalies around the time of the earthquake (Sadovski 1982; Sidorin 1992; Gokhberg et al. 1995; Rulenko 2000) their nature was unclear. In the present chapter we will try to fill up this gap and develop the physical mechanism of anomalous electric field variations before an earthquake. We will review the results of the electric field measurements in the boundary layer, will discuss the nature of the electrode effect and then will look at the microphysics of natural radioactivity interaction with molecules of atmospheric gases and the effects of this interaction.

Known measurements in seismoactive regions mainly consisted of observation of the vertical electric field *E* intensity. According to Gokhberg et al. (1995) A. Humboldt was the pioneer in studying the atmospheric electricity connection with earthquakes, observing in the middle of the nineteenth century with the help of an electroscope the “fast changes” of electric potential of air during earthquakes. Now the anomalous variations of *E* before earthquakes are registered in Central Asia (Cherniavski 1936, 1955; Bonchkovsky 1954; Cerfas 1971; Imankulov et al. 1988, 1990), at Northern Caucasus (Vorobiov and Remizov 1984) and in Transcaucasia (Matveyev et al. 1991; Morgunov and Matveyev 1991), at Kamchatka Peninsula (Rulenko et al. 1992, 1996; Vershinin et al. 1999) and Kuriles (Morgunov 1998), in China (Hao 1988; Jianguo 1989; Hao et al. 2000), in Japan (Kondo 1968), California (Bufe and Nanevicz 1976), in Europe (Nikiforova and Michnowski 1995). The most comprehensive review of the given phenomenon is submitted in the paper of Rulenko (2000). Besides the electric field *E* intensity observations, some cases of measurements in seismoactive areas of density of a vertical electric current j and specific electric conductivity of air σ are known (Rulenko 2000). Examples of the anomalous electric field registration before earthquakes are given in Fig. 1.10.

In all these observations the presence of a well defined connection between local deformation processes of rocks and anomalous changes of electric characteristics of near ground air is revealed. Considering the mechanisms of formation of *E* anomalies it is necessary to take into account that measurements of *E* were carried out at height 0 ÷ 4-5 m above a surface of the ground. In conditions of fair weather in this layer under action of the electrode effect there is a positive volumetric electric charge that provides positive and higher, than in the free atmosphere, values of *E*. Despite the action of the electrode effect, in near ground air in fair weather the negative volumetric charge is registered sometimes which in part or completely compensates the action of a positive charge at some heights. Its origin can be connected with natural radioactive emission of the ground. Using the model of the classical electrode effect (Hoppel 1962, 1967) it is theoretically shown, that the high intensity of ion production in air near the ground surface results in the occurrence of a negative volumetric charge due to upward outflow of negative ions from the area of increased ionization under the action of an electric field. Thus, there can be a situation when the field does not have the time to carry away all negative ions to the higher altitudes and they collect in some layer of air. The given phenomenon is named electrode effect reversion.

The basic natural ionizer of near ground air is radon (Roffman 1972). Having a large atomic mass, radon concentrates in the near surface layer of friable rock formations. Before earthquakes when the powerful process of deformation of rocks is activated, the concentration of radon in over surface gas increases what occurs most strongly at the final stage of earthquake preparation (Nishimura and Katsura 1990) and as a consequence, increases the density of the radon flux in the atmosphere. Radon ionizing action on the planetary boundary layer determines ion physics and the electricity balance in this layer. Small atmospheric ions, existing by virtue of a balance between ionization of the neutral gas and recombination and attachment to aerosol particles, cause the conductivity of the air. Yet, many facets of the nature and behavior of these particles are still poorly understood (Hoppel et al. 1986). One can find a very rare example of simultaneous measurements of radon daughter positive and negative ions concentration (upper panel), atmospheric electric field (middle panel) and radon concentration (lower panel) in Fig. 3.3.

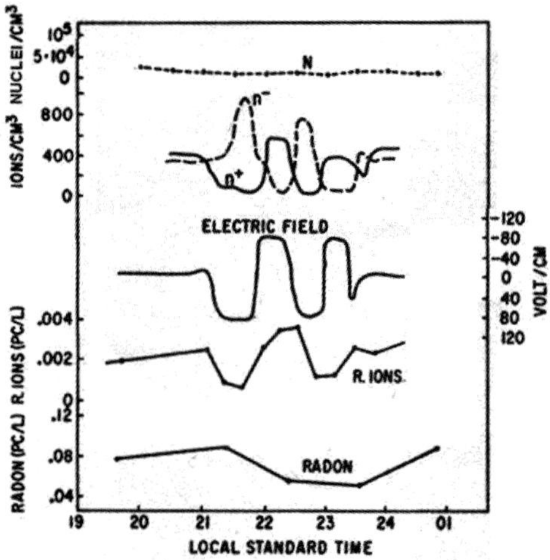

Fig. 3.3 Experimental behavior of the concentration of ^{222}Rn daughter ions in the thunderstorm environment of August 13th and 14th, 1969, at Langmuir Laboratory (Roffman 1972)

3.2 Atmospheric Plasma

All components of an atmosphere exist in continuous interaction, and though it is considered, that the bottom of the atmosphere consists only of neutral gases, but to not divide the atmosphere into different layers we will introduce the concept of *atmospheric plasma*. The concept "plasma" is usually not applied to the troposphere. At the same time looking at the encyclopedic definition, i.e. "ionized gas with equal concentration of positive and negative charges (quasi-neutrality)", there will not be any inconsistency. Moreover, for the troposphere a quite natural concept is that of a characteristic radius of ionic atmosphere, or Debye radius of shielding:

$$r_d = \left(\frac{\varepsilon_0 kT}{e^2 n_\pm} \right)^{1/2} \tag{3.1}$$

showing at what distance from the electric charge center immersed in the atmospheric plasma, its electric field will be effectively compensated by the mobile charges of the opposite sign. In a near-ground layer of an air under normal atmospheric conditions the concentration of light ions is close to $n_{\pm} = 10^2 - 10^3$ cm^{-3}, their thermal temperature $kT \cong 0.025$ eV; ions are single-charged, therefore $r_d \cong 10 \div 30$ cm (Smirnov 1992). The atmospheric plasma is strongly collisional: the collision frequency of thermalized electron with molecules of the air exceeds 10^{10} Hz. It is also multicomponent – alongside with electrons, the molecular and complex ions, there are present neutral and charged aerosol particles, cloud drops and precipitation drops, ice crystals, aerosols and other impurities distorting the classical structure of plasma.

The troposphere, as a rule, is only partially ionized, i.e. the condition $kT << \varepsilon_i$, is fulfilled where ε_i – effective potential of ionization of an air. At the same time during thunderstorms, with the presence of clouds and sources of artificial ionization the conditions could appear, when the atmospheric plasma will be essentially non-equilibrium locally, i.e. will have charged particles essentially distinguishable by electric mobility, magnitude and sign of electric charge. For example, such a situation is realized in seismoactive zones due to emanation of radon and various metal aerosols from the Earth's crust, as well as under conditions of radioactive contamination near enterprises of a nuclear industry in the case of emergencies. In these cases, as was mentioned above, the near ground air layer is formed within a height range of about 4–5 m over the ground, and is subjected to the influence of radioactive emissions, where the fast ion – molecular reactions take place resulting in turn in a change in structure and properties of a near ground atmosphere.

3.3 Atmosphere Ionization

In the first chapter the influence of ionizing radiation on the troposphere was shortly described. However in this chapter we shall go into more detail on this question. The basic sources of ionization in the low atmosphere are cosmic radiation and the radioactivity of air and ground. But the Earth is the generator of many other gases that are decay products of natural radioactive elements (uranium, thorium, potassium) and bear extremely important information on the planet's interior, on the status of massifs in the upper part of the Earth's crust and geo-ecological conditions (Utkin 1997). Ground based measurements register gaseous discharges of various substances in an atmosphere before the main shock of the earthquake including radioactive elements, in particular radon and the products of its decay. Their concentration grows by 4–8 times before the earthquake in relation to the undisturbed value. The growth of radon concentration starts several days prior to the activation of seismic activity, and the duration of these variations is several hours to several days.

Thus, more than 60–70% of the general productivity of natural sources of ionization is provided by the ground radiation. In the ground one can find all three basic sources of ionization: alpha, beta and gamma. At an altitude of 1–2 m from a surface the intensity of ion production from these ground sources on the average does not exceed 3–5 cm^{-3}s^{-1}. The radioactivity of air provides the contribution to the intensity of ion production in a ground layer of air of about 3–4 $cm^{-3}s^{-1}$. The sources of emanations in normal continental air includes the following isotopes: tritium, carbon-14, sodium-22, phosphorus-32, berillium-7, phosphorus-33, sulfur-35, cripton-85, radon, thoron and the products of decay of the soil radon and thoron. The greatest contribution is provided by radon and its daughter products. But due to the short half-life of radon isotopes (maximum 3.8 days), a significant concentration of air ions is observed above uranium containing rocks and over tectonic faults in the seismically active zones. Apart from these isotopes, especially over oceans and in polar areas, the basic source of ionization is cosmic radiation, dispersion of water drops and seawater electrization. Space radiation is the generator of ions with an average intensity of ion production in the troposphere of 1–2 $cm^{-3}s^{-1}$. At altitudes up to 20 km the ionization velocity increases up to 5 $cm^{-3}s^{-1}$. In general the intensity of ion production from the considered sources within the near ground layer of air and above continents is 7–10 $cm^{-3}s^{-1}$, and over the seas and oceans – 4–6 $cm^{-3}s^{-1}$.

3.4 The Characteristics of Radiative Sources

3.4.1 Alpha Emission

The alpha-emitters in the soil produce the particles with energies of the order of several megaelectronovolts and free pass of several centimeters. The main sources of alpha particles are presented in the Table 3.1.

Table 3.1 The mean energy of the main natural alpha-particle sources E_α and their ion production rate intensity ν

Isotope	E_α MeV	ν $cm^{-3}s^{-1}$
Radon-222	6	3
Thoron	6.28	0.655
Polonium-216	6.78	0.17
Action	6.81 (82%) 6.64 (13%) 6.41 (5%)	0.17
Totally		~4.0

The kinetic energy of alpha particles varies from 1.83 MeV (144 Nd) up to 11.65 MeV (isomer 212 mPo). The alpha particle of 6 MeV energy free pass l_α in air is near 5 cm under normal conditions and can be estimated in the following way:

$$\begin{aligned} &\text{if } E_\alpha \leq 4 \text{ MeV}; \ l_\alpha = 0.56\, E_\alpha; \\ &\text{if } E_\alpha = 4 - 8 \text{ MeV}; \ l_\alpha = 1.24\, E_\alpha - 2.62 \end{aligned} \tag{3.2}$$

where E_α is the alpha-particle energy.

The alpha-particles spectrum under decay of mother nuclei consist of several monoenergetic lines corresponding to different energetic levels of the daughter nuclei.

The largest ionization density from alpha-particles (κ_α) is observed at the end of their pass; κ_α may be estimated as $\kappa_\alpha = E_\alpha /(\varepsilon_\alpha\, l_\alpha)$. The ionization process parameters and the air products composition are similar as a result of ionization by electrons with energy 70–100 eV.

3.4.2 Beta Emission

The elements radiating the beta-emission generate the particles with mean energy 0.249 MeV and free pass in clear air near 1 m. This emission consists of single charge electrons with different energies. With acceptable accuracy for practice purposes the specific ionizing ability of beta particles κ_β (cm^{-1}) can be estimated:

$$\kappa_\beta = 2^{10} E_\beta^{-0.14} \tag{3.3}$$

If E_β=5.6 MeV the κ_β magnitude can be estimated as $\kappa_\beta \approx 1.6 \cdot 10^2$ cm^{-1}. This means that the ionizing ability of beta particles is smaller than alpha particles by two orders of magnitude, but the free pass in the same proportion is larger.

3.4.3 X-Rays and Gamma Emission

These emissions loose their energy in the atmosphere due to the photoeffect (if the quantum energy is less than 0.1 MeV) or due to Compton scattering (if quantum energy is 0.1–10 MeV). Their main action on the atmosphere as in the case of alpha particles is provided by the secondary or delta-electrons, the energy spectrum of which extends from the order of 0.025 eV up to the quantum energy. The gamma-emitters contribution to the air ionization prevails in the near ground air layer. For example, the ion formation velocity at the 1 m altitude from gamma-emission of radium and thorium is (~4–5 cm^{-3}s^{-1}), and from β- and γ-emission of radioactive aerosols and gases is (~0.2 cm cm^{-3}s^{-1}).

3.4.4 Radon

This colorless, insipid and radioactive gas, determines the radioactivity of air in a ground layer and inside buildings. Radon because of specific features is the optimum indicator for various geological and geotechnical areas of research. First, radon is a radiogenic gas and as the product of decay of the uranium – radium sequence is continuously generated in rocks during radioactive decay. It is always present in any rock massif, and reduction in its concentration, for example, at the expense of diffusion from a massif in air is constantly compensated for by new generation of radon. Second, diffusion of radon in a rock massif and its emanation from a surface of ground are defined by the effective diffusion coefficient, which depends on many factors. The most important of these factors are porosity, permeability and fissuring. These properties of the media essentially depend on the strain-deformation status of a massif. It is obvious, that under compression the massif's permeability is reduced, and at unloading increases. The effective factor of diffusion accordingly changes. Hence, dynamic changes of concentration of radon in the near surface layer of the ground will reflect dynamic changes in the strain-deformation state of rocks in significant volume. Recent studies of this problem led to the detection of a new effect in the distribution of radon concentration during the observed changes in the strain-deformation state of rocks (Barabas' 1991; Utkin et al. 1993; Bulashevich et al. 1996). The studies were carried out at a Northern Ural bauxite mine in connection to a problem of mine blasts, which can be characterized as local earthquakes of technogenic origin.

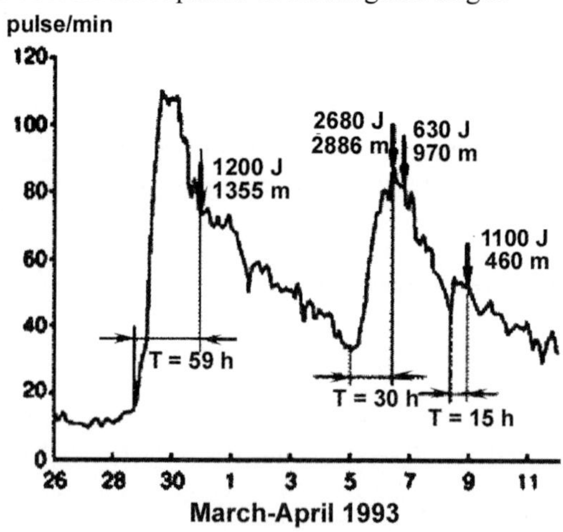

Fig. 3.4 Radon variation in observation well situated in far zone. Mine blasts are marked by *arrows* with indication of blast energy and distance to the well

In the far zone the radon variation is completely different than in the near vicinity: the blast is preceded not by a decrease but by a sharp increase of radon concentration (8–10 fold) in the well, and the blast is observed after passing the radon concentration maximum. Figure 3.4 shows the typical behavior of radon concentration for the isolated blasts, also for a swarm.

3.5 Ionized Atmosphere Models

Models of the ionized atmosphere were for a long time the subject of special attention. Basically they are connected with the study of cosmic radiation effects on the top and middle layers of the atmosphere (Mitra 1974; Zinn 1990) with the changes of ion-molecular structure of an ionosphere (Thomas 1971) and stratosphere (Ferguson 1979; Brasseur and Chatel 1983; Arijs 1983; Kawamoto and Ogawa 1984; Arijs 1992). However, compared to the number of papers devoted to research of the ionized troposphere, the number of papers is essentially less (Kawamoto and Ogawa 1986; Salm and Lutsb 1988; Smirnov 1992). The basic advantage of these models is that they account for the huge number of reactions and components of air, but at the same time this is their deficiency because they demand the use of significant computation resources. It is necessary to note also, that in all these papers the mean volume uniform ionization of air was taken into account only, though under certain conditions (for example, at track character of ionization) this consideration does not strictly obey the laws. To create a universal model of the ionized atmosphere, which takes into account the features of ionization of air in various, described above, situations is practically impossible. It is better to consider the separate processes applicable to specific situations, for example: the analysis of the processes occurring in the bottom layers of the atmosphere under the action of ionizing radiation (Boyarchuk and Svirko 1996; Boyarchuk 1997), to choose the most probable channels of ion kinetics (Boyarchuk 1999), or to create a model of the ionized atmosphere which takes into account the track character of ionization (Boyarchuk 2000).

3.5.1 Peculiarities of Ionization of Atmospheric Air – the Result of Track Structure

Usually during the analysis of the laws of radiochemical conversion of the atmospheric environment uniform volume ionization is supposed. Here we shall consider the influence of track structure on the general kinetic processes occurring in the ionized atmosphere.

We normally consider gaseous environments to be simpler then liquids or solid bodies. Owing to the low specific mass of gases, during the passage through them by high-energy particles, track effects are practically absent, and primary radiolysis products (electrons, ions, the free radicals and the excited molecules) are distributed in the volume at regular intervals. It essentially simplifies the theoretical analysis of the gas molecular-ion kinetics. However, under sufficiently high pressures (0.1–1 MPa) in air at room temperature micro areas with a high concentration of ionized particles may be formed, so-called tracks of high energy corpuscular radiation (protons, α- and β-particles, etc.) in which the products of radiolysis are located in most cases (Pikaev 1986). Hence, in studying the process of formation of negative and positive ions in the air environment with radioactive pollution it is necessary to distinguish between the volumetric ionization of X-rays or γ-rays, and local ionization by high-energy particles taking into account the track structure, caused, for example, by α-active radon.

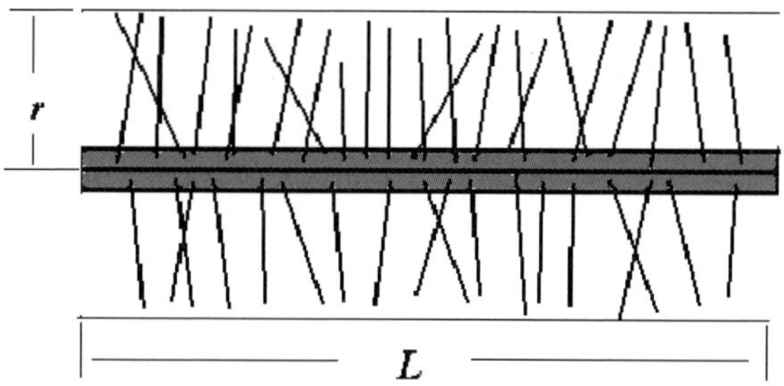

Fig. 3.5 The general schematic presentation of the charge particle (decay debris) track. r – track radius, L – track length

The main effect of ionizing radiation on the air environment is caused by the secondary electrons appearing under ionization of atoms and molecules. The energy of these electrons is in the range from thermal energy up to the energy of energetic primary particles. The structure of a track depends on the type and energy of the particle. Generally the energy of a particle is spent for the formation of the secondary electrons scattering in the sides from a trajectory of a particle (the main track) and forming spurs. In a case of α-particles or decay debris, the spurs are located so closely to each other, that around of the main track the continuous ionized area is formed. The resulting ion-molecular processes in such tracks can be of a different type, than in the case of the volumetric ionization that is caused by X-rays or γ-quantums.

Presently, there is no consistent theory describing processes on tracks of ionizing particles. Therefore, the experimental results are often interpreted with the help of the simplified theoretical models of Onsager (1938) and Jaffe (1913) for elementary processes in tracks. The first one was developed for the description of processes of a recombination on traces of high-energy particles in dense gas environments ($P \approx 20$ MPa). The second one has been intended for the description of processes of a recombination on traces of α- particles in gas at pressure ~ 1 MPa and consequently it is more convenient for the description of processes in the troposphere. The main provision of the Jaffe theory is the following: the ions formed as a result of the particles passage are concentrated in the cylinder formed around the particle trajectory.

Regarding the track to be axially-symmetric and neglecting the longitudinal heterogeneity, Budnik (1994) modeled the step-like spatial-temporal evolution of a track formed by the debris of decay in a He-Cd nuclear-excited plasma at various parameters of medium and different levels of pumping energy. The formation of a track was considered in two stages: just after the particle passage mainly there is a scattering of electrons and ions from an axis of a track and a relaxation of electrons in non-elastic collisions. Due to this process the concentration of ions and excited atoms changes (the duration of this stage exceeds the time of the relaxation of electrons in non-elastic collisions, but it is less than the characteristic time of their relaxation in elastic collisions). During this stage the ion – molecular reactions and radiating transitions were neglected. During the second stage of modeling the ion-molecular kinetics of the active medium are considered.

The conclusion is that with growth of pressure in the medium, at constant pumping energy, the influence of track structure on ion – molecular processes increases. Under a pumping energy increase with other parameters constant, the influence of track structure decreases. Depending on the parameters of the medium, the sizes of track structures will be ~1–10 microns, and the characteristic time of their existence ~10–100 nanoseconds.

3.5.2 The Model of the High-Energy Particle Track in the Lower Atmosphere

For the air environment of the bottom layers of the troposphere we shall consider, that positive and negative ions are equally distributed within the volume of the cylinder around the particle trajectory. Let us estimate the track cross-section size, considering, that it is defined by the average length of spurs – the ionized traces of secondary electrons forming the ionized area around the main track.

The free pass L of α-particles with energy E_α~5 MeV in tropospheric air does not exceed 4 cm, and the ionization cross-section of such a particle for air σ~$5.1 \cdot 10^{-16}$ cm^2 (Physical magnitudes 1991). Now we can determine the free pass length l of α-particles in air $l = \dfrac{1}{\sigma N_L} = 7.3 \cdot 10^{-5}$, where N_L – is the Loschmidt constant $2.69 \cdot 10^{19}$ cm^3), i.e. average distance between spurs. Knowing the length of free pass, it is possible to determine the average loss of energy ΔE on one act of ionization $\Delta E = \left(\dfrac{L}{l}\right)^{-1} E \approx 100\text{eV}$, it will correspond to the energy of secondary electrons. Using the Bethe-Born approximation (to use the Born approximation it is enough that the velocity of the approaching electron was large in comparison with the velocity of atomic electrons), we shall determine the ionization cross-section of air molecules by secondary electrons according to (Gudzenko and Yakovlenko 1978).

$$\sigma_2 = 4 \cdot 10^{-14} \xi \frac{\ln(\Delta E / I)}{\Delta E I} = 3 \cdot 10^{-16} \tag{3.4}$$

where I=15.58 eV and ξ=6 – ionization potential and number of equivalent electrons of a nitrogen molecule. As the minimal track radius we will consider the length of free pass of the secondary electron $r = \frac{1}{\sigma_2 N_L} \approx 1.3 \cdot 10^{-4}$ cm. The track volume will be consequently $V = \pi r^2 L \approx 2{\cdot}10^{-7}$ cm^3

Let us determine now the ion-electron pair concentration within this volume. From the kinetic energy formula one can determine the α-particle velocity in the air: $v = \sqrt{\frac{zE_\alpha}{uM}} = 1.5 \cdot 10^9$ cm/s, where M and z are the mass and the charge of α-particle, u – atomic mass unit, e – elementary electric charge. Averaged by the passage time power contribution $w = \frac{E_\alpha e}{Vt} \approx 3 \cdot 10^3$ W/cm^3, under condition that the characteristic pass time for the particle $t = \frac{L}{v} \approx 2 \cdot 10^{-9}$ s. The ionization frequency is determined in the following way:

$$\nu = \frac{w}{N_L E_i e} \approx 11 \text{ s}^{-1} \tag{3.5}$$

where E_i – the energy necessary to create the one ion-electron pair in the air (~34 eV). As a result the primary electron concentration within the α-particle track will be:

$$N_e = \nu N_L t \approx 10^{12} \text{ cm}^{-3} \tag{3.6}$$

Generally speaking the size and lifetime of the track is determined as with the ambipolar diffusion from the primary track volume, so by the primary ions recombination within this volume.

3.5.3 Numeric Modeling of the Molecular Processes within the Track

It should be noted, that the small volume of a track reduces practically to zero the probability of participation of atmospheric small gas constituents in ion-molecular processes; their concentration is $\leq 10^8$ cm^{-3}. It permits us to limit the number of considered molecules and ions. In our approach it is possible to consider the basic ion-molecular processes during the time when the track was not washed away with a sufficient degree of accuracy only for the ions of the main components of the air N_2 and O_2 whose concentration is stable and their concentration by several orders of magnitude exceeds the other possible small gaseous constituents of the atmosphere (Boyarchuk and Svirko 1996). It is also necessary to take into account the influence of water vapor molecules H_2O on ion – molecular processes and on the formation of ions.

Table 3.2 The basic ion-molecular reactions of the bottom atmosphere which are taken into account in our model

No	Reacton	Velocity α, cm^{3}s^{-1}
1	$O_2^+ + e \rightarrow O + O$	2.2×10^{-7}
2	$N_2^+ + e \rightarrow N + N$	2.9×10^{-7}
3	$e + O_2 + O_2 \rightarrow O_2^- + O_2^*$	2.5×10^{-30} (cm^{6}s^{-1})
4	$O_2^- + O_2^* \rightarrow O_2 + O_2 + e$	2.0×10^{-10}
5	$N_2^+ + O_2 \rightarrow O_2^+ + N_2$	2.0×10^{-11}
6	$A^+ + B^- \rightarrow A^* + B$	10^{-6}
7	$H_2O^+ + H_2O \rightarrow H_3O^+ + OH$	1.2×10^{-9}
8	$e + H_3O^+ \rightarrow H_2 + O + H$	1.3×10^{-6}

In Table 3.2 the elementary processes taken into account in our model within a track are submitted. Here we consider the simplified kinetic model of a track. In a more realistic situation of electron-ion recombination, the velocity will essentially depend on the electron temperature T_e and as the electron temperature within a track can be significant, the magnitude of recombination velocity may be lower.

The ion kinetic equations within a track, taking into account molecular diffusion and the drift of ions in the electric field, written down in a cylindrical system of coordinates, will look as follows:

$$\frac{\partial n_i}{\partial t} = -\frac{1}{r}\frac{\partial}{\partial r}\left[r\left(D_i \frac{\partial n_i}{\partial r} + \mu_i \frac{\partial}{\partial z}(En_i) \right)\right] + \sum_{i\neq j}^{N} \alpha_{ij} n_i n_j$$

$$\frac{1}{r}\frac{\partial}{\partial r}(rE) = 4\pi\rho \tag{3.7}$$

where n_i, D_i, μ_i – concentration, diffusion and mobility coefficients for electrons, positive and negative ions, and also the excited molecules [e], [O_2^+], [N_2^+] [O_2^-], [O_2^*], [H_2O^+], [H_3O^+]; α_{ij} – reactions velocities of ion – molecular reactions from Table 3.2; E – electric field, r – coordinate across a track.

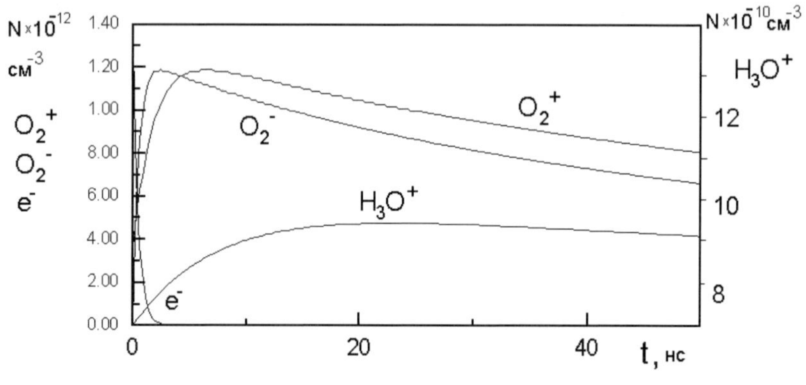

Fig. 3.6 Electron and main ions concentration variations with time within the track volume

In Fig. 3.6 the result of the numerical solution of the given system of the equations is depicted. It can be seen that the basic processes of a recombination occur between the positive and negative ions of oxygen, since all positive ions as a result of reaction of a recharging no. 5 (see Table 3.2) quickly pass into positive ions of oxygen. Electrons almost with the same speed attach to neutral molecules of oxygen in reaction no. 3. Formation of positive ions H_3O^+ occurs as a result of consecutive hydration according to reaction no. 7, but because of fast reaction no. 8 the resulting process will be by several orders slower and affects the basic processes a little. These processes occur during one nanosecond, then the concentration of negative ions decreases at the expense of a recombination and molecular diffusion, and ions are deduced from the reaction volume.

For comparison in Fig. 3.7 the three-dimensional dependence of concentration of negative ions of oxygen on time and distances from the center of a track is shown. From this figure it is possible to estimate the time of existence of an elementary volume of the homogeneous ionization created by a track of the energetic particle, on a level of ion concentration decrease in the center of a track in *e* times. For this time ($t_m \sim 125$ ηs) as a result of diffusion degradations the volume of area of homogeneous ionization will increase approximately by 4 times up to $V_m \sim 5{\cdot}10^{-7}$ cm^3. Such estimations can be carried out for β-emission also. In this case the volume of a track will increase approximately by 2 orders of magnitude at the expense of the greater free pas of β-particles in air under the same conditions.

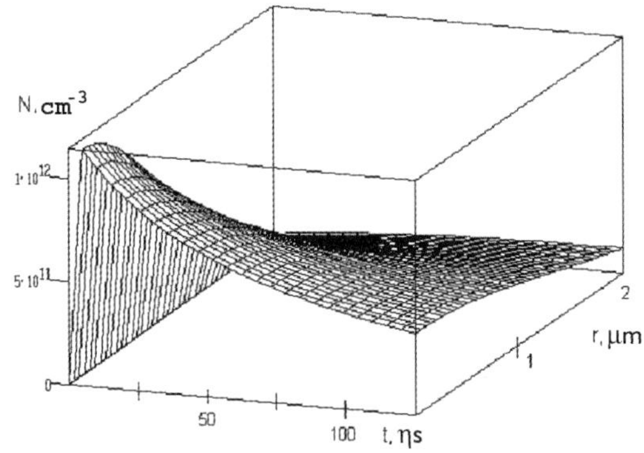

Fig. 3.7 Dependence of the molecular oxygen negative ions concentration N (cm^{-3}) on the time t (ηs) and distance from the track center r (μm)

3.5.4 The Model Application Criterion

The taking into account of the track structure of ionization of air or gases at atmospheric pressure is important for the solution of some specific problems. For example, during the analysis of formation of long-living ions in a near-ground atmosphere under significant emanation of radon. As was already mentioned above – radon is a α-active gas, hence in places of radon presence the ionization of air will have the track character.

As "rigid" criterion for the necessity to take into account the track structure of air ionization may be the moment of tracks overlapping. If to designate the value describing the number of tracks born in a unit of volume per second, i.e. volumetric density of nuclides activity as $a_{\alpha,\beta}$, the condition of tracks overlapping can be written down as $a_{\alpha,\beta}V_m t_m \ll 1$ – tracks do not overlap, and $a_{\alpha,\beta}V_m t_m \gg 1$ – tracks overlap; i.e. at substantial growth of a degree of ionization, the track centers can merge, and ionization occurs at regular intervals throughout the volume. So it is easy to estimate, that for a α-active source the boundary of criterion will be achieved at concentration of radioactive substances $a_{\alpha,\beta}$~500 Ki/cm^3 which corresponds to volumetric ionization velocity ~10^{18} cm^{-3}s^{-1}. Such levels of radioactive pollution can be observed only during large failures of nuclear objects (Isräel 1996). In nature, under natural conditions such levels are improbable. In the paper by Boyarchuk (2000) the various levels of radioactive pollution are analyzed and the conclusion is made, that practically always, except perhaps for the exception of catastrophic emergencies, during analysis of ion-molecular processes in air subjected to radioactive emission, it is necessary to take into account the track character of ionization.

Hence, according to rigid criterion, at normal levels of radioactive pollution, the track centers of non-uniform ionization in the troposphere will be always observed. It is necessary to take into account track heterogeneity, also during the calculation of ion kinetics. From this point of view, such a level of radioactive pollution of the atmosphere can be taken as a criterion, after which the changes in relaxation kinetic processes will be observed. According to the suggested track model of ionization, during the first moment after the beginning of the act of ionization, all ion-molecular processes with primary elementary ions occur with the basic components of air N_2 and O_2 in small areas around the tracks of high energy particles and continue during the time of a track's existence. Reactions of an electron-ion and ion-ion recombination prevail here. And after only 10^{-7} s the track is washed away, and the ions formed in it mix with the main volume of non-ionized air, entering in reaction with neutral molecules of small gas constituents of the atmosphere, for example, nitrogen or carbon containing compounds. Depending on the balance of concentration of the primary ions created in tracks and the small gas constituents present throughout the volume, processes of an ion-ion recombination or a reaction of ions with neutral molecules can prevail. In each of these cases the ion production velocity and final concentration of the basic ions of an atmosphere will be different.

The condition of such "soft" criterion can be written as:

$$X = N_i^2 \alpha_6 - \alpha N_i N_{NO} \tag{3.8}$$

where N_i – the ion concentration produced within the track in relation to the whole volume, N_{NO} – concentration of the main small gaseous constituents of the atmosphere, for example the typical concentration of nitrates in the troposphere ~10^{10} cm^{-3}, α_6 – coefficient of the ion-ion recombination (Table 3.2), α – velocity of reactions of ions with the neutral molecules ~10^{10} cm^{-3}s^{-1}.

If X>0, it is not necessary to take into account the track structure, since the mean volumetric concentration of ions is comparable with the concentration of small gas constituents, and reactions of a recombination between ions prevail. If X<0 track ionization influences the results of ionization because the main process of the ions destruction will be their reaction with neutral molecules, and as a result of it will be quickly formed, for example, steady and characteristic ions for the troposphere NO_3^-

In our model of a track we did not take into account the elementary ions hydration, which starts immediately after their formation and occurs with significant velocity. However, there is an assumption, that the hydrated ion will enter chemical reactions with surrounding molecules and ions differently, for example, with a slower recombination velocity. This hydration will be discussed in more detail below. There are experimental observations (Stozhkov et al. 1997) and the indirect data (Smirnov 1992), confirming this assumption.

Thus, it is necessary to distinguish two mechanisms of atmospheric air ionization (Boyarchuk et al. 2000):

- The first operates at a small degree of ionization when track effects of non-uniform ionization are significant. During the first moment after the beginning of the act of ionization all ion-molecular processes with primary elementary ions occur in small areas around the tracks of high-energy particles and continue during the existence of a track. And only then are they deduced in common reaction volume. In this limiting case a small degree of ionization allows us to neglect the change of a chemical composition of neutral components of air under the action of radiation, and to not consider the reaction of the charged particles with neutral ones, appearing in air as a result of an irradiation.

- The second mechanism operates at the high degree of ionization when the track centers of ionization merge. In this case the kinetics of ion-molecular processes it is enough to consider as mean volumetric from the very beginning of the act of ionization, but it is necessary to take into account the changes of a chemical composition of neutral components of the air.

3.5.5 The Low Degree of Ionization – the Model of Stable Ions Formation in the Low Layers of the Atmosphere

In a real situation the level of radioactive radiation seldom exceeds the ionization velocity $10^5 - 10^8$ $cm^{-3}s^{-1}$. A low level of ionization allows us to neglect the change of a chemical composition of neutral components of air under the action of radioactivity, i.e. to not consider reactions of the charged particles with neutral ones, appearing in air as a result of an irradiation (Medvedev et al. 1980). It is possible to considerably reduce the number of reactions that is necessary to take into account. Thus, there exists an opportunity for some ordering of the numerous ionizing-recombination processes and formation of the closed system of the equations describing the kinetics of ion-molecular processes in time. Under the low level of ionization the stable long-living positive and negative ion complexes – clusters are formed which strongly influence the electric characteristics of an atmosphere.

3.5.6 Analysis of the Most Probable Ion-Molecular Reactions of the Lower Atmosphere

In Table 3.3 the basic ion-molecular reactions of the bottom atmosphere, which are taken into account in our model, are submitted. As a result of these reactions the formation of stable ion structure occurs[1].

Table 3.3

№	Reaction	$\alpha_n(cm^3s^{-1})$
	Electron-ion reconbination	
1	$e + O^+ \rightarrow O$	$3.0 \cdot 10^{-12}$
2	$e + N^+ \rightarrow N$	$3.0 \cdot 10^{-12}$
3	$e + O_2^+ \rightarrow O + O$	$2.2 \cdot 10^{-7}$
4	$e + N_2^+ \rightarrow N + N$	$2.9 \cdot 10^{-7}$
5	$e + NO^+ \rightarrow N + O$	$4.1 \cdot 10^{-7}$
	Electron attachment to the oxygen molecules	
6	$e + O_2 + X \rightarrow O_2^- + X$	$2.5 \cdot 10^{-30}$ $[cm^6s^{-1}]$
7	$e + O_2 \rightarrow O^- + O$	$4.8 \cdot 10^{-14}$
8	$e + O_3 + O_2 \rightarrow O_3^- + O_2$	10^{-13}
	Ion-ion recombination	
9	$A^- + B^+ \rightarrow A + B^*$	$\sim 10^{-6}$
	Negative ions destruction	
10	$O^- + O_2^* \rightarrow O_3 + e$	$3.0 \cdot 10^{-10}$
11	$O_2^- + O_2^* \rightarrow O_2 + O_2 + e$	$2.0 \cdot 10^{-10}$
	Positive ions reactions with the neutral molecules	
12	$O^+ + O_2 \rightarrow O_2^+ + O$	$4.0 \cdot 10^{-11}$
13	$O^+ + N_2 \rightarrow NO^+ + N$	$4.0 \cdot 10^{-12}$
14	$N^+ + O_2 \rightarrow O_2^+ + N$	$5.0 \cdot 10^{-10}$

[1] The reactions velocities are taken from the papers Ferguson, 1979); (Arijs, 1983); (Kawamoto and Ogawa, 1984, 1986); (Medwedev and all, 1980); (Pshezhtsky and Dmitriev, 1978); (Willis and Boyd, 1976); (Kossy and all, 1994); (Atmosphere. hand book, 1991); (Pikaev. 1986

15	$N^+ + O_2 \rightarrow NO^+ + O$	$5.0 \cdot 10^{-10}$
16	$N_2^+ + O_2 \rightarrow O_2^+ + N_2$	$\sim 10^{-11}$
	Negative ions reactions with the neutral molecules	
17	$O^- + O_2 + M \rightarrow O_3^- + M$	$\sim 1.1 \cdot 10^{-30} (M = N_2, O_2)$
18	$O_2^- + O_3 \rightarrow O_3^- + O_2$	$6 \cdot 10^{-10}$
19	$O_2^- + NO_2 \rightarrow NO_2^- + O_2$	$7 \cdot 10^{-10}$
20	$O_2^- + N_2O \rightarrow NO_2^- + NO$	$2 \cdot 10^{-14}$
21-1	$O_3^- + NO \rightarrow NO_2^- + O_2$	$8 \cdot 10^{-10}$
21-2	$O_3^- + NO \rightarrow NO_3^- + O$	$\sim 10^{-11}$
22-1	$O_3^- + NO_2 \rightarrow NO_2^- + O_3$	$7 \cdot 10^{-10}$
22-2	$O_3^- + NO_2 \rightarrow NO_3^- + O_2$	$2.8 \cdot 10^{-10}$
23	$O_3^- + CO_2 \rightarrow CO_3^- + O_2$	$5.5 \cdot 10^{-10}$
24	$NO_2^- + O_3 \rightarrow NO_3^- + O_2$	$1.2 \cdot 10^{-10}$
25	$NO_2^- + NO_2 \rightarrow NO_3^- + NO$	$2 \cdot 10^{-13}$
26	$NO_3^- + O_3 \rightarrow NO_2^- + 2O_2$	$1 \cdot 10^{-13}$
27	$NO_3^- + NO \rightarrow NO_2^- + NO_2$	$1 \cdot 10^{-12}$
28	$NO_3^- + H_2SO_4 \rightarrow HSO_4^- + HNO_3$	$2.6 \cdot 10^{-9}$
29	$CO_3^- + NO \rightarrow NO_2^- + CO_2$	$1.1 \cdot 10^{-11}$
30	$CO_3^- + NO_2 \rightarrow NO_3^- + CO_2$	$2.0 \cdot 10^{-10}$

3.5.7 Analysis of Processes of the Main Negative Ions Formation in the Lower Atmosphere

According to our model the formation of the main negative elementary ions of the bottom layer of an atmosphere – ions of oxides of nitrogen and carbon – proceeds within three basic channels:

$$\begin{aligned} &1.\ O^- + O_2 \rightarrow O_3^- \rightarrow NO_2^- \rightarrow NO_3^- \\ &2.\ e + O_2 \rightarrow O_2^- \rightarrow O_3^- \rightarrow CO_3^- \rightarrow NO_2^- \rightarrow NO_3^- \\ &3.\ e + O_2 \rightarrow O_2^- \rightarrow O_4^- \rightarrow CO_4^- \rightarrow NO_3^- \end{aligned} \tag{3.9}$$

As a result of recharging reactions:

$$O_2^- + O_3 \rightarrow O_3^- + O \quad (\alpha \sim 6 \cdot 10^{-10}\ cm^3 s^{-1}) \tag{3.10}$$

$$O^- + O_2 + M \rightarrow O_3^- + M \tag{3.11}$$

($\alpha \sim 1{,}1 \cdot 10^{-30}$ cm^6s^{-1}, where M may be N_2 or O_2)

the considerable amount of molecules of ozone and oxygen will turn into negative ions O_3^- which represent very steady complexes (Raiser 1992); and the probability of these reactions is large, since the energy of affinity to an electron for ozone (2.89 eV) considerably exceeds the energy of affinity of O (1.47 eV) and for O_2 (0.46 eV). The negative ion O_3^- enters the numerous reactions with oxides of nitrogen:

$$O_3^- + NO \rightarrow \begin{bmatrix} NO_3^- + O \\ NO_2^- + O \end{bmatrix} \tag{3.12}$$

$$O_3^- + NO_2 \rightarrow \begin{bmatrix} NO_3^- + O_2 \\ NO_2^- + O_3 \end{bmatrix} \tag{3.13}$$

$$O_3^- + N_2O \rightarrow NO_2^- + NO \tag{3.14}$$

These reaction velocities are of the order of $\sim 10^{-10}$ cm^3s^{-1}. It is necessary to note, that the formation of complex ions on the basis of NO_3^- in the troposphere of the Earth can occur at the expense of oxidation of nitrogen during interaction with CO_2 [channel No 2 in Eq. (3.9)], for example in reactions:

$$O_3^- + CO_2 \rightarrow CO_3^- + O_2 \tag{3.15}$$

$$CO_3^- + NO_2 \rightarrow NO_3^- + CO_2 \tag{3.16}$$

Transition of an electron from one molecule to another in these reactions is promoted by the circumstance, that its energy of affinity to molecules NO_3, NO_2 and CO_2 is even higher (3.88 eV, 3.82 eV and 3.80 eV respectively). Generally speaking, in the air environment the dioxide of carbon plays a smaller role than oxygen. Because of the action of ionizing radiation in the atmosphere it appears the significant amount of ions of O_2^- and O_3^-, and the formation of NO_X^- goes basically according to the first channel in Eq. (3.9). In other circumstances, for example, under increased concentration of CO_2, the second channel of formation of ions can prevail.

The ion O_4^- can be considered as the ion-ligament $O_2^- \cdot O_2$ (similarly also the ion CO_4^- can be considered as the complex $O_2^- \cdot CO_2$). The reaction:

$$O_2^- + O_2 + M \Leftrightarrow O_2^- \cdot O_2 + M \tag{3.17}$$

is more characteristic of the upper stratosphere where the reversed reaction velocity ($2.7 \cdot 10^{-14}$ cm^3s^{-1}) and the three-particle reaction velocity ($3.4 \cdot 10^{-31}$ cm^6s^{-1}) are small. But for the troposphere where there is a high level of water vapor concentration direct hydration is more characteristic:

$$O_2^- + H_2O + M \Leftrightarrow O_2^- \cdot H_2O + M \qquad (3.18)$$

particularly because the direct reaction velocity ($2.2 \cdot 10^{-28}$ cm^6s^{-1}) is higher than the reversed one ($\sim 10^{-14}$ cm^3s^{-1}). When the essential concentration of the carbon dioxide prevails by velocity the reaction:

$$O_4^- + CO_2 \Leftrightarrow CO_4^- + O_2 \qquad (3.19)$$

because the direct velocity here ($4.3 \cdot 10^{-10} cm^6s^{-1}$) is essentially lower than the reversed one ($\sim 10^{-13}$) (Willis and Boyd 1976; Przerzecki and Dmitriev 1978).

Thus, as a result of action within any of the three channels of Eq. (3.9) mainly the elementary negative ions NO_X^- will be formed.

As a result of the processes of hydration, the more complex structures – complex ions, which represent connections of ions with atoms or molecules, are formed. As the bonds in such ions are caused by far-ranging interaction they are weaker than valent chemical bonds. Complex ions are much more stable than the molecular connections formed as a result of Van-der Vaals interaction of molecules, because one of the components of the complex ion is the charged particle.

Table 3.4. Negative ion complexes formation (Smirnov 1978)

Reaction	α, 10^{-30} cm^6/s
$O^- + H_2O + O_2 \rightarrow O^- \cdot H_2O + O_2$	100
$O_2^- + H_2O + O_2 \rightarrow O_2^- \cdot H_2O + O_2$	160
$O_3^- + H_2O + O_2 \rightarrow O_3^- \cdot H_2O + O_2$	210
$O_2^- \cdot H_2O + H_2O + O_2 \rightarrow O_2^- \cdot (H_2O)_2 + O_2$	540
$NO_2^- + H_2O + NO \rightarrow NO_2^- \cdot H_2O + NO$	150
$NO_3^- + HNO_3 + N_2 \rightarrow NO_3^- \cdot HNO_3 + N_2$	10000

3.5.8 Analysis of Processes of the Main Positive Ions Formation in the Lower Atmosphere

Formation of the most widespread positive ions of the troposphere occurs basically as a direct interaction of primary elementary ions O_2^+ and NO^+ with molecules of water according the following reactions:

1.
$$\left.\begin{array}{l} O_2^+ \xrightarrow{O_2} O_4^+ \xrightarrow{H_2O} O_2^+ \cdot H_2O \\ O_2^+ \xrightarrow{H_2O} O_2^+ \cdot H_2O \end{array}\right\} \xrightarrow{H_2O} H_3O^+$$

2.
$$\left.\begin{array}{r} NO^+ \xrightarrow{N_2} NO^+ \cdot N_2 \xrightarrow{CO_2} NO^+ \cdot CO_2 \xrightarrow{H_2O} NO^+ \cdot H_2O \\ NO^+ \xrightarrow{H_2O} NO^+ \cdot H_2O \end{array}\right\} \rightarrow \tag{3.20}$$

$$\xrightarrow{OH\forall HO_2} H_3O^+$$

3.
$$NO_2^+ \xrightarrow{H_2O} NO_2^+ \cdot H_2O \xrightarrow{H_2O} H_3O^+ \cdot H_2O$$

The ion O_4^+ can be also regarded as an ion-ligament O_2^+-O_2. Its formation is a result of the three-particle reaction:

$$O_2^+ + O_2 + M \rightarrow O_4^+ + M \tag{3.21}$$

with the velocity $2.6 \cdot 10^{-30}$ cm^{6}s^{-1} (Brasseur and Chatel 1983), and decay may occur both as a result of reaction:

$$O_4^+ + M \rightarrow O_2^+ + O_2 + M \tag{3.22}$$

with velocity $2.4 \cdot 10^{-6}$ cm^{6}s^{-1} (Brasseur and Chatel 1983), and as a result of the process:

$$O_4^+ + O \rightarrow O_2^+ + O_3 \tag{3.23}$$

which has a velocity $(3\pm2) \cdot 10^{-10}$ cm^{3}s^{-1} (Fehsenfeld and Ferguson 1969).

Due to the small concentration of the oxygen atoms in a near ground layer of air because almost all oxygen is in the O_2 state, the last reaction (3.23) can be neglected. However, the velocity of the other reaction (3.22) ion O_4^+ destruction is so high, that it brings to nothing the reaction of formation, and, hence, and efficiency of the first channel (3.20). Nevertheless, the experimental observations testify to the presence of this ion in an atmosphere. There are assumptions of formation $O_2^+ \cdot (H_2O)$, without using O_4^+ through unknown complex $O_2^+ \cdot X$, for example, $O_2^+ \cdot N_2$ (Smirnov 1992), but, generally speaking, this question remains open.

In the case of a dense near ground atmosphere with the significant contents of water vapor molecules it is possible to assume with a high degree of probability that will the following reaction will prevail:

$$O_2^+ + H_2O + M \rightarrow O_2^+ \cdot H_2O + M \tag{3.24}$$

Then through the consecutive reactions with the water vapor molecules the stable complex is formed on the basis of H_3O^+ (or $H^+ \cdot H_2O$). This is the work of channel 3 in Eq. (3.20). However, with the presence of a significant concentration of such atmospheric small gas components as CO_2 and NO, the H_3O^+ formation can occur more quickly using channel no. 2 [Eq. (3.20)], especially at low temperature (T<215 ° K) with a constant velocity ~10^{-9} cm^3s^{-1} (Kossy et al. 1994), in the following reactions:

$$\begin{aligned} &NO^+ + 2N_2 \rightarrow NO^+ \cdot N_2 + N_2, \\ &NO^+ \cdot N_2 + CO_2 \rightarrow NO^+ \cdot CO_2 + N_2, \\ &NO^+ \cdot CO_2 + H_2O \rightarrow NO^+ \cdot H_2O + CO_2. \end{aligned} \tag{3.25}$$

Usually it is enough for only one step of the hydration of NO^+. After that the reaction of a regrouping with free radicals OH and HO_2 takes place. Their occurrence in the atmosphere is typical to processes of ionization:

$$\begin{aligned} &NO^+ \cdot H_2O + OH \rightarrow H_3O^+ + NO_2, \\ &NO^+ \cdot H_2O + HO_2 \rightarrow H_3O^+ + NO + O_2, \\ &NO^+ \cdot H_2O + HO_2 \rightarrow H_3O^+ + NO_3 \end{aligned} \tag{3.26}$$

and, then, the twice hydrated positive nitrogen oxide ion in reactions with water will be transformed to a final steady ion of oxonium and a nitric acid (Fehsenfeld and Ferguson 1969):

$$NO^+ \cdot 2H_2O + H_2O \rightarrow H_3O^+ \cdot 2H_2O + HNO_2 \tag{3.27}$$

The third channel from the system (3.20) will be effective with the presence of a significant concentration of nitric dioxide NO_2 and is realized in reactions:

$$\begin{aligned} &NO_2^+ + H_2O + N_2 \rightarrow NO_2^+ \cdot H_2O + N_2, \\ &NO_2^+ \cdot H_2O + H_2O + N_2 \rightarrow NO_2^+ \cdot 2H_2O + N_2, \\ &NO_2^+ \cdot 2H_2O + H_2O \rightarrow H_3O^+ \cdot 2H_2O + HNO_3 \end{aligned} \tag{3.28}$$

with velocities $5 \cdot 10^{-28}$ cm^6s^{-1}, $2 \cdot 10^{-27}$ cm^6s^{-1} and $>10^{-10}$ cm^3s^{-1}, respectively (Fehsenfeld and Ferguson 1969). It can be regarded that $H_2NO_3^+ \equiv NO_2^+ \cdot H_2O$ (proton transition from H_3O^+ to HNO_3) which means the hydrated ions NO^+, NO_2^+ are in reality the protoned acids $H_2NO_3^+$ and $H_2NO_2^+$

It should be noted that as a result of the water molecules ionization the following ions are formed: H_2O^+, OH^+ and H^+, which, in turn participate in the following reactions:

$$\begin{aligned} &H_2O^+ + H_2O \rightarrow H_3O^+ + OH && (1.2 \cdot 10^{-9}\ cm^3s^{-1}), \\ &OH^+ + H_2O \rightarrow H_3O^+ + O && (1.5 \cdot 10^{-9}\ cm^3s^{-1}), \\ &H^+ + 2H_2O \rightarrow H_3O^+ + H_2O && (3.0 \cdot 10^{-9}\ cm^3s^{-1}) \end{aligned} \tag{3.29}$$

One can see that all of them lead to the formation of ion H_3O^+.

In Fig. 3.8 the provisional diagram of the ion structure formation near the surface of the Earth, according to the suggested model is submitted which takes into account the above proposed reactions. It is necessary to note, that the process of hydration begins immediately after the formation of the very first ions. In our model the assumption is made, that the basic processes of hydration will take place after the formation of the main ions NO_3^- and H_3O^+ in alpha-particle tracks.

3.5.9 Numeric Solution of the Kinetic Equations

Let W be the probability of molecular ion appearance, q [cm^{3}s^{-1}] – velocity of electrons birth. In the bottom layers of an atmosphere under the influence of ionizing radiation a direct ionization mainly takes place, leading to the formation of molecular ions of nitrogen or oxygen. That's why we will estimate the probability of a molecular ion appearance in the act of ionization as W=0.75. For molecular ions the birth velocities are equal to $A_i = n_i q W / N_L$, and for atomic ions they are $A_i = n_i q(1-W)/N_L$ where n_i - concentration of corresponding molecules [cm^{-3}], N_L = $2.687 \cdot 10^{19}$ cm^{-3} — Loshmidt constant.

The general kinetic equations can be written as (Boyarchuk 1997, 1999):

$$\frac{dn_e}{dt} \quad q \quad n_e \quad {}^{at}_{ij} n_j \quad {}^{rec}_{ij} n_j \quad {}^{det}_{ij} n^*_j$$

$$\frac{dn_i}{dt} \quad A_i \quad {}^{rec}_{ij} n_i n_j \quad {}^{n}_{ij} n_i n_j \quad {}^{rec}_{ij} n_i n_e$$

$$\frac{dn_i^-}{dt} \quad {}^{at}_{ij} n_e n_j \quad {}^{n}_{ij} n_i n_j \quad {}^{n}_{kl} n_k n_l \quad {}^{rec}_{ij} n_i n_j \quad {}^{det}_{ij} n_e n^*_j \qquad (3.30)$$

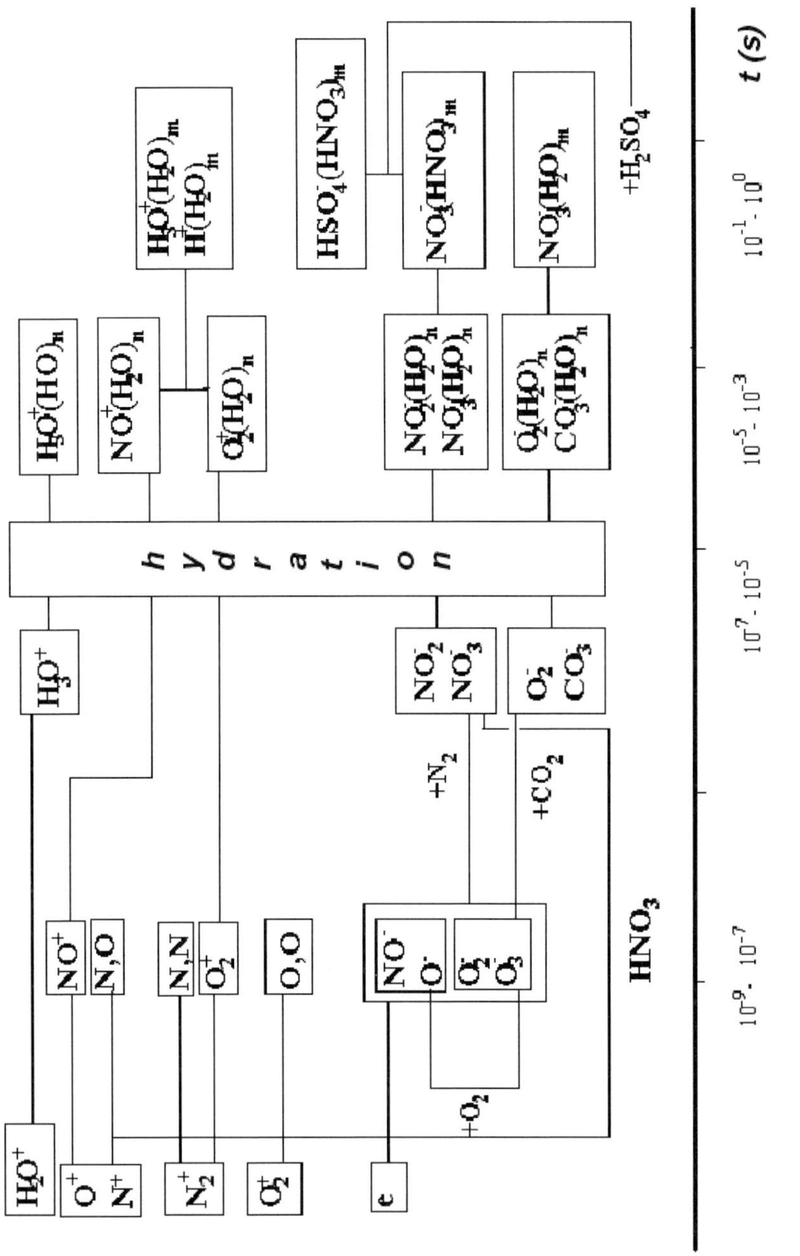

Fig. 3.8 Schematic presentation of the near ground kinetics with the basic ions formation

Here $n_i^{\pm}$ is the concentration of the positive and negative ions, n_e – electron concentration, n_i^* – concentration of excited molecules, α^{at}, α^{rec}, α^{det}, α^{n} – the velocities of attachment, recombination, detachment (negative ions destruction) reactions for electrons, and interaction with the neutral molecules respectively. It is not difficult to determine that this system satisfies the electric charge conservation law within the system: $n_e + \sum n_i^- = \sum n_j^+$

Let us consider the system (3.30) solution for the case of radionuclides concentration in the near ground layer of air ~ 10^{-4} Ki/cm^3. The natural nitrates concentration we put: O_3 – $5 \cdot 10^{11}$ cm^{-3}; NO – 10^9 cm^{-3}; NO_2 – $2 \cdot 10^{11}$ cm^{-3}; CO_2 – 10^{10} cm^{-3} (Smirnov 1992). The numerical solution of the system (3.30) by the Runge-Cutte technique within the time interval 0.1 s is shown in Figs. 3.9 and 3.10, which corresponds to normal and reduced concentrations of nitrates in the atmosphere; *t=0* corresponds to the beginning of ionization. For the initial period of time from 0 up to 10^{-7} s the ions kinetics within the tracks were considered, and since the moment of track degradation, the ion kinetics was considered in the total volume. Because of the low intensity of radioactive emission, according to the "soft" criterion accepted earlier, in kinetic processes the reactions of ions (basically negative ions) with neutral molecules of minor gas constituents of a near ground layer of an atmosphere will prevail.

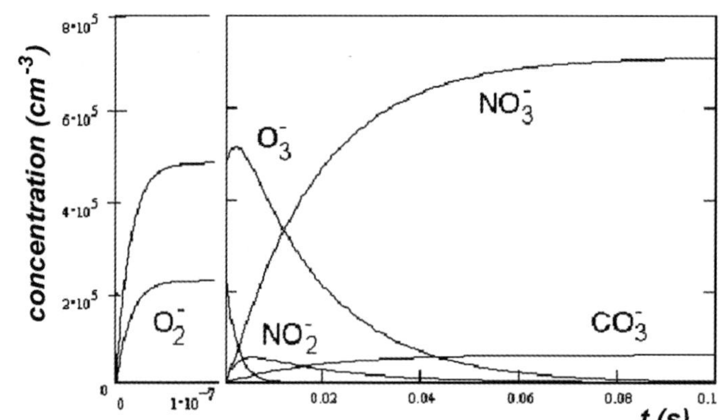

Fig. 3.9 Temporal variation of the concentration of ions *N* (cm^{-3}) for the case of the natural concentration of nitrates in the troposphere ([NO] ~ 10^9 cm^{-3}, [NO_2] ~ $2 \cdot 10^{11}$ cm^{-3})

One can clearly see from Figs. 3.9 and 3.10 how the formation of negative ions develops with time. The reaction velocity and the chosen reaction channel depends on the initial concentration of minor gas constituents of an atmosphere (Fig. 3.10).

As follows from the presented model, under the action of ionizing radiation the NO_3^- and H_3O^+ ions collect in the atmosphere. From our calculations it follows, that at the considered levels of ionization of an atmosphere, concentration of elementary ions is small in comparison with concentration of molecules of water vapor under usual conditions of relative humidity (70%~10^{17} cm^{-3} molecules of water vapor). Therefore with sufficient accuracy it is possible to consider that all negative elementary ions will be hydrated. This means that the result of the calculation of concentration of ions H_3O^+, NO_3^- and CO_3^-, submitted in Figs. 9 and 10, corresponds to the concentration of complex ions such as $H_3O^+\cdot(H_2O)_n$, $NO_3^-\cdot(H_2O)_n$ and $CO_3^-\cdot(H_2O)_n$.

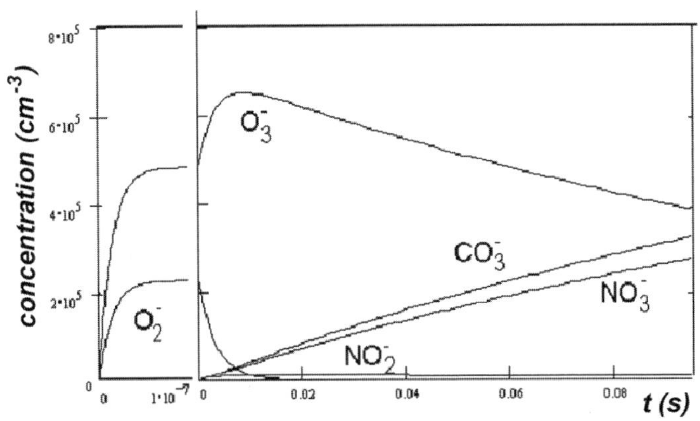

Fig. 3.10 Temporal variation of the concentration of ions N (cm^{-3}) for the case of reduced concentration of nitrites in a troposphere ([NO] ~ 10^8 cm^{-3}, [NO_2] ~ $2\cdot10^{10}$ cm^{-3})

3.6 On the Conception of Neutral Ion Clusters in Air

Summarizing the modeling described above one can conclude that under the action of an ionizing radiation in atmosphere in the initial stage a large amount of O_2^+ ions is formed both as a result of direct ionization, and as a result of charge exchange between an initial ion N_2^+ and electrons, which fast adhere to atoms of oxygen, since the oxygen has a significant energy of affinity to electrons, forming the negative ions O^- and O_2^-. As a result of fast ion-molecular reactions during an interval of the order 10^{-7} s the main elementary tropospheric ions will be formed: O^-, O_2^-, NO_2^-, NO_2^- NO_3^-, CO_3^- and O_2^+, NO^+, H_3O^+. The concentration of electrons is so insignificant, that they can be neglected. The large amount of water vapor molecules contained in the troposphere (~10^{17} cm^{-3}), having a noticeable dipole moment p=1.87 D, leads to hydration of elementary ions and the formation of ion complexes of a type $NO_2^- \cdot (H_2O)_n$ and $NO_3^- \cdot (H_2O)_n$, $NO_3^- \cdot (HNO_3)_n(H_2O)_m$ and $O_2^+ \cdot (H_2O)$, $NO^+ \cdot (H_2O)_n$, $H^+ \cdot (H_2O)_m$ and $H_3O^+ \cdot (H_2O)_n$ which happens rather fast.

The ions $NO_3^- \cdot (H_2O)_n$, $NO_3^- \cdot (HNO_3)_n(H_2O)_m$ and $H_3O^+ \cdot (H_2O)_n$ could be regarded as the main ions of the troposphere with a high level of probability. The average time of life of these ions reaches 30–40 min. and more under normal conditions (Smirnov 1992). However the physics of existence of such cluster ions and more complex constructions remains unexplored till now (Castleman 1982). In Fig. 3.11 one can find examples of similar structure formation. The models of the chemical composition of cluster ions are also controversial. In particular, mass-spectrometer and theoretical studies of ionization of mixtures like ozone-water vapor at pressure close to normal atmospheric pressure show, that even at the minimal concentration of water vapor, a significant amount of proton hydrates $H_3O^+ \cdot (H_2O)_n$ are observed (Smirnov 1992).

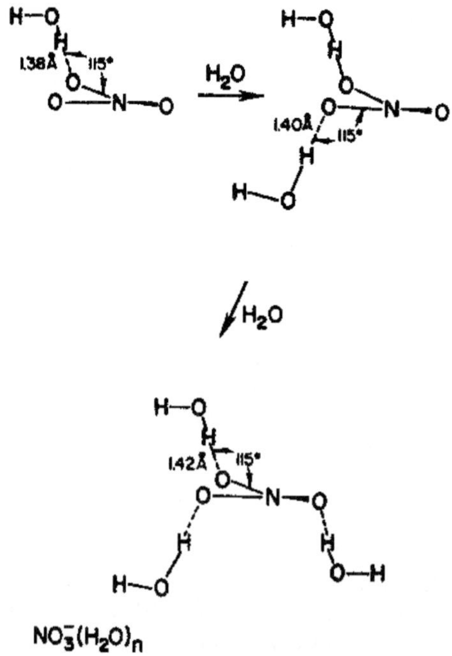

Fig. 3.11 An example of expected structure for two and three water molecules clustered around NO_3^- (Lee et al. 1980). Analogous chain-like structures are also predicted for the case of water molecules bound to the ion HCO_3^- (Lee et al. 1979)

In hydrated ions the cluster shell consisting of three up to five molecules of water, has a bond energy near 4 eV, it should not fail from impacts at a temperature of below 1,000 *K*. It is natural to assume, that such a shell can hinder recombination.

From times when the theory of electrolytic dissociation appeared, it is known, that if solvation heat surpasses the energy educed during the ion recombination, the shells around the ions will detain recombination. Such a situation is confirmed for electrolytes. It is natural to assume, that it is possible also for a gaseous phase, i.e. recombination detained when the energy of ion hydration surpasses the energy educed during their recombination (Stakhanov 1996). Table 5.3 gives the bond energies measured for water molecules in the hydrated ions.

Note, that for the reactions:

$$\begin{aligned} &H_3O^+(H_2O)_{n-1} + H_2O \rightarrow H_3O^+(H_2O)_n \\ &OH^-(H_2O)_{n-1} + H_2O \rightarrow OH^-(H_2O)_n \end{aligned} \quad (3.31)$$

the ion of oxonium H_3O^+ is in the reality the product of proton hydration by the single water molecule: $H^+ + H_2O \rightarrow H_3O^+$. But the bond energy educed after this reaction is so high (7.18 eV) in comparison with bond energies of subsequent water molecules that it is better to speak of the oxonium ion hydration than the proton hydration. One can see from Table 3.5 that the bond energy of the hydrating shell from seven water molecules of the ion $H_3O^+(H_2O)_7$ is 5.4 eV, and $OH^-H_2O)_7$ is 4.5 eV. This means that for the destruction of these two cluster ions to form H^+, OH, *e* and 15 water molecules it is necessary to spend 18.8 eV. At the same time during the proton-electron recombination only 13.5 eV is educed. Even if we take into account that the newly formed hydrogen atom can unite with the OH group, forming one more water molecule and educing near 5 eV, we will obtain as a total only 18.5 eV.

Table 3.4. Experimental values of the bond energey for water molecules (eV) in hydrate shells of H_3O^+ and OH^- (Kebarle et al., 1967; Arshadi and Kebarle, 1970).

ΔH	n-1; n 0; 1	n-1; n 1; 2	n-1; n 2; 3	n-1; n 3; 4	n-1; n 4; 5	n-1; n 5; 6	n-1; n 6; 7
$H_3O^+(H_2O)_n$	1.56	0.97	0.74	0.67	0.57	0.51	0.45
$OH^-(H_2O)_n$	1.1	0.71	0.66	0.62	0.61	-	-

Thus, the system of hydrated ions appears to be energetically stable. Experimentally are measured only the recombination cross-sections of the positive hydrated oxonium ions with a free electron (Biondi et al. 1972). Recombination coefficient with an electron appears to be large and grows with increase in number of molecules in the hydrate shell of clusters from 10^{-6} for H_3O^+ up to 10^{-5} cm^3s^{-1} for $H_3O^+(H_2O)_5$. On could expect this because of the increase in their geometrical sizes. However the recombination coefficients of positive and negative hydrated ions at n=3–5 have not been measured yet.

From the above it is possible to assume, in particular, that newly formed ions (such as N_2^+ and O_2^-) will interact in a different way to that which is characteristic for the usual air ions such as $O_3^-(H_2O)_n$ and $H_3O^+(H_2O)_m$. One may expect also the difference in the recombination coefficients of ions in α- particles tracks and during the volumetric ionization of air, for example, by X-ray radiation. Retardation of the recombination processes of ions in the bottom layers of tropospheric plasma has been noticed recently in papers of the group from FIAN (Stozhkov et al. 1997) at altitudes 0–35 km where experiments were conducted. It was shown that the ion balance is described by a linear equation, instead of the square-law.

Retardation of ion cluster recombination can lead to formation in an atmosphere of the electrically neutral associates, formed under collision and association of positive and negative light and middle-sized ions. In the dusty plasma this effect is called coagulation (Horanyi and Goertz 1990). A similar hypothesis was proposed earlier to explain a number of observable effects of the air artificial ionization, particularly, the formation of condensate particles (Smirnov 1983, 1987, 1980, Fig. 5.12). The numerical modeling of classic Coulomb plasma (Mayorov et al. 1994; Yakovlenko 1995; Stakhanov 1996) by multiple particle approximation has shown an essential (up to 12 orders of magnitude) delay in the recombination time of a system composed of heavy particles of identical mass, and the conditions are reviewed, in which the plasma of ions with envelopes of water molecules can be sustained in a supercooled ionized state for an anomalously long time period.

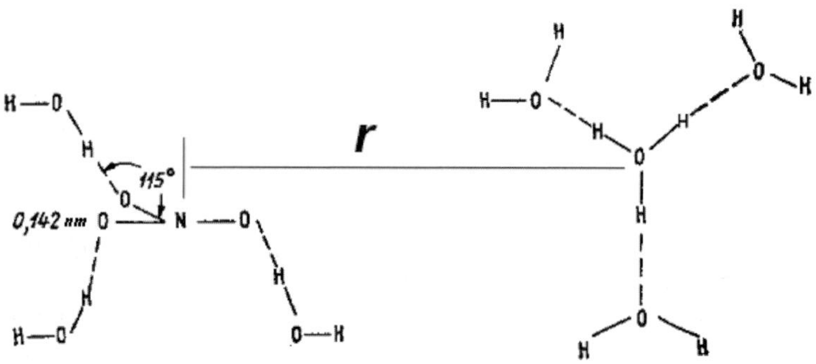

Fig. 3.12 The geometry of hydration clusters with saturated association on the basis of ions NO_3^- and H_3O^+

The basic conditions of existence of ion associates within the atmosphere are: firstly, the presence around each molecular ion (the nucleus) of shells from oriented molecules of water vapor, which prevents the transition of a free electron from one nucleus to another. There are always a lot of water molecules in the bottom layers of the troposphere, and this condition is usually fulfilled. As a second condition, it is necessary to fit the initial condition of the shell strength to prevent its destruction and neutralization of charges:

$$E_b + E_h \geq \varphi \cong \frac{q^2}{4\pi\varepsilon_0\varepsilon_*\rho_i} \tag{3.32}$$

where $E_b \cong 1$ eV – is the energy of the ion-molecular bond close to the total hydration energy; E_h – is the energy of the valent (hydrogen) bond between the adjoining hydrogen and oxygen ends of molecules of water (every bond is of order the 0.15±0.5 eV), $\varphi \cong 1.5$ eV, is a Coulomb potential between the molecules drawn together to the distance close to the sum of ions radius ($\cong 10^{-9}$ m); $\varepsilon_* \cong 2$–3 – dielectric permeability within the complex of ions.

The degree of how this non-equality will be fulfilled, depends on the ion size and on the width of the ion shell from the dipole molecules. To avoid the disruption of the system it is necessary to satisfy the condition:

$$E_{\text{thermal energy}} = E_{\text{electrostatic}} : r \leq \frac{q^2}{4\pi\varepsilon_0\varepsilon_* kT} \tag{3.33}$$

The example of such a complex is:

$NO_3^- \cdot (H_2O)_n$, $NO_3^- \cdot HNO_3H_2O$ and $H_3O^+ \cdot (H_2O)_m$

The reaction of association of such a complex is the record of the dissociated nitric acid formula:

$NO_3^- \cdot (H_2O)_T + H_3O^+ \cdot (H_2O)_m \Leftrightarrow HNO_3 + H_2O(n+m+1)$

As a result the neutral cluster is formed:

$NO_3^- \cdot (H_2O)_p H_3O^+$ where p = m + n

Its stability is determined both by ion and covalent bonds.

It is necessary to note, that the received estimate of the common energy of the internal bond in a neutral complex formed from two ions of different polarity, gives a rather high value ~1.5 eV, in comparison with the average thermal energy ~0.026 eV.

Thus, if the ion has no hydrate shell, after the collision of two ions of different polarity their recombination happens without formation of the neutral cluster. But as we mentioned above in the process of the air "aging" the complex structures are formed of the ion radical type: $H_3O^+(H_2O)$ and $NO_3^- \cdot (H_2O)_n$, $NO_3^- \cdot (HNO_3)_n(H_2O)_m$ etc. As a result of their association the neutral cluster is formed of the type: $NO_3^- \cdot (H_2O)_p\ H_3O^+$. Its stability is determined both by ion and covalent bonds.

If we assume, that all acts of ion collisions result in the formation of ion pairs and that light colliding ions are aged, i.e. have an optimum chemical composition, the equilibrium concentration of neutral associates (the loss mechanism – is the collision with each other) in an atmosphere at background ionization will be ~10^5 cm^{-3}, and in a zone of strong ionization ~10^9 cm^{-3}.

With the presence of an external influence on such atmospheric plasma, the neutral associates can be destroyed, and this process will locally add one more source of charged particles. For example, during the process of spatial charge separation in a near-ground layer of an atmosphere as a result of action of the so-called electrode effect, or during the drop charge separation in a cloud or a fog, the "priming" electric fields can be generated whose magnitude will be sufficient to destroy these neutral associates. The process of destruction of the neutral associates can be described within the frame of theory developed by Kikuchi (2001) of the particle acceleration in the electrostatic field. The last will add negative ions in this local volume of an atmosphere that will lead to the strengthening of effect of spatial charge separation and increase the electric field. Thus, even at background concentration of a source of ionization there can be conditions at which formation of a significant spatial charge is possible. With the presence of a source of artificial ionization the effect amplifies. In all these cases the neutral associate is the accumulator of the long-living ions, ready to "engage" under the first request.

The concept of neutral associates has an essential deficiency: there are no direct methods and devices to measure any characteristics of particles in an atmosphere in sizes around 1 nanometer with a zero electric charge. Nevertheless, all the indirect experimental facts regarding anomalous electric field generation before strong earthquakes testify to the abrupt appearance of the large quantity of ions for very short intervals which may be provided solely by the destruction of the preliminary "prepared" neutral associates.

3.7 Electric Field Generation

It is well known that the Earth degassing has many components (Toutain and Baurbon 1998). Among them most attention is attracted to radon because of its radiation properties. This is important due to ecological effects (Segovia et al. 2002). In the seismically active areas radon emanation can reach 25 eman (Allegri et al. 1983) which corresponds to the ion formation velocity of the order $Q \sim 16 \cdot 10^3 cm^{-3}s^{-1}$. As was shown above as a result of ionization long living ion complexes can be formed in the near ground layer of the atmosphere of opposite signs.

It was determined experimentally that the mobility of the negative ions is 1.3–1.4 times higher than the positive ones. This can probably be explained by the asymmetry of the removal energy of ions of different signs from the oxygen atom in the water molecule. This means that the negative ion possesses lower energy than the positive one, i.e. a lower number of associated water molecules (Briant and Burton 1976). According to the ion classification proposed in Smirnov (1992), we relate the considered ion complexes to the type of light or intermediate ions with mobility's 0.05–5 $cm^2/(V \cdot s)$. Therefore, due to the different mobility's, which ions of different sign have, under action of the natural atmospheric field a non-compensated charge can be formed at the ground surface. It will distort the atmospheric electric field near the ground. This distortion is named the "electrode effect" (Hoppel 1967).

3.7.1 Electrode Effect

Let us consider the simplified situation when under slight turbulent diffusion (for example, before sunrise) the radioactive cloud will propagate in the narrow near-ground layer where the ions will be earned (Fig. 3.12).

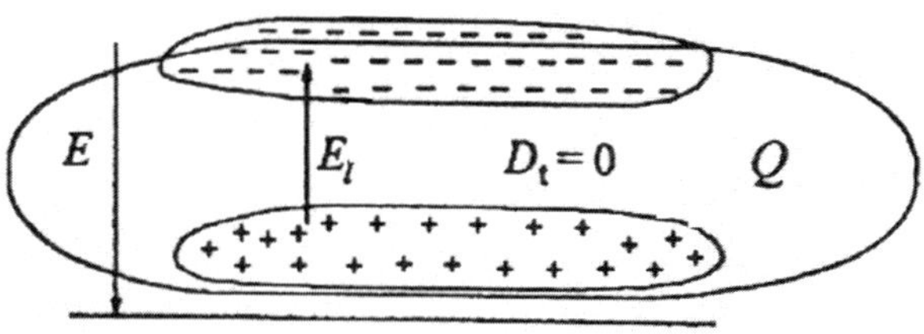

Fig. 3.13 Positive and negative space charges near the Earth's surface

Under action of the natural atmospheric electric field E the positive ions will move to a surface of the Earth where they will recombine, but because of their low mobility after some time the spatial layer of positive ions is formed at the surface. Negative ions will move vertically upwards (the electrons within the frame of this model may be neglected because of their low concentration at the surface of the Earth). In such a way the near ground "electrode layer" is formed with the local electric field E_l, which diminishes the natural atmospheric electric field E-E_l. This means that inside the electrode layer the electric field will be smaller than the natural one, and even sometimes can change its direction in comparison with the natural one. Over the electrode layer due to the presence of the non-compensated electric charge, the electric field will increase. We suppose in our model that the time of restoration of the ground surface charge is much less then all other characteristic times of the processes presented in the model, so we will consider the surface charge unchanged. So the main consequence of the electrode effect is the formation of a non-compensated electric charge over the ground surface. The turbulent movements and regular winds may convey this charge inside the near ground atmosphere creating an anomalous electrode layer over the large surfaces. In reality such situations may be maintained for a short period of time, because the further turbulent movements will destroy the electrode layer mixing all ions.

We can describe the process kinetics in the following form:

$$\frac{\partial n_+}{\partial t} = \frac{\partial}{\partial z}\left((D_T + D_+)\frac{\partial n_+}{\partial z}\right) - b_+\frac{\partial}{\partial z}(En_+) + Q_0 - \alpha^{rek} n_+ n_-,$$

$$\frac{\partial n_-}{\partial t} = \frac{\partial}{\partial z}\left((D_T + D_-)\frac{\partial n_-}{\partial z}\right) + b_-\frac{\partial}{\partial z}(En_-) + Q_0 - \alpha^{rek} n_+ n_-, \tag{3.34}$$

$$\frac{\partial E}{\partial z} = 4\pi e(n_+ - n_-).$$

where e – electron charge, z – vertical coordinate (we consider the "condenser" to be infinite in the horizontal plane and calculate for the one-dimensional case); $n_\pm$ – the positive and negative ions concentration; Q_0 – velocity of the complex ions formation. The spatial distribution of the ion formation velocity connected with the radon emanation, is worth to determine in exponential form Q=Q_0e$^{-z/h}$; where h is the ionization layer height; $b_\pm$ and $D_\pm$ – are the mobility and diffusion coefficients for the corresponding ions; D_T – the turbulent diffusion coefficient determined according to Hoppel (1962); α^{rek} – ion recombination coefficient ~ 10^{-6} – 10^{-7} sm^{3}s^{-1} (Boyarchuk 1997).

We will define the boundary conditions in the following way: electric field magnitude near the ground surface $E_{Z=0}$=100 V/m, and at the boundary of the considered layer – - $\left.\frac{\partial \mathrm{E}}{\partial \mathrm{z}}\right|_{z=\infty} = 0$.

The initial values of the positive and negative ions concentrations will be set equal to the background concentration ~450 cm^{-3}.

In Fig. 3.14 we present the results of calculations according to the model. The concentration of the positive and negative ions, and the electric field are calculated after 53 s from the start of ionization (radon exit). One can see well in the graph the process of the electrode layer formation where the electric field diminution is observed. Over the layer the electric field magnitude growth is shown. The increase of the positive and negative ions mobility's will increase the effect.

Different small fractures and admixtures will have an effect on the final ion composition, changing mainly the light negative ions content. Having a higher affinity to electrons then the oxides of nitrogen they can replace the latter from the base of the complex ions and to the central ion – $M^-\cdot(H_2O)_n$. The aerosols with a high metal content ("metallic aerosols") carried by the gas discharges from the Earth's crust, have very high affinity to electrons and may become a base for the negative ion complexes, and increase the negative ion concentration significantly. Their exists experimental indication that the exit of the metallic aerosols may increase several times before strong earthquakes (up to 1.5 order of magnitude, Alekseev and Alekseeva 1992; Alekseev et al. 1995).

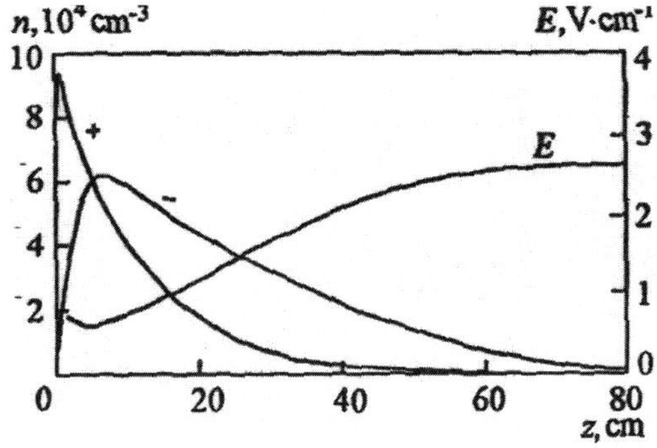

Fig. 3.14 Concentrations n of positive (+) and negative (–) ions and the electrostatic field E as functions of the altitude z near the surface 50 s after the radon emanation onset. The nobilities of negative and positive ions are $b_-=3.8\cdot10^{-1}\text{cm}^2\cdot(\text{V}\cdot\text{s})^{-1}$ and $b_+=2.4\cdot10^{-1}\text{cm}^2\cdot(\text{V}\cdot\text{s})^{-1}$, respectively; $D_+=2.8\cdot10^{-2}\text{cm}^2\cdot\text{s}^{-1}$ and $D_-=4.3\cdot10^{-2}\text{cm}^2\cdot\text{s}^{-1}$ at weak turbulent diffusion (K=0.1 $\text{m}^2\cdot\text{s}^{-1}$) and h=10 cm

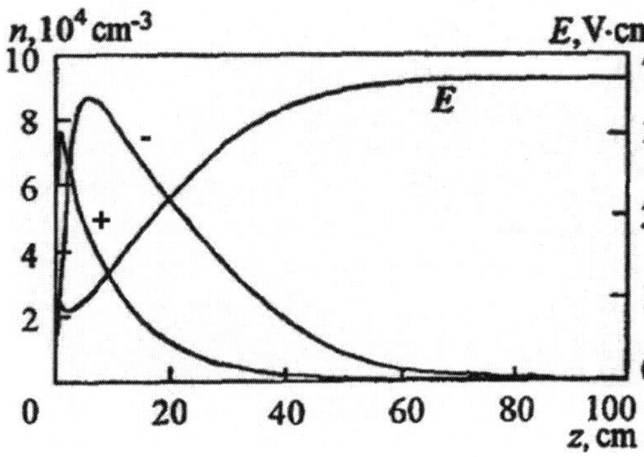

Fig. 3.15 Concentrations n of positive (+) and negative (–) ions and the electrostatic field E as functions of the altitude z near the surface 50 s after the radon emanation onset with the additional flow of metal aerosols from the Earth's surface

The situation with the exit of aerosols is modeled and shown in Fig. 3.15. One can see the essential increase of the volume charge at the expense of negative ions based on the metallic aerosols. This large negative charge leads to the increase of the electric field over it. But in this case the diminishing of the electric field within the electrode layer is lower. In Fig. 3.16a is presented the time dependence of the field evolution at the height of 60 cm (curve 1) over the negative electric charge and at the height of 3.5 cm (curve 2) within the electrode layer. One can see that after 40 s the process of anomalous field formation is stabilized.

One should take into account that the ions (of one or another sign) will be led out of the reaction volume by the external electric field E which, in turn will lead to the recombination rate slowing down because it is necessary for recombination to have the ions of both signs. It will be equivalent to the gain of the ions of one sign. This phenomenon also leads to amplification of the field generation effect. One can see in Fig. 5.16b the field temporal evolution at two heights as a result of ion generation fallout after 50 seconds. One can see that the natural field reconstruction process goes lower and weaker than the field generation and takes more than 200 s.

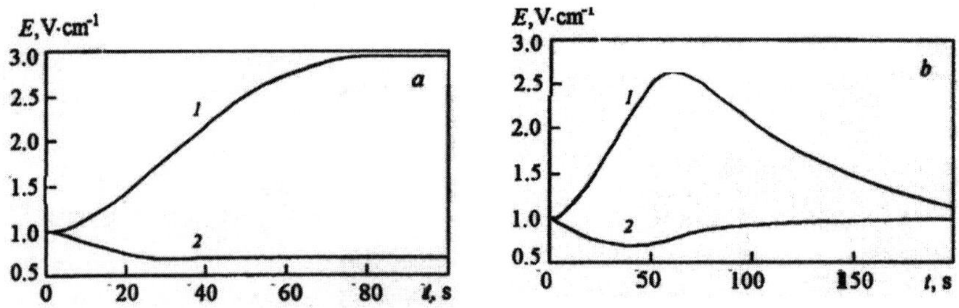

Fig. 3.16a,b Time evolution of the near-surface electrostatic field at an altitude of (*1*) 60 and (*2*) 3.5 cm: (**a**) permanent ion formation (**b**) time of ion formation is 50 s

3.7.2 Electric Field Reversals

In Fig. 1.10 three examples are presented of anomalous electric field registration before strong earthquakes. All examples demonstrate the phenomenon of the field reversal in comparison with the direction of natural atmospheric field (which is directed down). This phenomenon is not unique only for the cases of earthquakes. It is also observed during the passing of thunderstorm clouds over the observation site and for ground events such as fog or mist (see Fig. 5.17, Mühleisen 1958).

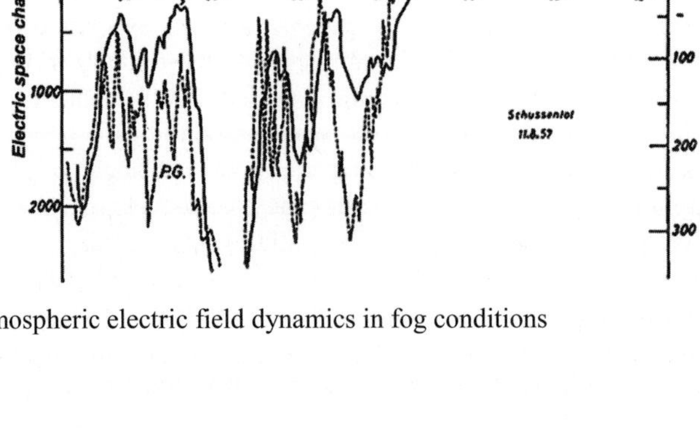

Fig. 3.17 Atmospheric electric field dynamics in fog conditions

The effect of the water drops on the generated electric field was calculated in Pulinets et al. (2000) and is presented in Fig. 3.18

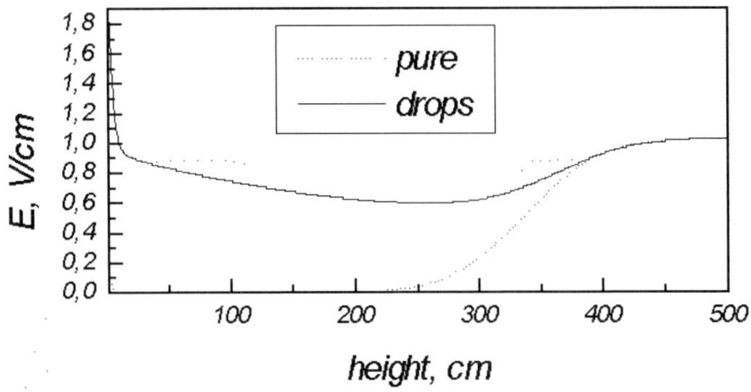

Fig. 3.18 Change of the electric field due to the negative charge transfer by the fog's drops

The major mechanism for charge separation here is the process of transfer of a negative charge downwards by falling drops. It can be seen in Fig. 3.18, that the negative charge transferred by drops of fog on the Earth's surface compensates the electrode effect. In this case the field decreases slowly.

However, if there is a significant source of positive ions above the surface, the situation will strongly change – the field over the electrode layer can change its sign. For example, there are experimental observations of fluxes of Fe_3^+ ions over the ocean surface in seismic activity regions before strong earthquakes or volcanic eruptions (Hata et al. 1998). It is known, that the ion production rate does not exceed a value of $cm^{-3}s^{-1}$ above a water surface under normal conditions. Therefore even with an insignificant flow of positive ions from a surface of water, the positive ions will not have time to recombine with negative ions, since the concentration of negative ions is not enough for neutralization of the positive ions. As a result, at the surface of water a positive charge layer will be formed, which can have an effect on the intensity of the local electrical field at the surface. The last case is presented in Fig. 3.19.

Positive charges can appear on the ground surface as well. A recent experimental study (Freund 2000) has indicated a direct connection between the stress applied to rocks and the generation of charge carriers, causing electric currents and affecting surface potentials. It was also shown theoretically (King and Freund 1984). Under the stress the rocks become a "positive hole" semiconductor due to existing peroxy links $O_3Si/^{OO}\backslash SiO_3$. A peroxy represents a positive hole pair (PHP), which is electrically inactive and dormant. During crack formation an acoustic wavelet will activate positive holes, which will propagate outward to the ground surface. The ground is expected to have a highly positive potential as a result of this phenomena. This potential will induce the flux of the positive ions into the atmosphere leading to the same effect as was described for the ocean surface. This means that over the ground surface we can expect the effect of the field overturn presented in Fig. 3.19, and registered experimentally before strong earthquakes.

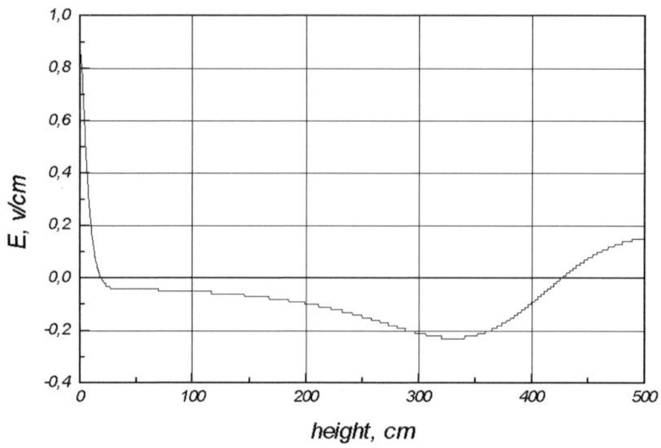

Fig. 3.19 Electric field overturn under the presence of positive ions flux

3.7.3 Active Experiments

The natural phenomenon of atmospheric precursors of earthquakes considered in the present chapter is well modeled by the effects of artificial ionization of an atmosphere (Solovyov 1941) and changes in an atmosphere under nuclear tests (Holzer 1972). In experiments on artificial ionization by *X*-ray emission, from the moment the X-ray radiators switch-on, the field dropped and was established during 0.5–1.0 minutes, Fig. 3.20. Reduction of a field was observed at significant distances up to 1 km from a radiation point. After deenergizing of the *X*-ray installation, electric field slowly, during 5–10, and sometimes 20–40 minutes came back to the normal state. The long lasting relaxation tells us of the long-term existence of a negative volumetric charge, the positive charge will be quickly neutralized by a terrestrial surface (Solovyov 1941).

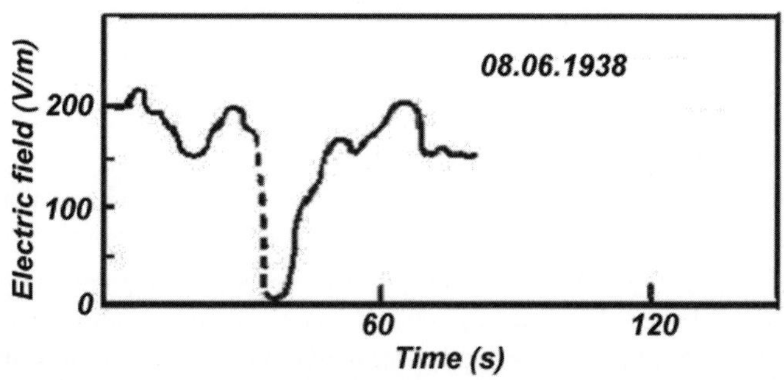

Fig. 3.20 Variation in the electric field *E* recorded by a Benndorf electrograph during the near-surface air ionization with *X*-rays (*dashed line*) and in the absence of *X*-rays (*solid line*)

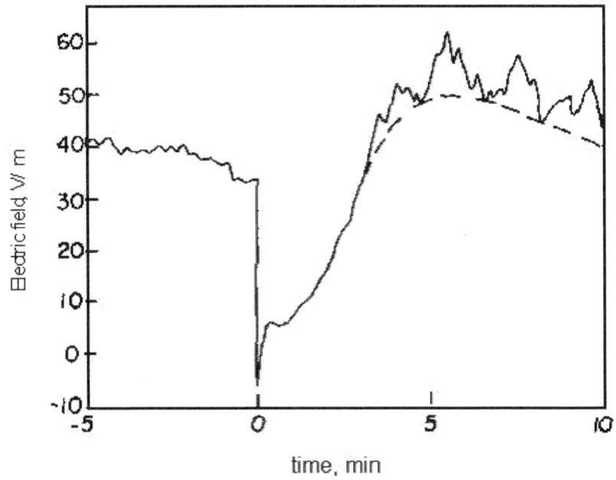

Fig. 3.21 Variation in the electric field E for the nuclear explosion at a distance of 7.8 km (Holzer 1972)

In the case of electric field measurements in the area of underground nuclear tests a similar situation is observed – the sharp falling of an electric field at the moment of the explosion, then followed by the slow restoration of a normal value for the field, please see Fig. 3.21. Ionization of a ground layer is achieved here by fission products leaving the ground.

The results presented here, change to a considerable extent the present view of the mechanisms of anomalous electric field generation in the vicinity of an impending earthquake and permit us to conclude:

- We obtained the theoretical support for the hypothesis of Pulinets et al. (1994, 1997) on the defining role of radon emanation and aerosols in the electric field generation within the area of earthquake preparation.
- The obtained results form the background to the calculations of anomalous electric field effects within the ionosphere and allow us to account for the mechanisms of generation of ionospheric precursors.

Chapter 4 Physics of Seismo-Ionospheric Coupling

4.1 Introduction

Prof. Seiya Uyeda, member of the Japan Academy of Sciences and Chairman of Inter Association Working Group on Electromagnetic Studies of Earthquakes and Volcanoes (EMSEV) of IAGA/IASPEI/IAVSEI, told me at one of the Commissions meetings that the electromagnetic phenomena associated with earthquakes would be recognized as real earthquake precursors only when the physical mechanism of their origin would be clarified. The purpose of the present chapter is to establish the real physical relationships between different parameters of atmosphere and near Earth plasma, which demonstrate their distinct variations before strong earthquakes. Several groups in the World are now working in this direction. Regardless of different approaches in their calculations, their conceptions have a lot in common, acknowledging the electromagnetic character of seismo-ionospheric coupling.

Actually, the problem of seismo-ionospheric coupling is one of particular cases of the more general problem of electromagnetic coupling of the atmosphere and ionosphere. This problem was studied for years but mainly in a traditional aspect where the Sun was regarded as the main source of energy, and the energy propagation direction, as well as the coupling itself was regarded only from top to bottom. Even recently developed documents and projects (Understanding Plasma Interactions 2001) look at the problem from this point of view. The NASA Space Mission GEC (Geospace Electrodynamic Connections) described in the cited report with 2008 as a planned launch date, regards the atmosphere only as a redistributor of the energy arriving from the magnetosphere. But in Fig. 1.3 of this document one can find a new element – the large red arrow indicating the action of acoustic gravity waves and tides from bottom to the top.

At the same time yet from the 30th another group of scientists developed the problem of the global electric circuit (The Earth's Electrical Environment 1986; Roble 1991). Regardless of some progress in the understanding of cloud charging processes (Tinsley 2000), and other problems concerning the details of thunderstorm discharges, the conception itself has not changed since the 30-th. Still the thunderstorm discharges were regarded as the main sources of potential difference between the ground and ionosphere, and closing the global electric circuit are a system of polar and auroral currents and a fair weather return current. The system is regarded as a condenser with a top plate – the ionosphere, and a bottom plate – the ground. This conception is fully described in Bering III et al. (1998).

Some agitation appeared after the discovery of the red sprites and the blue jets phenomena – electric discharges from the top of thunderstorm clouds up to the ionosphere (Sentman et al. 1995). It was a very strong and clear indication of the other direction of energy transfer: from the atmosphere into the ionosphere. It was a signal indicating that the global electric circuit is not as simple as perceived. International groups and projects appeared trying to reconsider the conception of the global electric circuit (SPECIAL, CAWSES). And the most important step made within the frame of these projects – is the attempt to unite the problems of solar-terrestrial relationships with the problems of the global electric circuit and climate. But again, the problem of penetration of the quasistationary electric fields of seismic origin into the ionosphere was not included in these projects. The red sprites and blue jets phenomena are transient and small-scale events in comparison with the large scale seismogenic electric fields.

The problem to be reviewed in the present chapter is closest in its statement to the problem of thunderstorm cloud electric field penetration into the ionosphere. Here we deal not with the transient effect on the ionosphere from below like red sprites but with the global system of thunderstorm activity with a time constant of the order of at least several hours. The first correct formulation of the problem was made by Park and Dejnakarintra (1973, 1977). Other approaches to the problem (Tzur and Roble 1985) did not put in question the results of Park and Dejnakarintra. Even more, this formulation was specially tested by Rodger et al. (1998) and was recognized to be correct. The formulation was modified by Hegai et al. (1990) what permitted an increase of the field penetration effectiveness by 10 times for daytime and by 100 times in nighttime conditions. The effect of ionosphere modification by the global system of thunderstorm activity was experimentally proved by Pulinets and Depuev (2003) using the global maps of electron concentration distribution from the Intercosmos-19 satellite topside sounding data and comparing them with the global distribution of the thunderstorm activity obtained with a satellite optical transient detector (OTD) device (Christian et al. 2003).

And finally, we should pay attention to the atmospheric wave and tide motions effect on the ionosphere. As one of the first publications on this subject we can mention Bowhill (1969) who introduced the term "coupling from below". But the most recent information and present state of the art one can find in the review papers Kazimirovsky (2002) and Kazimirovsky et al. (2003). In general, the coupling mechanism works due to the neutral atmosphere motions (in the form of internal waves, acoustic gravity waves, atmospheric fronts etc.), which dissipate in the thermosphere involving in their motion the ionized part of the thermosphere and "switch on" by this motion the electrodynamic forces causing the variations in the ionosphere.

Actually, we live now in the transient period when different disciplines which existed separately, unite to explain ionosphere variability as a result of forcing and coupling from above (solar-terrestrial connections), and from below (meteorological effects, electric fields, anthropogenic effects, etc.).

Concerning the effects from below, we can separate two channels of energy transfer: a wave (or acoustic) channel through wave motions of the neutral atmosphere and an electrodynamic channel through atmosphere conductivity changes and quasistationary electric field effects. During the final stage of earthquake preparation one may expect both effects: generation of acoustic gravity waves by gas discharges and large-scale electric field for which we have experimental evidences. Both these possibilities will be reviewed in the following paragraphs.

4.2 Wave Channel

The hypothesis on the possible internal gravity (IGW) or acoustic gravity waves (AGW) generation before earthquakes was proposed by many authors. For example, the excitation of atmospheric oscillations by seismic gravitational vibrations was proposed by Garmash et al. (1989) and Lin'kov et al. (1990). And Shalimov (1992) proposed the AGW effect on the atmosphere as a possible source of the disturbances observed in the ionosphere before strong earthquakes and as a mechanism of seismo-ionospheric coupling. These ideas are summarized in the paper of Shalimov and Gokhberg (1998). Among the different possibilities of excitation of the air oscillation the authors propose three: (1) "piston-like" motion of the Earth's crust having a block structure, (2) unsteady thermal anomalies, created by greenhouse lithospheric gas input into the atmosphere, (3) unsteady mass input of lithospheric gas into the atmosphere. Taking into account that up to now there is no experimental indication of the fast vibration-like crust movements before an earthquake, the gas discharges in the vicinity of an active tectonic fault become the prevailing source of the air oscillation (Voitov and Dobrovolsky 1994).

4.2.1 Source Description

Gokhberg et al. (1996) proposed the ground velocity description on the Earth's surface as a linear source of the temperature anomaly:

$$v_z(x) \approx \Delta z A \approx \frac{\Delta z}{gH} \frac{\omega k_b \Delta T}{m_n} ch^{-2}\left(\frac{x}{l}\right) \tag{4.1}$$

where ΔT~3 K is the magnitude of the temperature anomaly caused by green-effect gases, l~100 km is the lithospheric source width, Δz~3 m – the thickness of the near ground gas layer. This one-dimensional approach was modified by Mareev et al. (2002). The authors regard two types of sources: the strip-like with different distributions inside the strip, and a disk-like source area also using different distributions inside the disk. To avoid a complex formulation of the problem, which one can find in the cited paper, we provide only the picture, which allows a clearer understanding (Fig. 4.1). For the strip-like source distribution used by Gokhberg et al. (1996) the authors propose three types of gas source distribution inside the strip (Fig. 4.1a – left panel). (1) – ordinary distribution, (2) – large scale structuring, (3) – small scale structuring. Their calculations show that a small-scale structured source radiates internal gravity waves more effectively with a factor ($l/L>>1$) where L is the characteristic scale of the used source function. The right panel of Fig. 4.1 demonstrates the radiated power calculated for three types of distribution of the gas sources inside the bar.

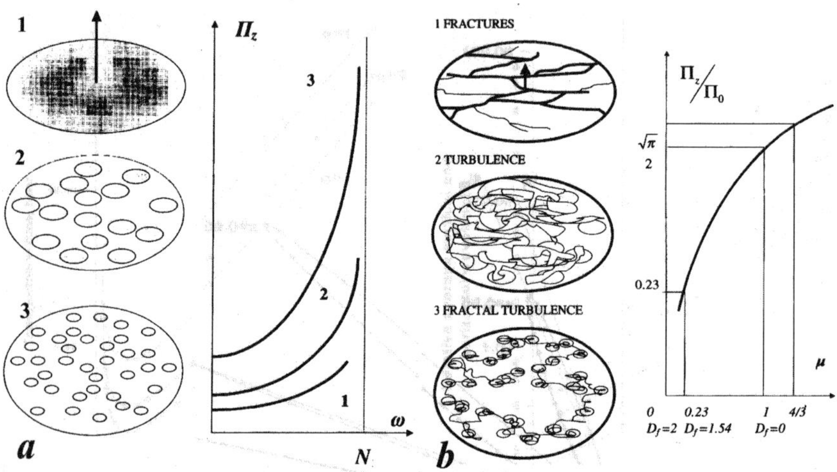

Fig. 4.1a,ba *Left panel* – 3 types of the gas sources distribution: 1 – ordinary distribution, 2 – large scale structuring, 3- small scale structuring. *Right panel* – calculated radiated power for the correspondent source distribution. **b** *Left panel* – 3 types of gas sources distribution within the disk-like source area. *Right panel* – dependence of radiated power of the fractal size. (After Mareev et al. 2002)

It seems that the more realistic is the disk-like distribution proposed by Mareev et al. (2002). Inside this disk (by which is meant the earthquake preparation area) three types of random sources are regarded which are described by different fractal dimension of the turbulence.

The common conclusion for both types of sources is that the total radiated power increases proportionally to the relation between the size of the area occupied by gas or temperature inhomogeneities, and their characteristic scales both for one-dimensional (strip-like) and two dimensional (disk-like) regions.

4.2.2 Wave Propagation and Penetration into the Ionosphere

The majority of authors calculating the gravity wave propagation from the ground up to the atmosphere and ionosphere use the classic approach of Hines (1960) with some minor modifications in terms of equations set 4.2–4.4. The consideration is always conducted in quiet undisturbed isothermal atmosphere without dissipation, which brings too optimistic and sometimes misleading conclusions.

Equations with sources of heat $A(\vec{\mathbf{r}},t)$ and mass $q(\vec{\mathbf{r}},t)$ (Mareev et al. 2002) will be:

$$\frac{\partial \rho}{\partial t} + div\rho \vec{V} = \rho q(\vec{r}) \tag{4.2}$$

$$\rho \frac{\partial \vec{V}}{\partial t} = \rho \vec{g} - \nabla p \tag{4.3}$$

$$\frac{dp}{dt} - \gamma \frac{p}{\rho}\frac{d\rho}{dt} = A(\vec{\mathbf{r}},t)\rho \tag{4.4}$$

were ρ is the air density, p is the air pressure, and $\vec{V}$ is the amplitude of air velocity in the wave. With the z-axis directed up the relation of air pressure and density will be:

$$\frac{dp}{dz} = -g\rho(z) \tag{4.5}$$

The air density changes in isothermal atmosphere exponentially:

$$\rho = \rho_0 \exp(-\alpha z) \tag{4.6}$$

where α is the reversed atmosphere scale height $H=g/RT$.

Even without any calculations the estimations show that in such an approach the vertical component of the air velocity will grow exponentially in accordance with (4.6):

$$u \sim \frac{\rho \exp(-\alpha z)}{\rho_0} \tag{4.7}$$

Shalimov (1992) shows that if on the ground surface the relation $\rho/\rho_0 \sim 10^{-6}$ at altitude 100 km under H=5 km will be $\rho/\rho_0 \sim 0.1$. So one obtains incredible amplification of the vertical air velocity component, which then due to interaction with the charged part of the atmosphere (ionosphere) through ion-neutral collisions causes variations of plasma density within the ionosphere.

4.2.3 AGW Effects on the Ionosphere

For the ionospheric conditions in the E and F layers of the ionosphere (Liperovsky et al. 1992) the ion-neutral collision frequency ν_{in} and ion gyrofrequency ω_{Hi} will be much higher than the frequency of the gravity wave ν_{in}, $\omega_{Hi} >> \omega$. For the cold plasma approximation the ionosphere response will be described by the continuity equation:

$$\partial n / \partial t + n_0 div\mathbf{v} + \mathbf{v}\nabla n_0 = 0 \tag{4.8}$$

Assuming $n_0 = n_0(z)$, $k(z) = \nabla_z n_0/n_0$, and $n \sim \exp(-i\omega t + ikx + ik_z z + z/2H)$ for gravity wave with $\omega << \omega_b$ where ω_b – the Brunt-Väisälä frequency $\omega_b = \dfrac{g\sqrt{\gamma-1}}{c_s}$

(γ-adiabatic constant) for $k_z >> k,\ 1/H$ for two ultimate cases one obtains:

1. $\omega_{Hi} << \nu_{in}$, – E-layer

$$\frac{n}{n_0} \approx -\left[\frac{\rho}{\rho_0}\right]\frac{\gamma}{\gamma-1}(kH+1+\gamma) \tag{4.9}$$

2. $\omega_{Hi} >> \nu_{in}$, – F-layer

$$\frac{n}{n_0} \approx \frac{\nu_{in}}{\omega_{Hi}}\left[\frac{\rho}{\rho_0}\right]\frac{k_z}{k}\frac{\gamma}{\gamma-1}(kH+1) \tag{4.10}$$

From Eqs. (4.9) and (4.10) one can see that the sharp gradients in k lead to ionosphere response amplification on the gravity wave passing by $kH\gamma/(\gamma-1)$ times. For $k>0$ and $\rho/\rho_0 \sim 0.1$ (H=5 km) for E-layers with $k^{-1} \sim 3$ km the sharp decrease of electron concentration should be observed of the order $n/n_0 \approx 0.5$.

The effect of the acoustic branch of AGW on the ionosphere was studied by Blaunstein (2000). He considers two cases: quasielectrostatic approximation valid for the lower ionosphere (D and E layers) when the charged particles motion has no influence on the change of the plasma concentration and an electrodynamic case for the altitudes h>130–140 km when the plasma becomes magnetoactive. For the latter case the amplitude of the traveling plasma disturbances depends on the orientation of the wave vector $\boldsymbol{k}$ of the acoustic wave relative to the geomagnetic field direction $\boldsymbol{B}_0$. Studying the evolution of traveling plasma disturbances he obtained the large-scale stratification of the plasma density in the ionosphere presented in Fig. 4.2 with the scales a few km across the geomagnetic field direction and tens of km along the field direction.

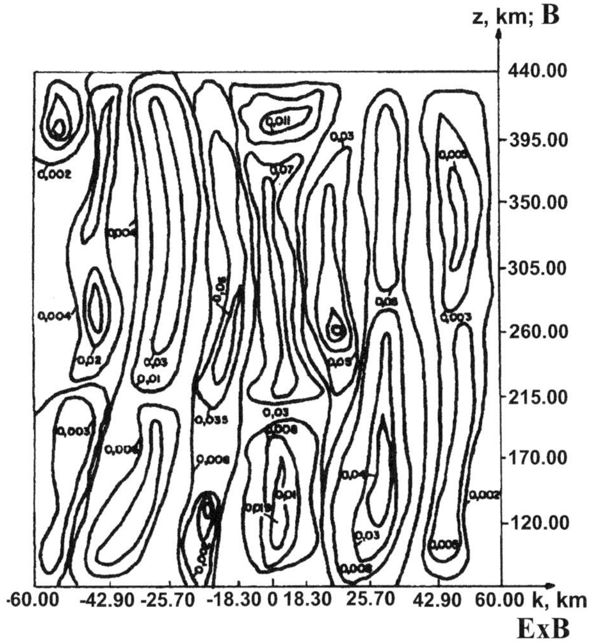

Fig. 4.2 Stratification of the ionosphere caused by the action of AGW according to calculation of Blaunstein (2000)

Some ideas of AGW action on the ionosphere in the area of the equatorial anomaly based on the satellite data were depicted by Molchanov et al. (2002a, 2002b) but they look at the present time more like hypothesis and rough estimations than real theory.

4.2.4 Limitations and Contradictions

There are two main problems with the AGW as concerns their precursory effects on the ionosphere.

1. As was mentioned earlier, all calculations existing up to now are made in linear isothermal atmosphere without dissipation. Playing with scale height of the neutral atmosphere (from 5 km to 45 km) the authors with the simple exponential dependence obtain at the altitude 300 km an amplification factor from several hundreds up to 10^{13} (depending on the author's conscience). But the atmosphere is essentially not isothermal (see Fig. 1.21). The negative temperature gradient on the mesopause altitudes will lead to the capture of the gravity waves due to refraction or at least to essential attenuation. The other factor affecting the acoustic gravity waves is winds, especially in the zonal direction. Didebulidze (1997) provides an estimation of the horizontal wind effects on AGW.

Due to this effect in some periods when the prevailing direction of winds is "working" on the AGW damping, the AGW activity observed by different experimental techniques is very weak (Gavrilov and Fukao 1999). And finally, the dissipative processes, in particular, diffusion. Swenson et al. (1995) within the frame of complex experiment ALOHA-93 detected by airglow imagers the limiting altitude for penetration of AGW near 130 km. They presented the expression which defines the upper altitude limit to which AGW can propagate given the vertical diffusion constrains of the atmosphere. The detailed discussion on the problem of dissipative processes connected with acoustic gravity waves one can find in Imamura and Ogawa (1995). So, in estimations of the amplification coefficient of AGW rather modest values should be taken ($\sim10^2$–10^3). For reasonable AGW vertical velocity amplitudes able to have some effect in the ionosphere (~50 m/s) one will have on the ground surface for a 10^3 amplification factor the value of 5 cm/s. To have such displacements BEFORE earthquakes is absolutely unreal. At least such oscillations should be easily registered by a barograph, which has never been reported.
We can also mention the factor of practical applicability. The velocity of the gravity branch of AGW has a large horizontal component. Depending on the wavelength, the direction of wave energy propagation will change but will be essentially horizontal (see Fig. 5 in Mareev et al. 2002). This result is confirmed by the calculations of Rapoport et al. (2004), see Fig. 4.3 where the calculated effect of the gravity wave on the electron density in the ionosphere is shown. Due to the large horizontal component of the group velocity the AGW reaches the ionospheric heights quite far from the vertical projection of the epicenter: 3,000–3,500 km as one can see from the figure. Taking into account the necessity for real earthquake prediction the determination of the future epicenter position, it is very doubtful this precursor can be used in earthquake prediction.

2. The second very important problem for the AGW hypothesis (which is a natural consequence of the first one) is that there is not any direct indication on the AGW effects on the ionosphere and their possible penetration into the ionosphere from the ground level regardless of its theoretical popularity. Of course there are registered some oscillatory variations in the ionosphere before earthquakes (references cited in Chapter 2). But all of them are indirect. There exists another possibility of acoustic gravity waves excitation directly in the ionosphere (Hegai et al. 1997) which will be discussed in the next chapter, and it is not necessary for these waves to penetrate from the ground level. On the other hand, there no any experimental evidences of oscillatory atmosphere pressure changes before earthquakes, which might be associated with the acoustic gravity wave excitation.

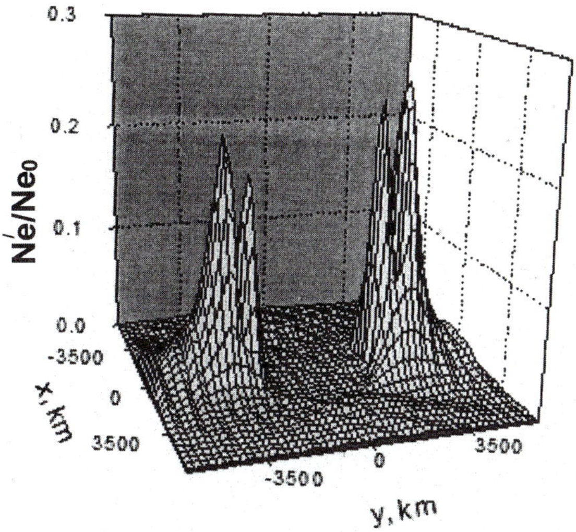

Fig. 4.3 Relative perturbations of electron concentration in the ionosphere under AGW action according to the calculation of Rapoport et al. (2004)

Most experimental results indicate the large scale and persistent character of the ionospheric irregularities observed as seismo-ionospheric precursors (Pulinets and Legen'ka 2003) which AGW theory cannot explain. And even if there should exist some irregularities connected with AGW of pre-seismic origin, there arises a severe problem with their identification and registration. The traveling ionospheric disturbances generated by the auroral processes are very active. As indicated by long duration (1997–2001) optical and ionosonde measurements (Aushev et al. 2002), in 70% of observations the acoustic gravity waves propagating from top to bottom are observed, and they cannot be associated with earthquakes. As concerns the AGW propagating from the bottom, their natural activity connected with meteorological factors is so strong (Boška and Šaul 2001; Manson et al. 2002) that it is very problematic to detect the acoustic gravity waves associated with the seismic activity on their background taking into account that these waves would be much weaker than the waves of meteorological origin. At least, there exists a rich body of literature describing direct observation of AGW penetration into the ionosphere by radars, Doppler installations and optical techniques. But there are no publications reporting the observation of gravity waves associated with seismic activity. So, the AGW hypothesis needs strong experimental support in the form of direct observation of atmospheric movements that can be interpreted as AGW generated by preseismic activity.

To conclude the discussion on the possibility of a precursory effect associated with acoustic gravity waves we will refer to the paper of Davies and Archambeau (1998), which has the most bearing on this subject. The authors of the paper solve in the most accurate (known up to now) form, including nonlinear effects, the problem of excitation of ionospheric perturbations by acoustic gravity waves generated by shallow earthquakes. Everybody should agree that the earthquake itself generates much more intense air oscillations than any possible precursory phenomena. Figure 4.4 presents their calculation of the relative electron density perturbations created by AGW generated by a magnitude 4 shallow earthquake at the altitude 250 km (near the F2-layer peak). One can see that the relative change of the electron concentration in the maximum phase at least reaches 0.3% what is two orders of magnitude lower than the ordinary value of the day-to-day ionosphere variability and practically undetectable.

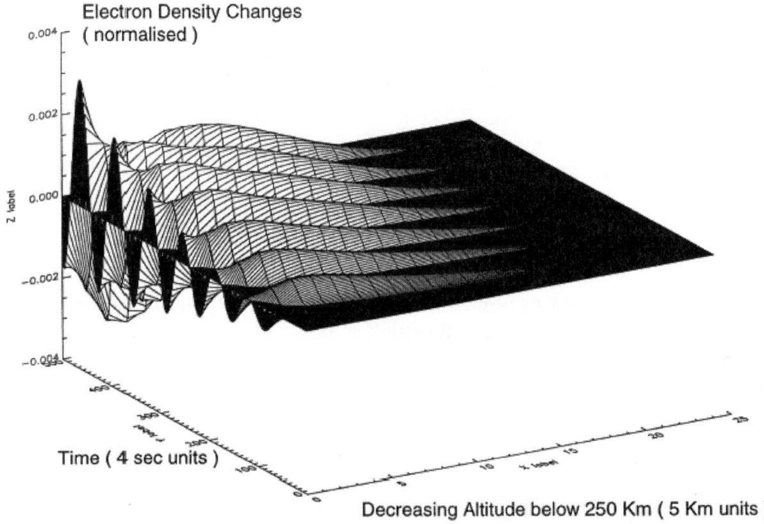

Fig. 4.4 Electron density changes (normalized) for altitudes between 125 km and 250 km after the onset of a gravity wave generated by an earthquake with magnitude 4 at the depth of 5 km (after Davies and Archambeau 1998)

The calculations of Davies and Archambeau (1998) are supported by the experimental results of Calais and Minster (1998) who measured experimentally by the GPS TEC technique the ionospheric effect from the Northridge earthquake of 17 January 1994, M=6.7. The TEC variations associated with AGW were 2–2.5 orders of magnitude lower than the background one. At this point we can probably stop this discussion of the precursory abilities of acoustic gravity waves with following conclusion: *regardless of the possibility of AGW generation near the seismic source before an earthquake, their penetration into the ionosphere is quite improbable, if possible at all.*

4.3 Electric Field Channel

Chapter 3 describes the physical mechanism of the substantial anomalous electric field generation in the area of earthquake preparation. Experimental data confirm this and give the values from several hundreds up to several kilovolts per meter of the vertical electric field. The purpose of the present paragraph is to estimate how strong (if any) the effect of this field may be on the ionosphere. The history of quasistationary electric field penetration into the ionosphere is very long and complicated. Firstly, it is connected with the problems of the global electric circuit and penetration of the electric fields of the thunderstorms (Roble and Tzur 1986; Roble 1991). There were different approaches to the calculation of the electric field penetration into the ionosphere (Park and Dejnakarintra 1973, 1977; Tzur and Roble 1985). According to the ionospheric effects the approach of Park and Dejnakarintra is now commonly accepted (Rodger et al. 1998).

The problem of the electric field of seismic origin does not differ too much from the problem of the thunderstorm cloud except some change of the electric field source altitude. But taking into account the Earth's orography the difference in several kilometers is negligible. That's why the first paper calculating the seismogenic electric field penetration into the ionosphere was based on the Park and Dejnakarintra formulation (Kim et al. 1994). Taking into account that Kim et al. (1994) changed the boundary conditions of Park and Dejnakarintra, we will discuss this problem more carefully below.

4.3.1 Penetration of Electric Field of Seismic Origin into the Ionosphere

Up to the present several papers have appeared with calculations of electric field of seismic origin penetration into the ionosphere. Some ideas of the possibility of electric field of seismic origin action on the ionosphere were expressed by Gokhberg et al. (1983b). These ideas were supported by some estimations in Gokhberg et al. (1995). But the problem was solved with the assumption of isotropic air conductivity, which is not the case (see Fig. 1.23). Kim et al. (1994) solved the problem in the formulation of Park and Dejnakarintra and their approach will be discussed in detail below. Sorokin et al. published several papers reviewing the electrodynamic model of seismo-ionospheric coupling. Taking into account that there is some repetition between the papers we will use the paper by Sorokin et al. (2001), which contains the main elements. The authors deal with the modification of air conductivity by radon emanation in the area of earthquake preparation. After obtaining the modified vertical profiles of atmosphere conductivity the authors under the condition of the vertical current $j_s(0) = 3 \cdot 10^{-9}$ A/m^{2}obtain the electric field at altitude 80 km $E(h)$=3 mV/m. These values rise some doubts in their reality from both sides: ionosphere and near ground processes. Taking from the paper the near ground air conductivity $\sigma(0) = 2 \cdot 10^{-14}$ S/m we obtain the value of the electric field near the ground surface $j_s(0)/ \sigma(0) = 3 \cdot 10^{-}$

$^9/2\cdot10^{-14}$ = 150 000 V/m!!! On the other hand, if we go to the ionosphere and try to calculate the electric field at least at the altitude 100 km using the value given in the paper of 3 mV/m at the altitude 80 km, we obtain (using the conductivity value 10^{-3} S/m):

$E(h) \approx -E_0(h) + j_s/\sigma(h) \approx 0.003$ mV/m – it is a negligible value which cannot create any effect in the ionosphere. So we probably will not use the results of the present paper in the following discussion.

Grimalsky et al. (2003) calculated the penetration of the electrostatic field of seismic origin into the lower ionosphere. The authors take into account the geomagnetic field inclination and possible asymmetry of the electric field source on the ground. They claim that they use quite new boundary conditions (contrary to Park and Dejnakarintra 1973 and Kim and Hegai 1999). Let us look carefully at these conditions by direct citing from the paper: *"Consider some altitude well above 90 km, say in the F2 region of the ionosphere, where the conductivity is very large in all directions including horizontal. In this region we should set Φ=const, for potential in a horizontal plane."* This statement does not conform to any elementary knowledge of ionospheric physics. It well known (see, for example Schunk and Nagy 2000) that ionosphere conductivity is several orders of magnitude higher along geomagnetic field lines than across them. So even on the geomagnetic equator where the geomagnetic field lines are horizontal, the zonal conductivity will be very low because of a direction perpendicular to geomagnetic field lines. This means that there is no place in the World where the ionosphere conductivity was the same in all directions in the horizontal plane. Yes, due to magnetic field inclination the conductivity in the horizontal direction may be high, but this does not mean that it is the same in all directions in this plane, due to the arguments given above, and consequently it is impossible to obtain Φ=const in the horizontal plane. Then the authors project this erroneous configuration to the lower altitudes and make their calculations. Their "pioneer" approach neglects all existing ionosphere electrodynamics. If to assume Φ=const at the altitude 200 km, no electric fields in the ionosphere of middle and low latitudes due to the dynamo effect of neutral winds in the *E*-region (100–150 km) can exist which is quite surprising. According to the above discussion, this paper will be out of the scope of further consideration.

The approach presented in Kim et al. (1994), and Kim and Hegai (1999) looks quite reasonable besides their approximation of vertical geomagnetic field lines (which limits the theory application to the high and middle latitudes). Taking into account that for calculations in the F-region the magnetic field inclination is taken into account, we will follow their calculations.

4.3.2 Calculation of the Electric Field in the E-Region (Cylindrical Source)

Assuming the circular shape of the earthquake preparation area the problem will be solved in a cylindrical system of co-ordinates (r, φ, z). The origin of the coordinate system is put in the epicenter and the z-axis is directed up. Let us suppose that the electric field perturbation ΔE_z on the ground surface connected with the earthquake preparation is Gaussian-like:

$$\Delta E_z = E_0 \cdot exp\{-d \cdot (r/a)^2\}, \tag{4.11}$$

where E_0 – is the magnitude of ΔE_z in the epicenter, and a – is the characteristic size of the field localization area, $d=4 \cdot ln(10)$ – is a factor providing tenfold decrease of electric field perturbation at the distance $a/2$ from the epicenter of the anticipated earthquake.

The Park and Dejnakarintra (1973) approach for the thunderstorm cloud electric field penetration at the ionosphere heights will be used. The electric potential ϕ distribution of the electrostatic field $\boldsymbol{E}$ at the ionospheric heights could be determined in a general case from the continuity equation for the electric field density $\boldsymbol{J}$, which in a stationary case could be expressed as:

$$\mathrm{div}(\boldsymbol{J}) = 0, \qquad \boldsymbol{J} = \boldsymbol{\sigma} \cdot \boldsymbol{E}, \qquad \boldsymbol{E} = -\nabla \phi \tag{4.12}$$

where σ – is the tensor of electric conductivity in the Cartesian coordinate system, which could be expressed as follows (Park and Dejnakarintra 1973):

$$\sigma = \begin{bmatrix} \sigma_1 & \sigma_2 \sin I & \sigma_2 \cos I \\ -\sigma_2 \sin I & \sigma_1 \sin^2 I + \sigma_0 \cos^2 I & (\sigma_1 - \sigma_0) \cos I \sin I \\ -\sigma_2 \cos I & (\sigma_1 - \sigma_0) \cos I \sin I & \sigma_1 \cos^2 I + \sigma_0 \sin^2 I \end{bmatrix} \tag{4.13}$$

where σ_1 and σ_2 – Pedersen and Hall conductivities respectively, σ_0 – conductivity along the geomagnetic field lines, I –magnetic field inclination.

In the selected cylindrical coordinate assuming a horizontal stratification of the medium and a vertical geomagnetic field, it is easy to deduce the following equation for electric potential ϕ from the continuity equation:

$$\frac{\partial^2}{\partial r^2}\phi + \frac{1}{r}\frac{\partial}{\partial r}\phi + \frac{1}{\sigma_1}\frac{\partial}{\partial z}\left(\sigma_0 \frac{\partial}{\partial z}\phi\right) = 0 \tag{4.14}$$

The distribution of the conductivities with altitude can be given as follows: $\sigma_0 = \sigma_1 = b \cdot \exp\{z/h\}$ for $0 \le z \le z_1$, $\sigma_{0,1} = b_1 \cdot \exp\{(z-z_1)/h_{0,1}\}$ for $z_1 \le z \le z_2$, where $z_2 = 90$ km, $z_1 = 50$ and 65 km for day and night conditions, respectively; $h = 6.5$ km; $h_0 = 3$ km; $h_1 = 4.5$ km; $b = 2 \times 10^{-13}$ Mo/m; $b_1 = b \cdot \exp\{z_1 / h\}$. Such a distribution of $\sigma_{0,1}$ roughly corresponds to the empirical model for conductivity (Cole and Pierce 1965). In this case the general solution to the equation for the electric potential ϕ can be represented in the form:

$$\phi(0 \le z \le z_1) = \int_0^\infty J_0(kr) \cdot \{A_1(k)\exp(c_1 z) + B_1(k)\exp(c_2 z)\} dk,$$
$$\phi(z_1 \le z \le z_2) = \int_0^\infty J_0(kr) \cdot \{A_2(k) I_\nu(kf) + B_2(k) K_\nu(kf)\} f^\nu dk, \tag{4.15}$$

where

$$c_1 = -\frac{1}{2h} - \left(\frac{1}{4h^2} + k^2\right)^{1/2}; c_2 = -\frac{1}{2h} + \left(\frac{1}{4h^2} + k^2\right)^{1/2}; \nu = \frac{h_1}{h_1 - h_0};$$
$$f = \frac{2h_1 h_0}{h_1 - h_0} \exp\left\{-\frac{h_1 - h_0}{2h_1 h_0}(z - z_1)\right\}$$

Here J_0 is the Bessel function of zero order; I_ν and K_ν are the modified Bessel functions of the first and second kind, respectively, and A_1, B_1, A_2, B_2 are numerical coefficients. Above the altitude $z = z_2 = 90$ km the geomagnetic force lines can be considered equipotential, so the distribution of ϕ for $z \ge z_2$ is the same as at $z = z_2$. From here we obtain that the electric field of the source of the localized field in the ionosphere at altitudes $z \ge 90$ km is determined by the relation

$$E_r = -\frac{\partial}{\partial r}\phi|_{z=z_2} = \int_0^\infty J_1(kr)\left\{A_2(k) I_\nu(k f|_{z=z_2}) + B_2(k) K_\nu(k f|_{z=z_2})\right\} f^\nu |_{z=z_2} \, kdk \tag{4.16}$$

where J_1 is the Bessel function of the first order, and the coefficients A_1, B_1, A_2, B_2 can be found from the boundary conditions of the problem:

a) $$-\frac{\partial}{\partial r}\phi_{|z=0} = E_0 \exp\{-d(r/a)^2\};$$

b) b) ϕ is continuous at $z = z_1$;

c) c) $$\frac{\partial}{\partial z}\phi_{|z=z_2} = 0.$$

The results of calculations of the ionospheric distribution for the horizontal field strength E_r normalized by E_0 for various sizes of the field localization region E_z during the day and night are shown in Fig. 4.5. It is seen that during the day the degree of field penetration into the ionosphere is much smaller than during the night. The field strength therewith critically depends on the characteristic size a. For example, at a=100 km the maximal field strength E_r^{max} is more than an order of magnitude larger than E_r^{max} at a=20 km and nearly three times smaller than E_r^{max} at a=200 km, both for day and night conditions. The dependence of E_r on r is characterized by a fast initial increase and slow decrease after reaching maximum. The absolute value of E_r^{max} even for a=200 km at night is only 0.7 mV/m for E_0 =1,000 V/m, i.e. the efficiency of electric field penetration into the ionosphere is low.

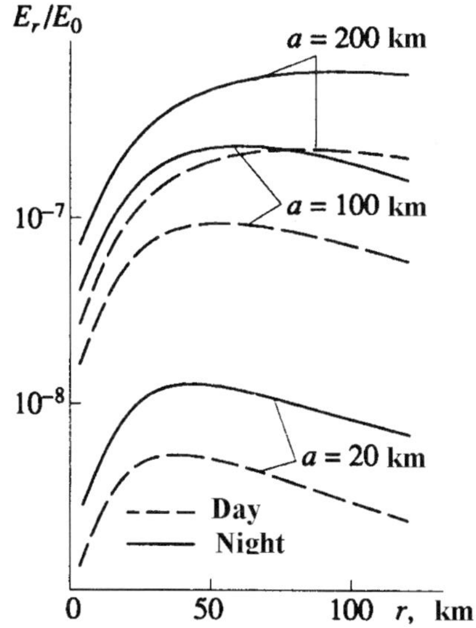

Fig. 4.5 Electric field strength E_r, normalized by E_0 in the ionosphere at the altitude z=90 km as a function of radial distance r from the electric field's source center

Thus, the electric field strength produced by an atmospheric source at ionospheric altitudes is noticeable only when the region of horizontal localization of the field E_z near the Earth's surface is sufficiently wide ($a \geq 100$ km) and $E_0 \geq$ 500–1,000 V/m. Our model calculations of electric field generation mechanism presented in the previous chapter as well as experimental data confirm the possibility of the existence of such a source. Some idea of the possible size of the area occupied by the anomalous electric field generated before strong earthquakes is given by the result of electric field measurements at Swider geophysical observatory in Poland before the Carpathian earthquake of 30.08.86, M=7.2 (Nikiforova and Michnowski 1995), Fig. 4.6. The amplitude of electric field variation exceeded 1,000 V/m (from –250 V/m up to 750 V/m at the distance near 700 km from the epicenter. These results conform well with the earthquake preparation zone concept.

4.3.3 Calculation of the Electric Field in the E-Region (Elongated Source)

It is well known that all seismic events are associated with some tectonic fault regardless of the source depth (Rundle et al. 2000). So one can expect that the source of the electric field may be not cylindrically symmetrical but elongated along the tectonic fault. Leaving the same conditions of calculation as in the previous paragraph, let us describe the source in the form (Kim et al. 1995) with two peaks of positive and negative electric field, Fig. 4.7a:

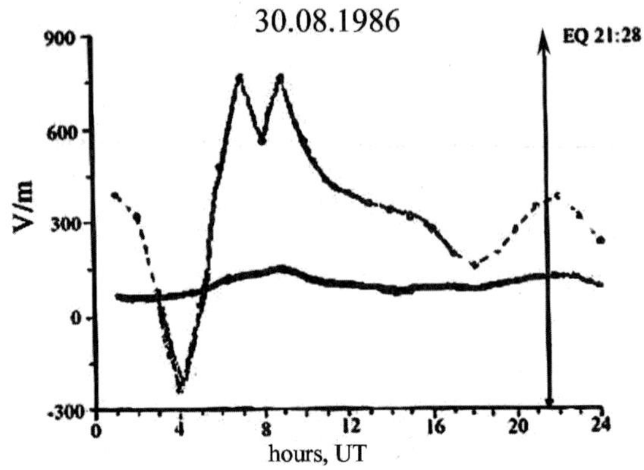

Fig. 4.6 Atmospheric electric field anomaly before the Carpathian earthquake of 30.08.86, *M*=7.2 registered at Swider observatory (Poland) 700 km away from the epicenter. *Bold curve* shows the monthly median fair weather daily variations. *Vertical line* indicates the earthquake time

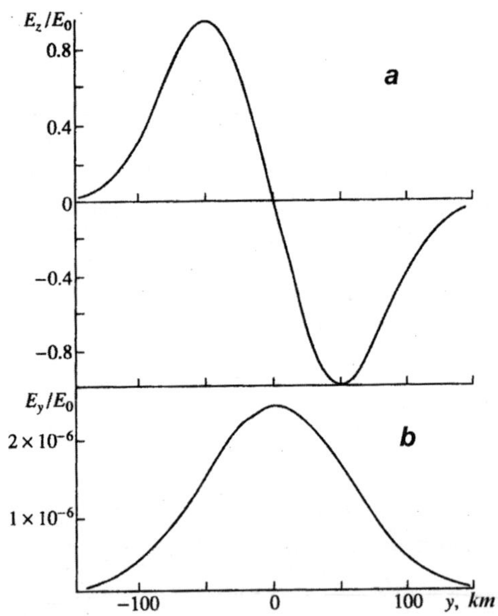

Fig. 4.7a,b ***a*** – Model vertical electric field distribution on the ground surface along the tectonic fault, ***b*** – horizontal (perpendicular to geomagnetic field) distribution of the electric field in the *E*-region of the ionosphere according to Kim et al. (1995)

$$E_z = E_0(\exp\{-4[(y+b/2)/a]^2\} - \exp\{-4[(y-b/2)/a]^2\}) \qquad (4.17)$$

Here E_0 is the electric field magnitude, a determines the peak width, and b – the distance between peaks. Let a=b=100 km. The results of field calculation in the E-region are shown in Fig. 4.7 b. For the value of E_0 = 300 V/m, the electric field in the E-region for nighttime conditions will be 0.75 mV/m. The electric field at the ionospheric heights is localized in the area of 300 km width with the maximum over the tectonic fault center. We can conclude that the effectiveness of the electric field penetration from the elongated fault is three times higher than from the Gauss-like source under the same dimensions of the field source on the ground surface.

Calculations show also the strong horizontal gradients of electron concentration across the fault projection as well as electron density profile distortions in the E-region of the ionosphere.

4.3.4 Effects of the Anomalous Electric Field in the E-Region

Let us look at the effects of the seismogenic electric field on the E-region of the ionosphere. Figure 4.8 shows the calculated electron concentrations N_e as a function of distance r from the vertical projection of the ground source for three altitudes z=115, 125, and 135 km. For a positive direction of E_z near the Earth's surface (electric field directed up) N_e decreases above the localization region of E_z (Fig. 4.8 a) and the minimum value occurs at r=0. With E_z <0, i.e. when the field near the surface of the Earth is directed downward, the electron concentration above the localization area increases reaching a maximum at r=0 (Fig. 4.8b). Thus, the effect of the electrostatic field is most pronounced in the nocturnal E-region directly over the source of the electric field.

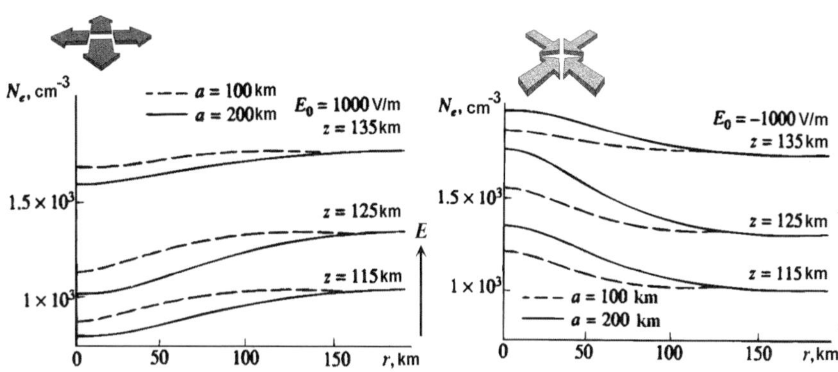

Fig. 4.8a,b Electron density modification in the E-region: (**a**) electric field at the ground surface is directed up, (**b**) electric field at the ground surface is directed down

4.3.5 Formation of Sporadic Layers in the E-Region

It is well known that sporadic layers forming under some conditions in the *E*-layer consist of metallic ions (mainly of meteoric origin). One can find the direct Lidar measurements of the metallic ion layers in the ionosphere (Gong et al. 2002). Kim et al. (1993) considered the possibility of taking into account the metallic ions M^+ in the ionosphere in the presence of the anomalous electric field. Let us consider the effect of the electric field E_r in this case, i.e. assuming that, in addition to molecular ions NO^+, N_2^+, and O_2^+ which are ordinarily the main ions in the *E*-region, the ionospheric plasma may also contain metallic ions M^+ with a mean mass close to that of molecular ions. In the case when the vertical electrostatic field E_z is directed downward the radial field component E_r at ionospheric levels and correspondingly the Pedersen ion drift are directed towards the *z*-axis, which provides the condition of ionospheric plasma convergence above the electrostatic field generation region. The results of calculations for E_0 =-1,000 V/m and a=200 km are shown in Fig. 4.9. Before the moment of "turning-on" the electric field, as well as at the boundary r=1,500 km, the molecular and metallic ion concentrations are determined using a photochemical approximation. As seen from the figure, 2 h after the beginning of the action of the electric field the metallic ion concentration above the field generation region becomes higher than for molecular ions concentration in the altitude range 112–134 km, and after 4 h the metallic ions dominate at altitudes of 107–146 km. Then at altitudes near z=120 km a maximum of the metallic ion concentration occurs reaching $2.5 \cdot 10^4$ cm^{-3} within 4 h, which is approximately 60 times as high as the corresponding initial unperturbed value $N(M^+)$. In contrast, the molecular ion concentration at the same altitudes decreases, and by the moment of "switching-off" the field a deep minimum of N $N(XY^+)$ forms there with a value of $1.0 \cdot 10^2$ cm^{-3}, i.e., $N(XY^+)$ drops by an order of magnitude relative to its initial level.

The presented calculations have not only academic interest. Looking at the Figs. 2.18–2.20 one can find a direct relation with the theoretical calculations that explain the sporadic layers appearance before strong earthquakes. It is just due to the anomalous electric field.

But the most interesting thing is that the anomalous electric field can itself transport the metallic ions from the ground surface or from the clouds containing the large amount of aerosols, especially during a volcanic eruption.

Volcanic eruptions, seismoactive zones and tectonic faults, anthropogenic activity, and dust storms enrich the tropospheric layer (0–15 km) with metallic ions. The diffusion cannot provide the vertical transport to the level of order (60–90 km) for time intervals of the order of days. Moreover, the inverse temperature height intervals where the temperature lowers with the height make such transport even more problematic. But quasistatic electric fields of the thunderstorm clouds or the seismogenic electric field described in the previous section can make such transport of metallic ions essentially more effective.

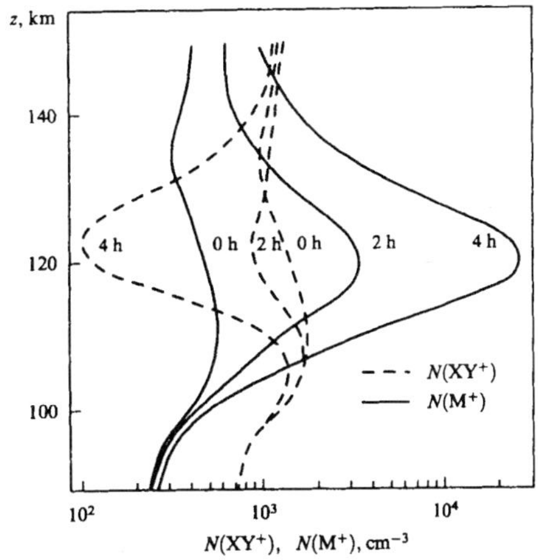

Fig. 4.9 Calculated altitude profiles of the molecular (*dashed lines*) and metallic (*solid lines*) ion concentrations above the electric field's source center (r=0) at t=0, 2 and 4 h after "turning-on" the electrostatic field

It was shown by Park and Dejnakarintra, (1973) that the vertical electric field of the thunderstorm cloud which at the heights of 15 km is of order 100 kV/m could provide the electric field at the height of 90 km of the order of 1 mV/m. To simplify calculations we will suggest that the troposphere height h=15 km corresponds to z = z_{tr} = 0 km. In such conditions the ionosphere height h==90 km will correspond to z=z_i =75 km. Taking at z=0 the electric field value $E_0 = 10^5$ V/m we will describe its exponential drop with height as:

$$E_z(z) = E_0 \cdot \exp\{-z/H_E\} \tag{4.18}$$

where H_E=3 km – characteristic scale of electric field change in the giant thunderstorm cloud providing electric field of 1 μV/m at z_i according to Park and Dejnakarintra (1973).

The ions velocity V under action of electric field E could be expressed as

$$V = \mu E \tag{4.19}$$

where $\mu = q/(m_i \nu_i)$ – correspondent ion mobility, q – ion charge, m_i – ion mass, and ν_i – ion-neutral collision frequency which could be well approximated by exponential dependence:

$$\nu_i = \nu_0 \cdot \exp\{ -z/H_\nu\} \tag{4.20}$$

In the given case the rising time T from z_{tr} = 0 to z could be expressed by the integral:

$$T = \int_0^{z_i} \frac{dz}{V_z} = \frac{m_i v_0}{qE_0} \int_0^{z_i} \exp\{z / H_E - z / H_v\} dz \tag{4.21}$$

or after integration

$$T = \frac{m_i v_0}{qE_0} \cdot \frac{H_v H_E}{H_v - H_E} \cdot \{\exp\left(\frac{H_v - H_E}{H_v H_E} \cdot z_i \right) - 1\} \tag{4.22}$$

When $H_E = H_v$, we obtain:

$$T = \frac{m_i v_0}{qE_0} \cdot z_i \tag{4.23}$$

It is obvious that the transport time is proportional to the ion mass and collisional frequency and reverse proportional to the ion charge and electric field intensity. Let us calculate the T for Fe^+ and Mg^+ ions. Our calculations show that the collision frequencies for Fe^+ and Mg^+ could be expressed for the height interval 15–90 km as:

$$\nu(Fe^+) = \nu_0(Fe^+) \cdot \exp(-z/H_\nu), \tag{4.24}$$

$$\nu(Mg^+) = \nu_0(Mg^+) \cdot \exp(-z/H_\nu),$$

where $\nu_0(Fe^+) = 7{,}91 \cdot 10^8$ s^{-1}, $\nu_0(Mg^+) = 1{,}53 \cdot 10^9$ s^{-1}, and $H_\nu = 7$ km. Substituting the correspondent values in (9), we will obtain for the rising time:

$$T(Fe^+)_{[hours]} = 6.69 \cdot 10^{-6} \cdot (\exp\{4z/21\} - 1)$$
$$T(Mg^+)_{[hours]} = 5.54 \cdot 10^{-6} \cdot (\exp\{4z/21\} - 1) \tag{4.25}$$

if z is expressed in km. These dependencies are shown in Fig. 4.10.

One can see that for the living time of the giant thunderstorm cloud, which is of the order of 2 hours, the effective rising time for both kinds of ions is more than 75 km. We can claim therefore that the thunderstorm cloud can transport the metallic ions into the ionosphere. We can also see that if we diminish the value of the electric field by an order of magnitude which corresponds to the electric field of a normal thunderstorm cloud, the rising time up to a level of 65 km will be nearer 1 hour, which does not exceed the living time of an ordinary thunderstorm cloud. The same conclusion could be attributed to the seismogenic electric field. So we can conclude in general that the troposphere could be an effective source of the metallic ions for the ionosphere.

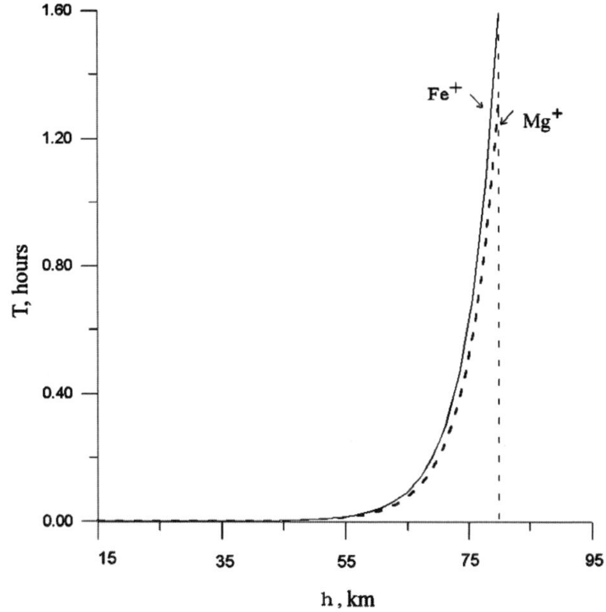

Fig. 4.10 Metallic ions rise-time dependence on height

These calculations also have some support in experimental data. Figure 4.11a presents the clear evidence of the strong electric field presence over the erupting volcano (Tolbachik, Kamchatka), and Fig. 4.11b demonstrates the sporadic E-layer formed after eruption of the Karymsky volcano at Kamchatka peninsula.

4.3.6 Electric Field Effects in the *F*-Region

Using the results of electric field calculation in the *E*-region we can make further calculations with altitude. Notice that at this height the electric field ***E*** is radial with respect to the *z*-axis and axially symmetrical. Above the altitude z=90 km the Earth's magnetic field lines are taken to be equipotentials.

From this point the inclination of geomagnetic field lines will be taken into account (see Fig. 4.12). The effect of the field will be particles ***ExB*** drift in the crossed seismogenic electric field and geomagnetic field. The ionospheric plasma in the F_2-region will move along quasi-circular trajectories around the geomagnetic line force z' cutting the z-axis at the height z=90 km. The z'-axis is a line of zero electric field strength.

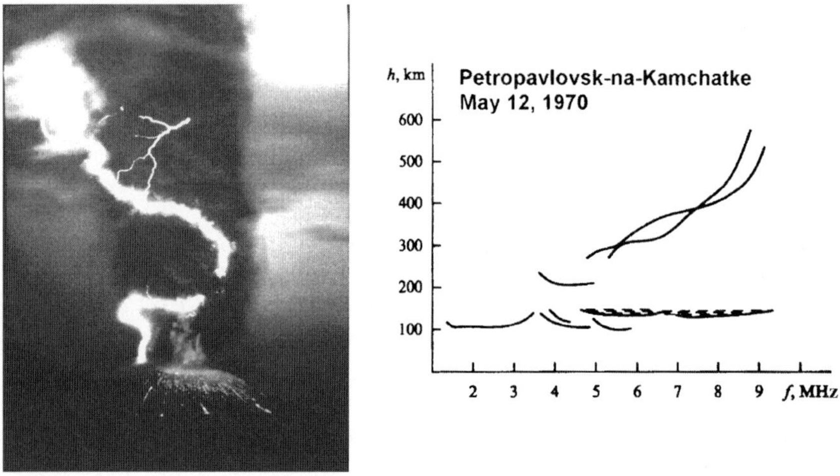

Fig. 4.11 *Left panel* – discharge initiated by the strong electric field generated over the erupting Tolbachik volcano. Strong vertical smog column indicates that particle rising is not only due to turbulence but to the vertical electric field (Photo Dr. Yu. Gordeev). *Right panel* – intensive sporadic E-layer registered after the Karymsky volcanic eruption at Kamchatka Peninsula (after Kolokolov and Shalagina 1978)

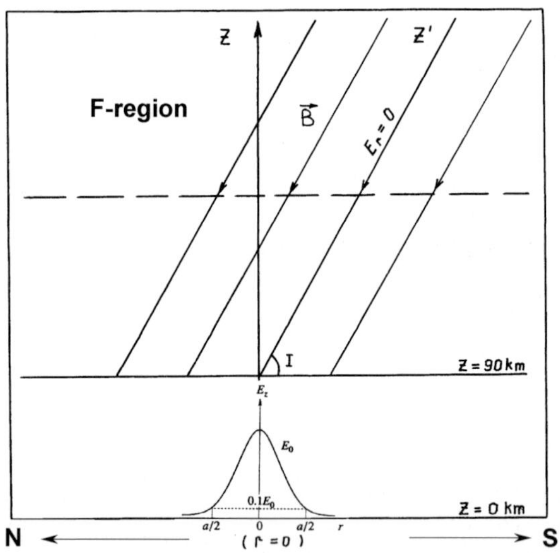

Fig. 4.12 Geometry of calculations of seismogenic electric field effects in the F-region of the ionosphere

The F_2 region plasma density is governed by the O^+ ion continuity equation because O^+ is the dominant ion in the F_2 region, so that $N_i(O^+) \approx N_e$ where N_e is the electron concentration of the F-layer.

For nighttime conditions the continuity equation can be written as:

$$\frac{\partial N_i}{\partial t} + \nabla\{N_i(\mathbf{V}_D + \mathbf{W})\} + \beta N_i = 0 \tag{4.26}$$

where V_D is the ion diffusion velocity along geomagnetic field lines, $\boldsymbol{W}=\boldsymbol{ExB}/\boldsymbol{B}^2$ is the $\boldsymbol{ExB}$ drift velocity, $\boldsymbol{B}$ is the geomagnetic induction and β is the linear recombination coefficient. The electric field is assumed to be "switched on" in steps at the time t=0, i.e., E_0=const≠0 for t≥0 and E_0=0 for t<0 (Kim and Hegai 1999). The expressions for V_D and β are adopted following Schunk (1988).

Equation 4.26 was numerically solved for the case of the upward vertical electric field E_z with $T_e = T_i = T_n = 800°$ K (T_n is the neutral atmosphere temperature) and the magnetic inclination $I=60°$. The initial ion density distribution is assumed to be given by:

$$\nabla\{N_i\mathbf{V}_D\} + \beta N_i = 0 \tag{4.27}$$

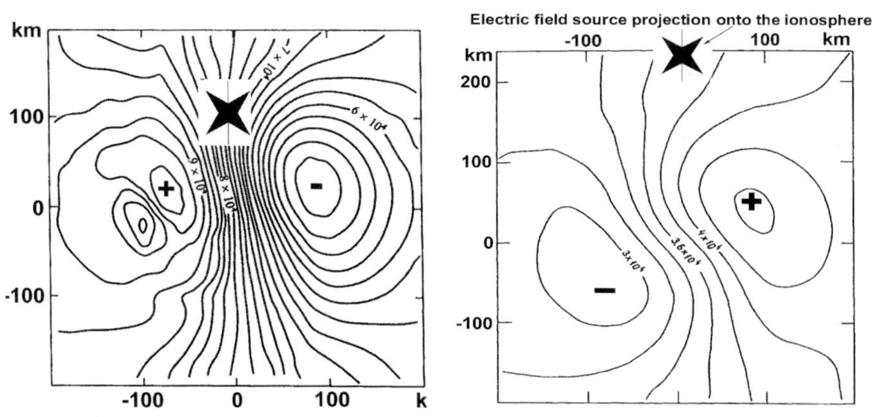

Fig. 4.13 Calculated electron density contours in the horizontal plane z=250 km (a) and z=500 km (b) for two hours after "turning-on" the electric field. The coordinate system (x,y) is centered at the point where the geomagnetic line z' intersects the plane (z' is a line of zero electric field). The x-axis points to the equator and the y-axis points eastward. The *asterisk* indicates the vertical projection of the electric field's source center onto the plane

For the upper boundary condition at z=700 km the downward plasma flux of $1.5 \cdot 10^8$ cm^{-2}s^{-1}is specified. At the lower boundary z–200 km the ion density is obtained from the equation of local ionization production and loss equilibrium.

The results of calculation are presented in Fig. 4.13.

The picture of the horizontal distribution of the ionospheric plasma concentration from a cylindrically symmetrical source is characterized by two foci of positive and negative perturbations of E_0. With altitude, the perturbation region of the ionospheric plasma concentration shifts as a whole from the center of the vertical field localization region on the base plane (z_0=0) towards the equator. In this way in the bottom part of the *F2* ionospheric region, the plasma concentration west and east of the magnetic meridian going through the epicenter increases and decreases, respectively, while in the upper part of the *F2*-region and at altitudes near the main ionospheric maximum the situation is the opposite: the focus of the positive perturbation of N_e lies east of the geomagnetic meridian, and west for the negative perturbation. And here we should make one very important remark: *the lower latitude of the earthquake epicenter, the larger displacement of the modified area in the ionosphere towards equator due to geomagnetic lines inclination.* Such bifocal distribution one can find experimentally, which is demonstrated in Fig. 4.14 where spatial distribution of the critical frequency deviation built from the European ionosonde network, which was registered before one of the earthquakes in Central Italy.

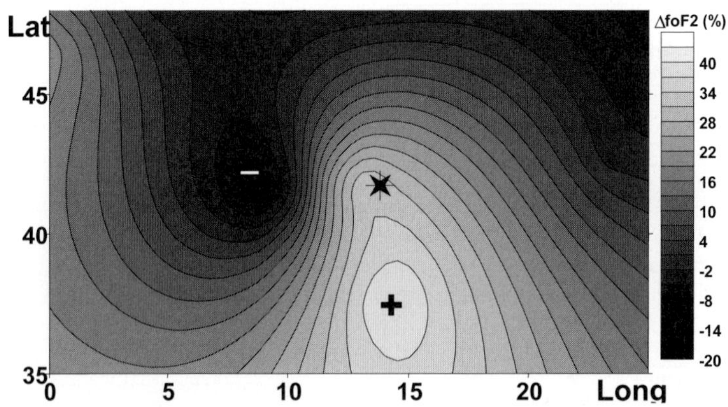

Fig. 4.14 Critical frequency deviation distribution 5 days before the strong earthquake in Central Italy (Abruzzo, 07.05.84, *M*-5.8) from the data of European ground-based ionospheric stations network

4.3.7 Effects in the F2-Region from Large Tectonic Faults

Effects from prolonged sources like tectonic faults were calculated as well (Pulinets et al. 1998c). This work was performed to interpret the data of the topside sounder (Pulinets, 1989) onboard the Intercosmos-19 satellite which registered strong longitudinal variations while passing over the Andes area. The Intercosmos-19 satellite had the elliptical orbit (480 – 950 km) with the high inclination (~ 74°) and the orbit period duration of ~ 90 min. In the middle latitudes a satellite with a high inclination orbit finds itself in the fixed sector of local time (within the range of 1 h), this allows us to obtain ionospheric parameters for a certain moment of local time. The data above the Andes area were collected when the satellite was in the sector of 05 – 06 LT for quiet geomagnetic periods (K_p <2 , D_{st}<10 γ) in July 1980 when the level of solar activity was high ($F_{10.7}$=152). On the basis of this dataset, the longitudinal profiles of the critical frequency and electron density at the height of 400 km were derived at three fixed latitudes φ=35°, 40° and 45°S for 05 LT. The critical frequency longitudinal distribution is presented in Fig. 4.15 for 25° S latitude, and profiles at 400 km altitude – in Fig. 4.16. For comparison, the appropriate electron density longitudinal profiles given by the ionospheric model IRI-85 are also shown in Fig. 4.16. It can be seen that the measured electron density as a function of longitude decreases by a factor of ~3 with the increase of longitude from 280° to 295°E. The observed longitudinal feature of electron density distribution is far from being reproduced by the IRI-85 model and it is difficult to explain it by helio-magnetospheric factors. Hence, it is reasonable to assume that the revealed longitudinal modification of the F2 region ionosphere is associated with some peculiarities of the considered geographical region (35°–45°S, 278°–296°E), which is the zone of high seismic and tectonic activity.

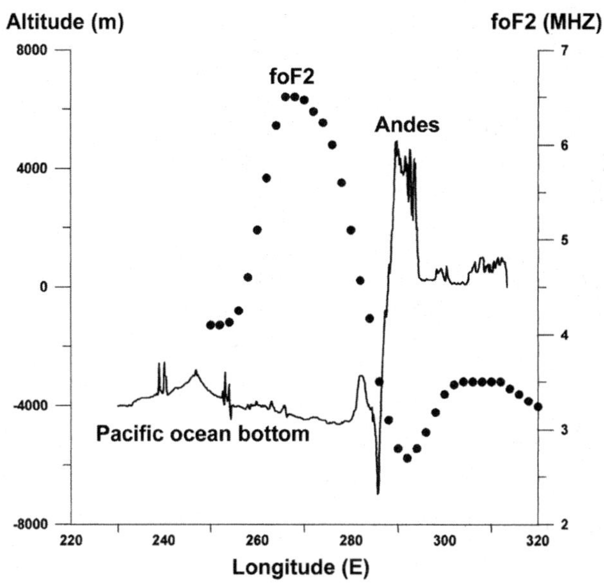

Fig. 4.15 Longitudinal distribution of the critical frequency *foF2* (*dots, right axis*) at the longitude 25°S from the data of the Intercosmos-19 satellite (15th–16th July 1980, 05LT) and the ground level (*continuous curve, left axis*)

For an explanation of these strong changes of electron density with longitude, it is necessary to note that they were observed in the longitudinal sector, which incorporates the global system of active tectonic faults extended meridionally along the Andes over thousands of kilometers. As was considered in the Introduction, appreciable perturbations of the vertical electric field E_z on the Earth's surface can take place in the zone of a tectonic fault. The perturbations of E_z can manifest themselves up to ionospheric altitudes and cause changes of ionospheric electron density. In this connection, we have assumed that the observed longitudinal changes of electron density in the *F2* region are caused by a large-scale perturbation of the vertical atmospheric electrostatic field E_z on the Earth's surface in the zone of the global system of tectonic faults in the Andes' area and in the following subsections concrete calculations supporting the propounded hypothesis are presented. First of all, we shall calculate the distribution of the penetration into the ionosphere electric field associated with the E_z perturbation localized in the region, which extends in the meridional direction on the Earth's surface.

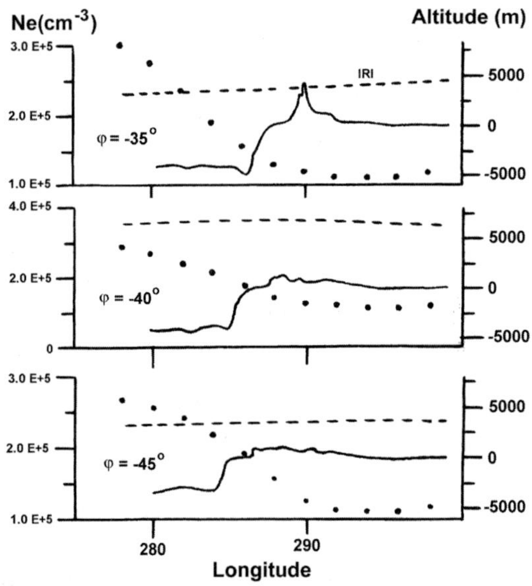

Fig. 4.16 Longitudinal profiles of the electron density observed in July 1980 by the Intercosmos-19 satellite (*dotted lines*) and given by the IRI-85 ionospheric model (*dashed lines*) at altitude z=400 km and 05 LT for three selected latitudes : 35°, 40°, 45°S. The *solid lines* show the Earth's surface relief

For the calculation we will use the same equations and conditions (4.12, 4.13) but with different geometry of the source in Cartesian coordinates (Fig. 4.17). We assume that the perturbation of the vertical electric field E_z on the Earth's surface is localized in an infinitely long stripe stretched out in the meridional direction, i.e., along the x-axis. The distribution of the electric field perturbation ΔE_z across the stripe is symmetric relative to the x-axis and given by:

$$\Delta E_z = E_0 \cdot \exp\{-[y/(b/2)]^2\} \tag{4.28}$$

where E_0 is an amplitude of ΔE_z which does not depend on time, and b is a transverse size of the electric field perturbation zone. The continuity Eq. (4.26) is solved for O^+ as the dominant ion in the ionosphere with the initial distribution presented by Eq. (4.27). We integrate Eq. (4.26) in the assumption that the vector of vertical electric field perturbation on the Earth's surface ΔE_z is directed downward and E_0=-65 V/m, b=800 km, the x-axis coincides with the longitude of 286°. The calculated electric field distribution in the E-region one can find in (Pulinets et al. 1998c), and here we present only the final result of calculations for longitudinal distribution of the electron concentration at the altitude 400 km presented in Fig. 4.18.

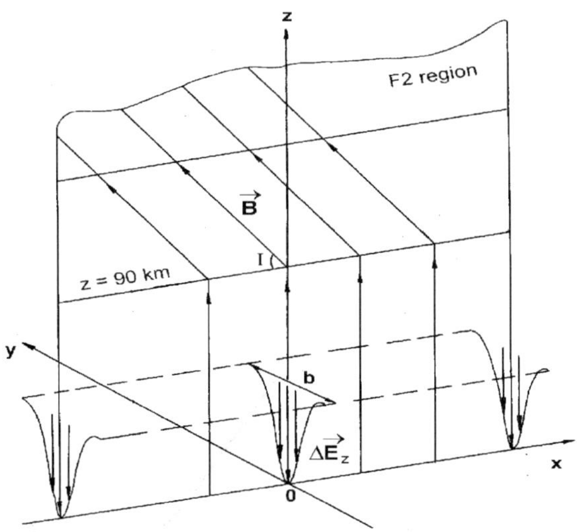

Fig. 4.17 Geometry of calculations

The comparison shows good agreement between the calculated and observed longitudinal variations of electron density. Thus, the results of the performed simulation support the proposed mechanism of strong longitudinal changes of electron density over the Andes region.

This result also has empirical support in another set of data. The tectonic plate boundaries present a giant set of tectonic faults that "lives" continuously. This means, that perturbations in the ionosphere should also be created continuously. The global network of GPS receivers can monitor the small-scale perturbations in the ionosphere. Figure 4.19 presents a comparison between the global seismic activity concentrated along the tectonic plates boundaries, and the global distribution of ionospheric scintillations (phase fluctuations) from the GPS receivers network.

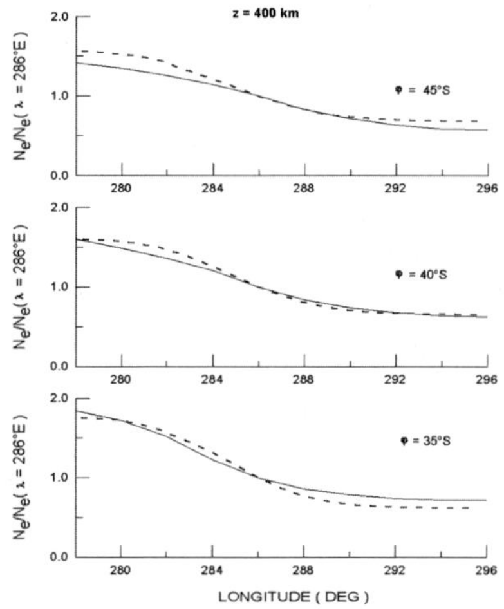

Fig. 4.18 Comparison of calculated (*dashed lines*) and observed (*solid lines*) longitudinal distributions at altitude 400 km for selected latitudes

Looking at the picture (Fig. 4.19) we obtain more support for the conclusion that seismic activity is one of the factors of ionosphere variability.

4.3.8 Generation of Acoustic Gravity Waves in the Ionosphere and their Possible Effects

It is natural to expect that the appearance of additional electric field in the ionosphere will cause the electric currents and turbulence. Hegai et al. (1997) solving the system of Eqs. (4.2–4.4) and using the formalism of Chimonas and Hines (1970) demonstrate the possibility of generation of acoustic gravity waves in the *F*-layer of the ionosphere. Taking the same conditions as were used in previous paragraphs (i.e. electric field strength in the ionosphere E_0=1 mV/m, and size of the area occupied by the field 200 km) for the electric field action as a step function during 60 min they obtain the unstable variations of the air pressure with period's characteristic for AGW. The atmosphere parameters were taken from the CIRA-72 model with H=32 km, $\rho_0(z_0=280\text{ km}) = 3.48\cdot10^{-12}$ kg/m^3, $g_{280\text{ km}}$= 8.94 m/s^2. Figure 4.20 presents the relative pressure P_L for the distance r=1,000 km at different altitude levels of the *F2*-region of the ionosphere.

The AGW causes the motion of neutral and ionized components of the atmosphere, but the motion of charged particles is restricted because of the presence of the geomagnetic field (magnetized plasma), and therefore, their velocity differs from the velocity vector of neutral particles. The charged particles cannot move across the geomagnetic field lines so they have only vertical and meridional components. This difference of the movements of neutral and charged particles can be expressed in the form:

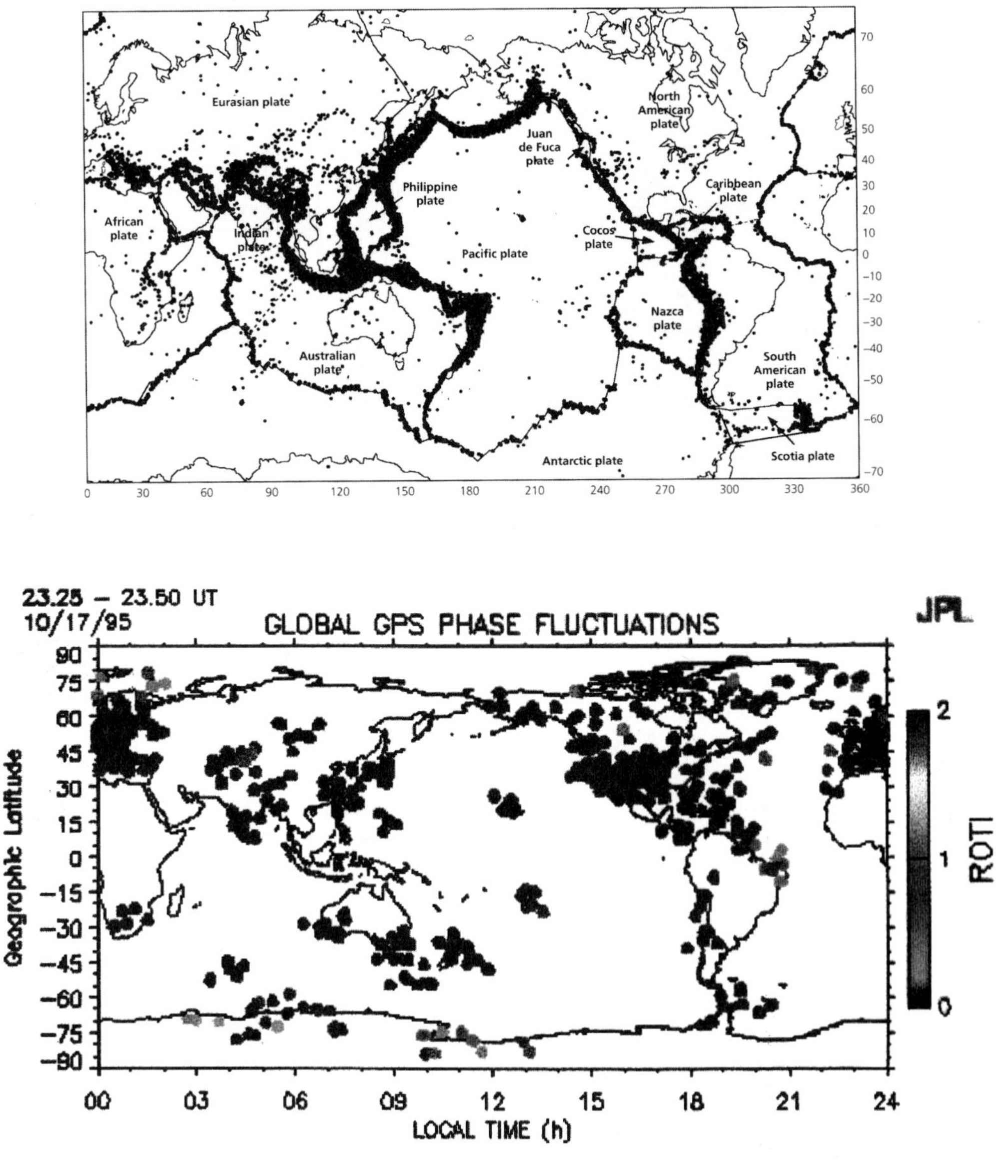

Fig. 4.19 *Upper panel* – global distribution of seismic activity, *lower panel* – global distribution of GPS receivers signal phase fluctuations caused by the small scale irregularities in the ionosphere

$$V_z = - U_x \sin I \cos I + U_z \sin^2 I \qquad (4.29)$$

$$V_x = U_x \cos^2 I - U_z \sin I \cos I$$

where U_x and U_z are the meridional and vertical velocity components of the neutral particles, and V_x and V_z – correspondent components of the charged particles, I – magnetic inclination.

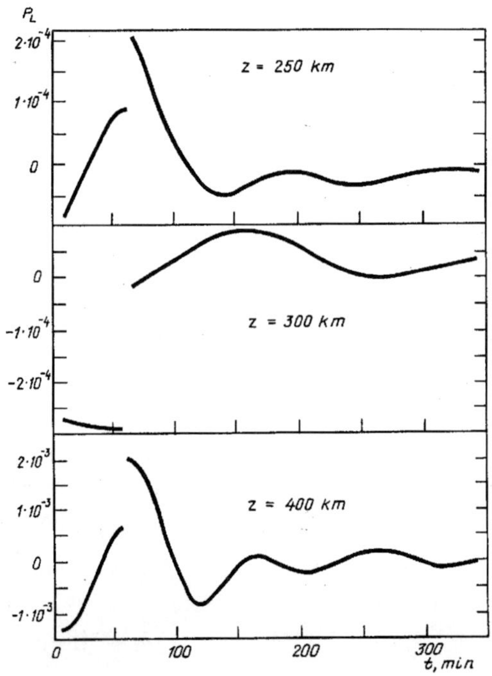

Fig. 4.20 Variations of the relative pressure P_L at a distance r=1,000 km from the epicenter of impending earthquake at different altitude levels of the ionosphere

The calculations of electron density disturbances were carried out for sufficiently strong electric field (20 mV/m) of short duration (15 min). The epicenter was localized at the point with I=45°. Results of the calculation are presented in Fig. 4.21.

Conclusions which can be made from the calculation results are very interesting: the disturbances created by the AGW are very anisotropic. They are maximal in an equatorward direction, notable in the poleward direction and practically insignificant in both zonal directions. Disturbances have a clearly wavelike character in the equatorward direction and a smoother character in the poleward direction. Their duration is much longer than the primary pulse of the electric field. These calculations can explain the satellite measurements of small-scale ionospheric irregularities before the seismic shock presented in Fig. 2.26 as well as the optical measurements before earthquakes described in the Chapter 2.

It is very interesting that for both the quasistationary effectand the AGW effect in the ionosphere the result is similar in the sense that the effect in the ionosphere is displaced from the vertical epicenter projection in an equatorward direction which should be taken into account in experimental data interpretation.

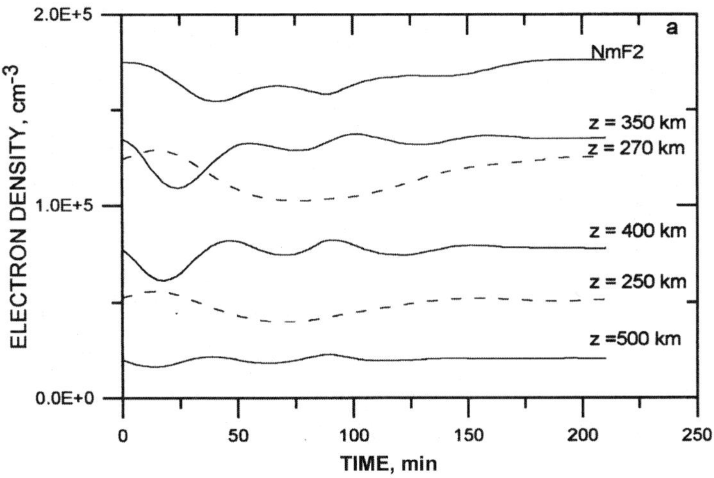

Fig. 4.21 Variations of the peak electron density N_mF2 and N_e at selected altitudes as a function of time at the horizontal distance r=1,000 km in an equatorial direction from the impending earthquake epicenter

4.3.9 Electric Field Effect in Upper Ionosphere and Magnetosphere

Taking into account the high conductivity along geomagnetic field lines they can be regarded as equipotential. In this case one can boldly project the electric field from the *E*-region along the geomagnetic field lines in the upper ionosphere. It is illustrated in Fig. 4.22 (Kim and Hegai 1997).

The continuity equation is solved along the field line with the assumption of temperature to be unchanged with the altitude. The calculations show formation of field aligned irregularities with specific parameters. Namely, a zonal spreading of the irregularities three times larger than the meridional one. This is a very remarkable fact because this latitude-longitude relationship repeats the relationship of the statistically obtained distribution of the VLF noises registered on the satellites before earthquakes (see Fig. 2.25). The magnitude of irregularities is higher for the periods of low solar activity due to changes of atmosphere scale height (from 5% during periods of high solar activity up to 15% for low solar activity). So, the formed plasma irregularities in the magnetosphere and pre-seismic VLF noises registered onboard artificial satellites (Larkina et al. 1989), have the same spatial coverage and position in relation to the future epicenter. It is worthwhile to look for the connection between them.

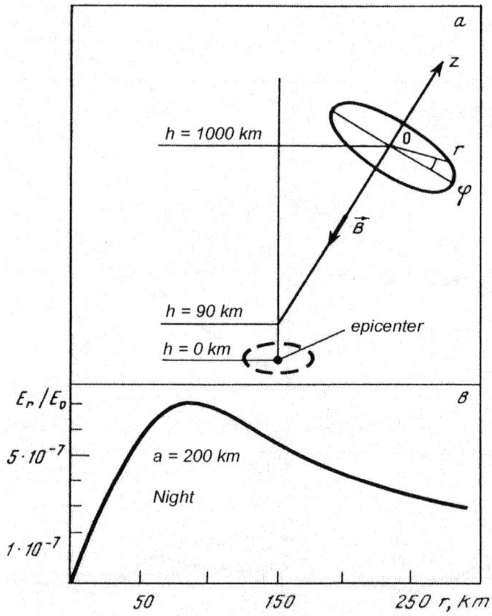

Fig. 4.22 Geometry of calculations of electric field effects in the upper ionosphere and magnetosphere

To understand whether there is any relation between the registered VLF noises and formation of plasma irregularities we use the results of (Shklyar and Nagano 1998). Deriving the set of equations for the vector potential of electromagnetic field in plasma with density irregularities the authors analytically solve the problem of wave scattering on plasma irregularities. As an application of their theory, the authors regard the scattering of whistler waves in the magnetosphere. The scattering process has two components – scattering in k-space and scattering of quasi-electrostatic waves into quasi-longitudinal ones. Both processes contribute to an increase of the VLF emission level in the lower ionosphere. In the first one all the VLF noises of different origins will be effectively scattered on the density irregularities within the modified magnetospheric tube, which will increase the VLF emission energy within the tube, which will be registered by the satellite. In the second case the inter-mode scattering will be registered in the lower ionosphere as VLF emission generated in the ionosphere. In both cases the spatial distribution of the noises will be the same as the irregularities. Here we should mention that the plasma ducts can be formed not only by a seismogenic field but by the quasistationary electric field of a thunderstorm cloud, which means that physically the situation is less confined and is not only connected with the seismic activity (McCormick et al. 2002).

From the discussion above one can conclude that *VLF noises registered on the satellites before the strong earthquakes are not generated in the seismically active area and propagated into the ionosphere* (*as was thought earlier, Parrot and*

Mogilevsky 1989) but are the process of scattering of the natural VLF noises always present in the magnetosphere on the plasma irregularities formed due to the action of anomalous electric field.

In some way this process is equivalent to the trapping of the VLF noises in the modified magnetospheric tube that is presented in Fig. 4.23.

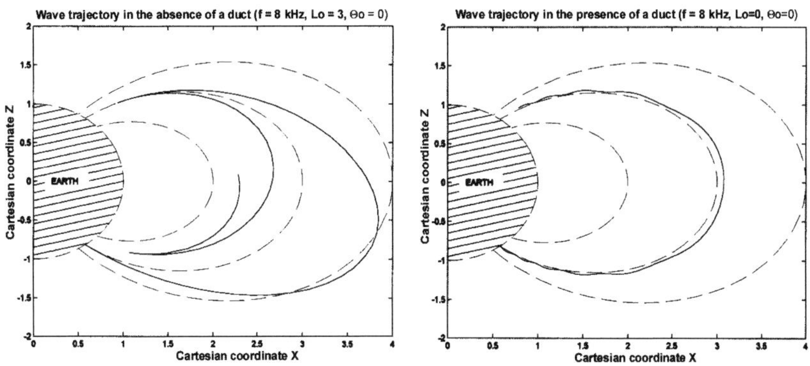

Fig. 4.23 Wave trajectories for VLF emission 8 kHz in the absence of irregularities (*left panel*), and in the presence of irregularities (*right panel*)

The concentration of VLF noises in some *L*-shell leads to the next process in the long chain of anomalous electric field effects – stimulated particle precipitation from radiation belts. The typical size of the irregularity formed by the seismogenic electric field across the geomagnetic field is about 300 km at altitudes around 2,000 km. Together with VLF noises of natural origin which have a sporadic character, the coherent VLF radio waves emitted by powerful ground-based radio stations, including the transmitters of the "Omega" navigation system, should be efficiently channeled in such ducts and should propagate along them in the plasmasphere penetrating in the region of trapped high-energy particles of the Earth's radiation belt. The resonance cyclotron interaction of the propagating VLF wave and the energetic-electron population can take place in the radiation belt, which leads to the pitch-angle scattering of the trapped particles. As a result, some energetic electrons will be scattered into the loss cone and precipitate in the lower atmosphere. The resonance cyclotron interaction of VLF waves with energetic charged particles is the subject of many publications (Dungey 1963; Inan et al. 1978).

Natural incoherent wideband VLF radiation trapped in a duct should lead to chaotic variations in the pitch angle of energetic electrons. Mathematically, such a pitch-angle scattering can be described as diffusion in the pitch-angle space (Roberts 1969). In this case, natural VLF waves can be efficiently amplified in a duct, which, in turn, should result in intense pitch-angle scattering of the energetic electrons and in their enhanced precipitation (Brice 1964; Nunn 1971; Ashour-Abdalla 1972). Coherent VLF radio waves have a much stronger effect on the pitch-angle distribution of energetic charged particles and on their precipitation from the radiation belt since, in this case, variations in the pitch-angle during the bounce period are no longer chaotic, but regular. The pitch-angle scattering of energetic electrons interacting with coherent VLF radio waves were calculated in Dysthe (1971), Gendrin (1975) and Inan et al. (1978) under various conditions.

Energetic electron precipitation from the radiation belt was observed at the middle and lower latitudes using different methods (Tulinov 1967; Paulikas 1975; Vampola and Kuck 1978; Voss and Smith 1980; Vampola and Gorney 1983). These publications do not report a connection with earthquake activity, but earlier (Chapter 2) we cited the results of satellite registration of particle precipitation associated with earthquakes (Voronov et al. 1989, 1990; Galper et al. 1989, 1995; Aleshina e al. 1992, 1993; Galperin et al 1992; Pustovetov and Malyshev 1993).

Thus we can conclude that *precipitation of energetic particles observed on satellites before strong earthquakes is caused by effective interaction of VLF noises trapped in irregularity ducts, which are created by anomalous electric field penetrating from the ground into the ionosphere and magnetosphere, with energetic particles of radiation belts.*

The collisional ionization produced by precipitating electrons in the lower ionosphere should give rise to an increase in the electron density in the nighttime mid-latitude ionospheric *D*-region. So starting from the ground with anomalous electric field we will finish our excursion by returning from the magnetosphere to the lower ionosphere.

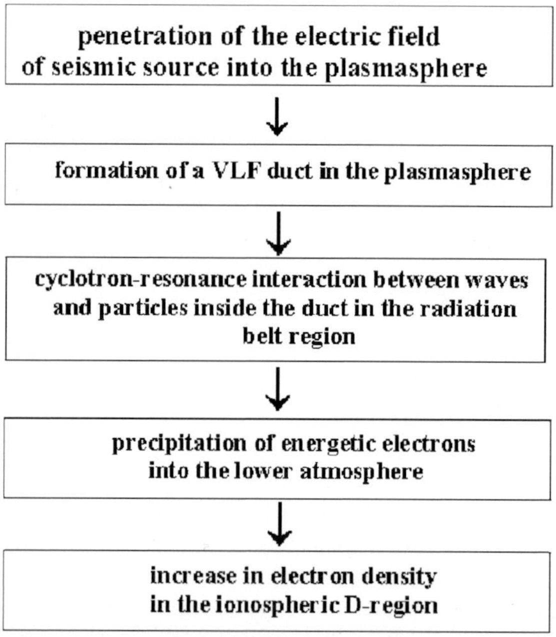

Fig. 4.24 Schematic presentation of the chain of physical processes from the anomalous electric field penetration into the ionosphere up to the increased ionization of the D-region

4.3.10 From the Magnetosphere to the *D*-Region

For clarity, we present in Fig. 4.24 a general sketch of the physical mechanism for increasing the electron density in the nighttime mid-latitude ionospheric *D*-region before a strong earthquake. It is worthwhile to mention here a recent paper (Rodger et al. 2002) where modification of the *D*-region due to whistler-induced electron precipitation has been studied. But we will follow the calculations of Kim et al. (2002) where the case was considered especially for seismic activity.

To calculate the rate of collisional ionization of the neutral atmospheric components by precipitating energetic electrons we apply the calculation technique developed in Rees (1963). According to this paper, the rate q of creation of ion pairs in a unit volume, normalized to the monoenergetic flux F of electrons with energy E_0 is given by the formula:

$$q / F = q_{E_0}(e) = \frac{E_0}{r_0 \Delta E_{ion}} \cdot \lambda(z/R) \cdot \frac{n(M)_z}{n(M)_R} \tag{4.30}$$

where ΔE_{ion} =0.035 keV is the ionization energy per ion pair, $r_0 = R/\rho$, ρ is the air density at the maximum depth R to which electrons with a given energy can penetrate, $\lambda(z/R)$ is the dimensionless distribution function of energy loss.

$$z = \int_{h}^{h_{top}} \rho dh' \tag{4.31}$$

h is the height in the atmosphere at which the ionization rate is calculated, h_{top} =1,000 km is the adopted altitude of the upper boundary of the atmosphere, and $n(M)_z$; and $n(M)_R$ are the number densities of atoms and molecules being ionized at depths z and R, respectively. In this case, the maximum depth of penetration into the atmosphere and the initial electron energy 0.4 keV$\leq E_0 \leq$500 keV at the upper boundary of the atmosphere are related as follows: R [g/cm^2] = $4.57 \cdot 10^{-6} (E_0 \text{ [keV]})^{1.75}$. If the flux of precipitating electrons having an arbitrary differential energy spectrum $g(E_0)$ in the energy range from $E_o^{\min}$ to $E_o^{\max}$ is isotropic over the upper hemisphere, then the ionization rate Q per unit volume can be presented in the following way in accordance with (4.29):

$$Q = \pi \int_{E_{0\min}}^{E_{0\max}} g(E_0) q_{E_0}(e) dE_0 \tag{4.32}$$

The experimental spectra of precipitating electrons registered on satellites were selected for calculation (Vampola and Kuck 1978).

We omit the detail of calculation which one can find in the paper (Kim et al. 2002) and show the final modified profiles of electron concentration in the *D*-region (Fig. 4.25).

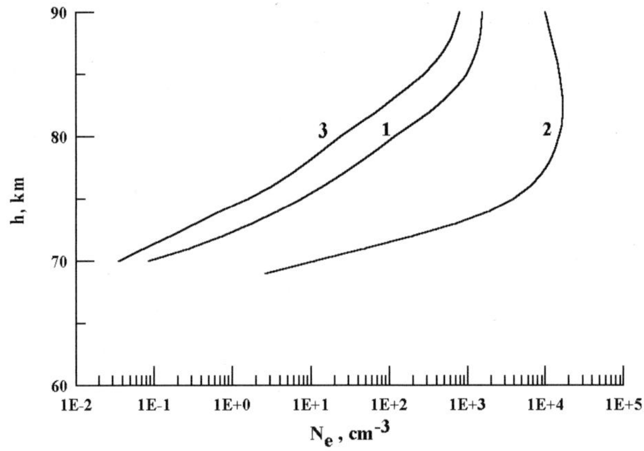

Fig. 4.25 Vertical profiles of electron concentration in the *D*-region of the ionosphere. 1 – the modified profile as a result of particle precipitation, 2 – modified profiles under action of coherent emission of a powerful VLF transmitter, 3-background profile

The vertical profiles of electron concentration N_e modified by the precipitating particles are shown in Fig. 4.25 (curves 1 and 2). For comparison, the background vertical profile is also plotted in the figure (curve 3) for the case of absence of precipitation. It is seen that the electron density in the nighttime ionospheric *D*-region before a severe earthquake can significantly increase. The effect of an increase in N_e is most pronounced if a powerful VLF radio station operates in the vicinity of the epicentral zone of an earthquake under preparation. In this case, a fairly pronounced, dense ionized layer with maximum electron density about $2{\cdot}10^4$ cm^{-3} is formed in the nighttime upper mesosphere at an altitude of about 85 km. Thus, the parameters of such a layer are close to the parameters of the N_e distribution in the daytime *D*-region for the middle and low latitudes. The disturbed electron density exceeds the background values of N_e by more than one order of magnitude at altitudes 85–90 km and by 2 and 3 orders of magnitude at altitudes of 75 and 85 km, respectively. And concentration still remains higher than 10^3 cm^{-3} for h>73 km. In regions far away from the operating VLF radio stations, the earthquake precursors in the nighttime *D*-region are less pronounced. However, variations in N_e are quite significant in these regions, too. The electron density at altitudes 80–85 km increases by about a factor of 8 compared with the background level, and the relative increase in N_e below 80 km exceeds one order of magnitude. Although the electron density reaches its maximum at the upper boundary of the *D*-region, at an altitude of 87 km, where N_e~$2.5{\cdot}10^3$ cm^{-3}, the value of N_e is nevertheless more than 10^3 cm^{-3} down to altitudes $h \approx 82$ km.

The effect of increasing concentration in the *D*-region is equivalent to the effective lowering of the ionosphere. Such lowering was detected also as a result of particle precipitation due to thunderstorm activity in California (Füllekrug 2003).

What consequences have the modification of the *D*-region before earthquakes? As we know, the radiowave propagation in different frequency bands is due to reflection of the radio waves from the ionosphere (Davies 1990). So almost all reported anomalies of the radio wave propagation before earthquakes: in the HF band (Davies and Baker 1965), LF band (Biagi et al. 2001), VLF band (Gufeld et al. 1992; Molchanov and Hayakawa 1998b) and even the VHF band over-horizon propagation (Fukumoto et al. 2002; Kushida and Kushida 2002) may be explained by the *D*-layer modification by precipitating particles. The lowering of the reflection point of the radio wave causes the distortions in the time delay, phase and amplitude, as well as point of arrival, which happens in the case of HF and VHF propagation.

So from this part we can conclude that *observed anomalies of radio wave propagation in VLF, LF, HF, and VHF bands before strong earthquakes are associated with ionosphere modification by the precipitating particles in increased ionization in the D-layer of the ionosphere.*

There exists one more possibility for radio wave propagation anomalies to be discussed in Chapter 7.

4.3.11 Ion Mass Changes in the Ionosphere as a Precursory Effect

In Chapter 2 we reported the results of Boškova et al. (1993, 1994) on the observation of the increased concentration of the light ions before the earthquakes. In a stationary state that assumes, in particular, the absence of particle drift, the concentration of different plasma species along the geomagnetic field line is determined by the partial balance of the pressure, gravity and the longitudinal (along the ambient geomagnetic field) polarization electric field forces. It is possible to seek for the effects of the seismogenic electric field on the ion balance along the magnetic tube solving the continuity Eq. (4.25). This problem was treated by Shklyar and Truhlik (1998). In the simplified model approach (only two ion species, isothermal ionosphere, neglecting of the geomagnetic field line curvature) they analyzed the ion balance within the magnetospheric tube affected by the electric field of seismic origin perpendicular to the geomagnetic field line. This arrangement exactly coincides with the field configuration regarded by Kim and Hegai (1997, 1999) for the upper ionosphere. Shklyar and Truhlic (1998) demonstrated the high sensitivity of the light ions to the changes in the background ionosphere on which the modified magnetic tube is loaned, especially in the transitions region where concentration of the light and heavy ions is comparable. This means that the slight local changes in the base of the magnetic tube caused by the particle drift under the action of electric field lead to the strong changes of light ion concentration, especially in the transition region.

4.4 The Model Generalization

In Chapter 2 we presented some examples of anomalous variations of different parameters registered by ground based and satellite techniques that appear before strong earthquakes and were regarded as short-term earthquake precursors. To be really accepted as precursors they should be explained, firstly, from the point of view of the physical mechanism responsible for them. The data analysis led us to the conclusion that most of the observed effects may be explained as a result of action of the anomalous electric field appearing some time (days or hours) before the seismic shock in the area of earthquake preparation. In turn, the other experimental measurements show that this anomalous (or *seismogenic*) electric field appears as a result of a complex chain of plasma-chemical processes and ion molecular reactions in the boundary layer of the atmosphere over the area of earthquake preparation. Due to this the following two chapters (3 and 4) were devoted to a detailed discussion of the physical mechanism of anomalous electric field generation, its penetration from the ground level to the ionosphere and magnetosphere, and effects which this field creates on its way at all levels of the near-Earth plasma.

Taking into account that the problem is very complex, involving many spheres of near-Earth environment (from underground up to the Earth's magnetosphere), is multidisciplinary (from seismology to plasma and quantum physics) let us try to generalize the model verbally with the help of a schematic chart presenting the main blocks of the model (Fig. 4.26) and an artistic view (Fig. 4.27).

In the area of earthquake preparation (the size of which is determined by the magnitude of the future earthquake) besides the mechanical transformations take place active geochemical processes including emanation of radon and several types of gaseous components including noble and greenhouse gases. The well-known effect is the electrode effect – modification of the near ground vertical profile of the atmospheric electric field by the ionization of the near ground layer of the atmosphere by radon. It is a global effect. The preparatory stage is the formation of near ground plasma in the form of long living ion clusters which are the result of ion-molecular reactions (after ionization by radon) in the near ground layer of the atmosphere with water molecule attachment to the finally formed ions. The water molecules high dipole moment prevents the formed ion clusters from recombination. The important feature is that the mobility of the positive and negative particles of this atmospheric plasma is different which will help in the further process of charge separation. Another very important process – is the formation of quasineutral clusters due to Coulomb attraction of the positive and negative ion clusters. So, as an end of the preparatory stage, we have the near ground layer of the atmosphere in the earthquake preparation area rich with latent ions masked by the formed neutral clusters.

The second step – is generation of the anomalous field itself. It is known that before the earthquake the intensive gas discharges take place in the area of earthquake preparation. These discharges play a twofold role. By generation of the air motions maybe in the form of acoustic gravity waves, they destroy the neutral clusters because of the weakness of the Coulomb interaction. And suddenly, in a short time interval the near ground layer of the atmosphere becomes rich with ions (the estimated concentration is $10^5 - 10^6$ cm^{-3}. The charge separation process described in Chapter 2 leads to the generation of anomalously strong vertical electric field in comparison with fair weather electric field (~1 kV/m and ~100 V/m respectively). This is the final stage of the first act – anomalous electric field generation. It should be noted that under different geophysical conditions (for example the presence of mist) the anomalous electric field might have a downward direction (coinciding with the natural direction of atmospheric electric field), or upward.

Before we move up into the ionosphere there is one additional remark. The gas discharges besides their destructive role for the neutral clusters, may carry with them submicron aerosols which (it is well known) will increase the intensity of the electric field.

And other remarks related to the observed precursors. As one can see from the discussion, the near ground layer of the atmosphere becomes the real plasma with particle concentration comparable with some regions of the Earth's ionosphere. In addition, this is a plasma in the strong electric field, where we should expect particle acceleration and excitation of plasma instabilities. We believe (and it is the purpose of our future work) that just these instabilities are responsible for different kinds of electromagnetic emissions from ULF to VHF measured in near earthquake areas (besides some underground processes by which these emissions tried to be explained up to now).

The process of water molecules attachment, and further interaction of the generated strong electric field with the water vapor in the atmosphere involves some latent processes changing the latent heat flux in the earthquake preparation area and the water vapor balance in it. This question will be discussed in more detail in Chapter 7. Nevertheless, here we should mention that with the same time scale as ionospheric disturbances initiated by the anomalous electric field, the anomalies of the latent heat flux and water vapor are observed by the remote sensing satellites over the earthquake preparation area.

Now we start to move up into the ionosphere. The electric field penetrating into the *E*- region of the ionosphere, depending on its sign may create positive or negative deviations in the *E*-region of the ionosphere which was registered experimentally. Depending on the shape of the area of the electric field generation (circular or elongated) the shape of the irregularity within the ionosphere will be different, but in all cases only the perpendicular (to the geomagnetic field lines) component of the anomalous electric field penetrates into the ionosphere. In the case of the anomalous electric field directed down on the ground surface, the sporadic layers will be formed in the ionosphere over the area of earthquake preparation, which is also registered experimentally.

Due to equipotentiality of geomagnetic field lines the electric field practically without any decay penetrates to the higher levels of the ionosphere. In the *F*-region we will mark two main effects. In the area of maximal conductivity due to Joule heating the acoustic gravity waves will be generated giving rise to the small-scale density irregularities within the ionosphere. These processes experimentally are manifested in periodic oscillations registered on different ionospheric heights by techniques of radiophysical and optical monitoring of the ionosphere and well supported by the experimental data. The other, probably main and well-documented effect – is the formation of the large-scale irregularities of electron concentration in the *F2*-layer of the ionosphere. They were registered by satellites, and from ground-based ionosondes and the ground network of GPS receivers. Due to the complex character of particle drift in the *F*-region in crossed electric and geomagnetic fields, the large scale anomalies in the *F*-region, as well as anomalies connected with AGW propagation will be registered not just over the impending earthquake epicenter, but shifted equatorward which one should keep in mind in practical applications.

At the higher levels we will have the following effects. The small-scale irregularities spread along the geomagnetic field lines into the magnetosphere creating the field aligned ducts where the VLF emission of different origins will be scattered. This will lead to an increased level of VLF emission within the magnetic tubes loaned on the areas of anomalous electric field generation. Due to plasma drift processes it will be not exactly the same shape as on the Earth's surface, but elongated in the zonal direction with a relation approximately 1:3 for meridional and longitudinal sizes of the modified volume of the magnetosphere. Just VLF emissions were the first to be registered by satellites in the early years of the study of electromagnetic precursors. The other effect of the ducts formation – is the cyclotron interaction of the VLF emissions with the particles of radiation belts leading to their loss cone precipitation. The precipitating particles associated with the earthquake preparation were also registered on many satellites.

And finally, the precipitating particles will produce ionization of the lower ionosphere leading to an increase of the electron concentration in the *D*-region of the ionosphere which is equivalent to the ionosphere lowering. This lowering changes the condition of radio wave propagation in different frequency bands (from VLF up to VHF) which also was registered experimentally.

This very long and complex description of the chain of the processes in seismo-ionospheric coupling is accompanied by the table where are shown the different types of precursors observed experimentally and the corresponding physical processes responsible for their generation.

4.4.1 Block Diagram of the Seismo-Ionospheric Coupling Model

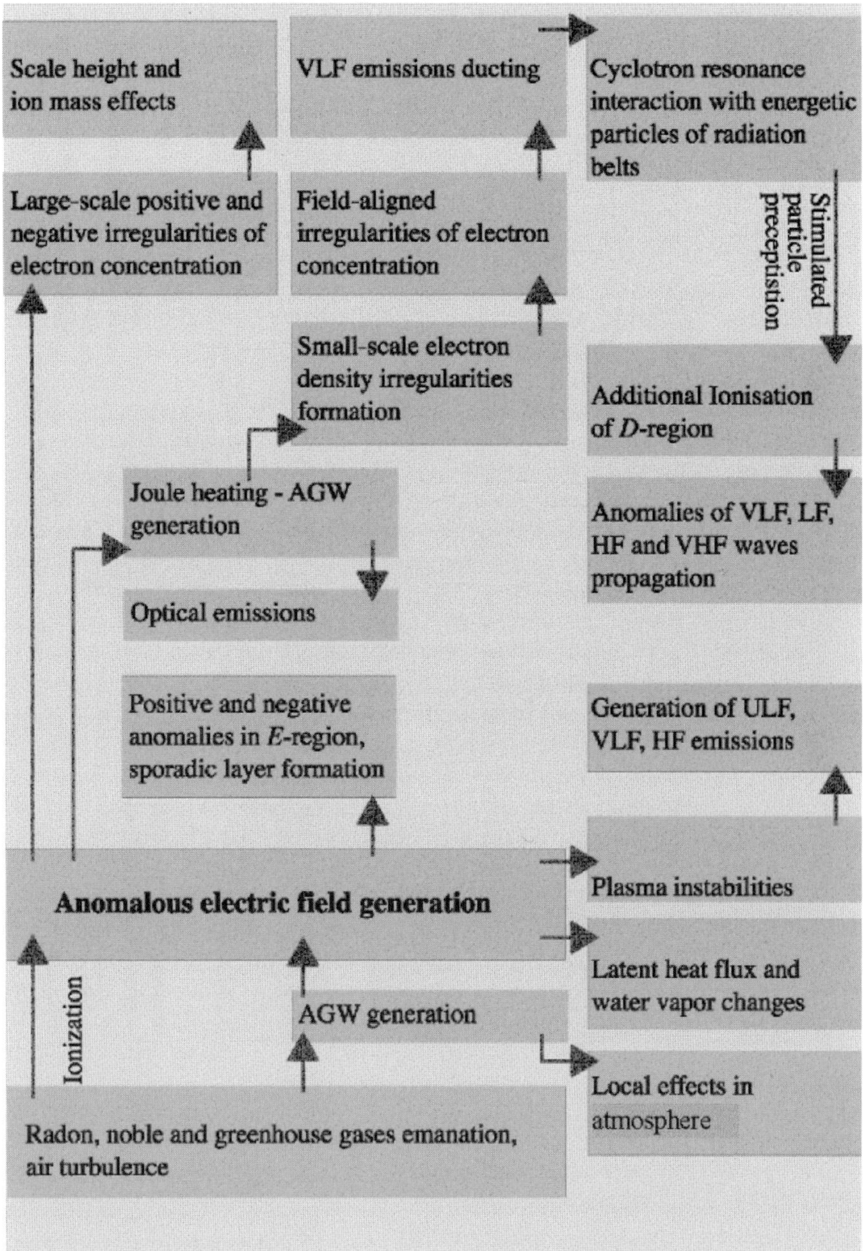

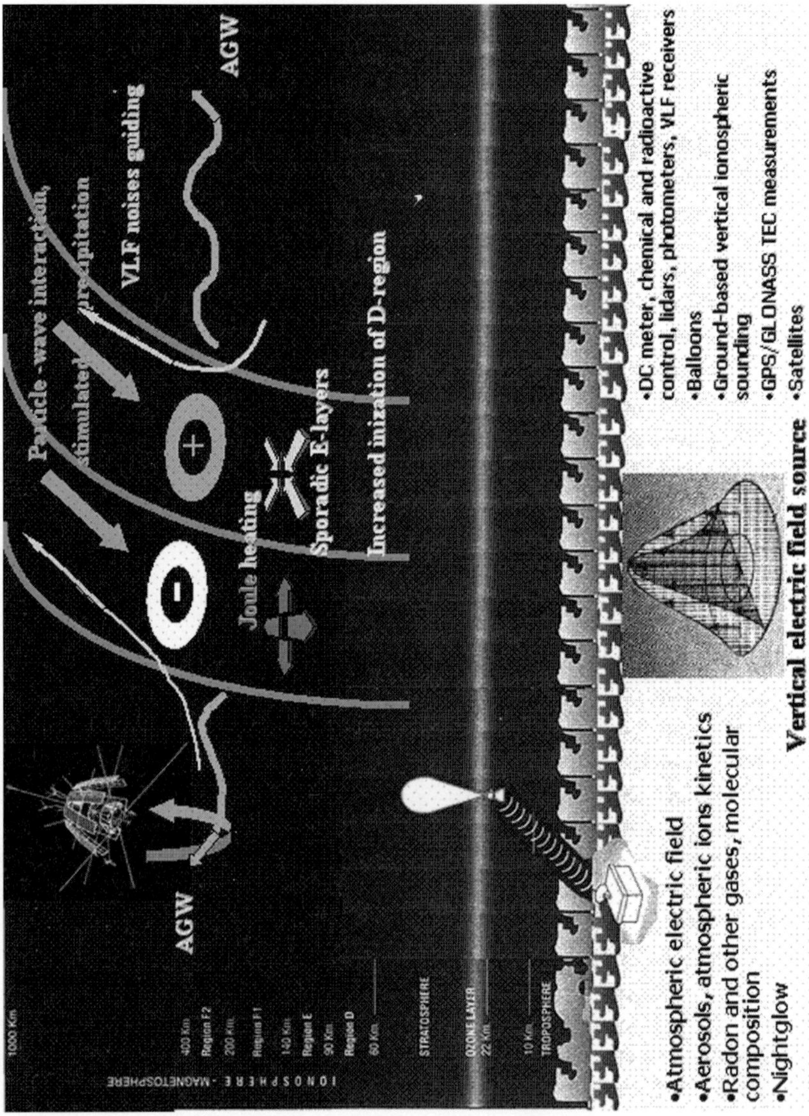

Fig. 4.26. Artistic view of the seismo-ionospheric coupling model

Table 4.1 Electromagnetic and plasma precursors and their physical mechanisms

Precursor	References	Physical mechanism
ULF, VLF, HF emissions registered on the ground level	Sobolev and Husamiddinov 1985; Fujinawa and Takahashi 1998; Vallianatos and Nomicos 1998; Mayeda 1999	Instabilities of plasma in strong quasistationary electric field
Latent heat flux, water vapor content	Tramutoli et al. 2001, Tronin et al. 2002; Dey and Singh 2004	Water molecules attachment in ion clusters, changes of water evaporation properties under action of strong electric field
Variations of electron concentration in *E*-region, formation of sporadic *E* layers	Liperovsky et al. 2000; Ondoh 2000	Penetration of anomalous electric field from the ground to E-region, particle drift
Optical emissions, oscillatory variations of electron density in ionosphere, small-scale ionospheric irregularities	Fishkova et al. 1985; Zelenova and Legen'ka 1989; Liperovsky et al. 1991; Chmyrev et al. 1997	Acoustic gravity waves generated in F-region due to Joule heating by anomalous electric field
Large scale ionospheric irregularities in F-region	Pulinets et al. 1991; Liu et al. 2002; Pulinets and Legen'ka 2003	Ionospheric particles *ExB* drift in anomalous electric field penetrating from the ground and geomagnetic field
Ions mass and scale height changes	Bošková et al. 1994; Pulinets et al. 2003 b	Interaction of anomalous electric field with polarization electric field of plasmasphere
Seismogenic VLF emissions registered by satellites	Larkina et al. 1983, 1989; Parrot et al. 1985; Parrot and Mogilevsky 1989; Molchanov et al. 1993	Trapping of VLF noises of different origin in the plasma ducts created by irregularities under action of anomalous electric field
Energetic particle precipitation	Galper et al 1983, 1995; Galperin et al. 1992	Cyclotron resonance interaction of radiation belts particles with VLF emissions trapped in magnetic tube modified by seismogenic electric field
Anomalies of VLF, LF, HF, VHF radio waves propagation over the area of earthquake preparation	Davies and Baker 1965; Gokhberg et al. 1989b; Gufeld et al. 1992; Molchanov and Haya-kawa 1998b; Biagi et al. 2001; Kushida and Kushida 2002	Particle precipitation-additional ionization of D-region, effective lowering of the ionosphere

Chapter 5 Main Phenomenological Features of Ionospheric Precursors of Earthquakes

5.1 Introduction

It is well known that between science and practical application of our knowledge there is a precipice. To understand the physics of the process is very important but not enough as in the case of the earthquake precursors. Real prediction is a great responsibility being based on precise calculation and system organization. One component of this system is continuous monitoring of the parameters used as earthquake precursors. In previous chapters we described several physical processes and ionospheric plasma parameters variation. To understand which of them could be the real precursor, at which the monitoring system should be aimed, we should know the behavior of these parameters in different geophysical conditions, in local time, their statistical characteristics, and how they look when registered by different techniques. This is what will name the morphological characteristics of the process. To move forward to the practical application one cannot avoid this step of studying the process – candidate to precursors. More than 10 years spent by the authors in studies of variations in the ionosphere observed over seismically active regions several days or hours before strong earthquakes permits us to derive at least the most important characteristics of their behavior and to come to a conclusion on how to use them in practice. Data obtained by ground-based ionosondes and by topside vertical sounding from satellites are used in conjunction with data from local probe satellite measurements.

Those who want to study the ionospheric variability associated with seismic activity face many difficulties, but the most challenging one is that of identification of this kind of variability. In our study we attempted to categorize the main features of ionospheric variations associated with pre-seismic activity. This classification is based on hundreds of cases processed using ground based and satellite measurements. It is also based on knowledge of the ionosphere's behavior both under quiet and geomagnetically disturbed conditions as a background for seismo-ionospheric variability separation.

Multiparameter analysis is also very important. Parameters that identify the precursor as univocal should be chosen so as to not confuse the observed variation with variability caused by another source.

So we regard this to be as important (maybe more important) than the description of the physical mechanism.

It is necessary to mention that morphological properties can be divided into two categories: objective and subjective. Objective ones are, for example, the absolute size of the modified area within the ionosphere, scale height or ion mass changes. The subjective parameters are those that are connected with the kind of observation technique with limited possibilities (in time or in space), so these characteristics are specific to the given type of measurement but not to the precursor itself. It relates especially to the one-point measurements where the device is able to register only one part of the process, and the amplitude and sign of the variation depends on the relative position of the measurement point and the future epicenter as it relates to ground based ionosondes.

5.2 When the Ionosphere Starts to feel Earthquake Preparation

5.2.1 Sensitivity to Earthquake Intensity

Theoretical calculations (Pulinets et al. 1998a) show that the vertical ground electric field starts to penetrate effectively into the ionosphere and create irregularities of electron concentration when the size of the area on the ground surface occupied by the anomalous field exceeds 200 km in diameter. This area could be identified with the zone of earthquake preparation (Dobrovolsky et al. 1979) or by another determination the precursory seismic activation zone (Bowman et al. 1998). Here the usual precursors used in seismology are borne in mind, foreshocks, deformations, seismic waves velocity anomalies, etc. Using the Dobrovolsky formula (1.8) $\rho = 10^{0.43M}$ km where ρ is the radius of the earthquake preparation zone, and M – earthquake magnitude, for ρ=100 km we obtain M=4.65. In the paper of Garavaglia et al. (2000) a good correlation between observed deformations and radon emanations is reported, so we can identify the size of the earthquake preparation zone used in seismology with our scale parameter used for the zone occupied by the anomalous electric field. The changes in radon concentration are due to the deformations and crack formation in the Earth's crust, so they should be observed within the zone of earthquake preparation. And in our model we use the radon as the source of ionization for the electric field generation mechanism, so we can identify these zones. Using the estimate above we can expect the ionosphere reaction from earthquakes with a magnitude higher than 4.65.

Another factor for estimation is simple scaling. The first most conducting layer of the ionosphere, the E-layer, is at the altitude ~100 km, and the electric field starts to penetrate into the ionosphere when its spatial scale reaches the size of the height of the ionospheric layer. We can say that a quantity between 100 and 200 km is a transition size for the electric field's area and the correspondent ionosphere reaction, which is shown by the results of statistical studies (Chen et al. 1999). The ionospheric precursors within the interval 5 days before the seismic shock were registered in 73% of cases for earthquakes with a magnitude of 5, and in 100% cases for earthquakes with a magnitude of 6. Therefore we can regard the magnitude of 5 as some threshold of ionosphere sensitivity for earthquake preparation.

5.2.2 Temporal Parameters of Ionospheric Precursors

Using the word "precursor", we mean that the observed variation appears before the seismic shock. But is there any statistical estimation of the words "in advance"? We will now discuss the variations of the critical frequency *foF2*, which reflects the variations of electron concentration on the main density peak of the ionosphere and is determined by formula (1.12). The main source of this parameter is the network of ground-based ionosondes. Usually only one ionosonde happens to fall within the area of strong seismic activity such as the Chung-Li ionosonde in Taiwan or the Rome ionosonde in Central Italy. So we encounter a case of one-point measurements when a site within a seismoactive zone is instrumented with only one sensor, the only way to identify seismo-ionospheric variations is through good knowledge of their temporal characteristics. Analysis of ionospheric data from many stations all over the world shows that tentatively the time-dependent evolution can be divided into two parts: one with a time-scale of days and another one with a time-scale of hours. We believe this reflects the different parts of the process, days – evolution of the seismic source, hours – combination of electric field and ionosphere variability.

The first question which arises in temporal characteristics analysis is how early prior to a shock, precursors appear and can be identified. Empirically it was found that precursors appear as far as 5 days before the shock (Pulinets 1998b). This result is displayed in Fig. 2.11 and the process of data processing is explained in Chapter 2. But Fig. 2.11 demonstrates only the case study. Is their any statistical confirmation of this fact? Unfortunately, the statistical data for ground-based ionosondes are still not published yet in their full extent regardless of their existence for at least two years, so we will refer to them as a private communication with permission of the authors (Liu et al. 2003). The robust statistical processing of the Chung-Li ionosonde (Taiwan) ionospheric data for the period of 1994–1999 of continuous observations in the seismically active area was conducted. This period of study includes 169 earthquakes with a magnitude higher than 5. To make a robust monitoring of the seismo-ionospheric variation, the 16th daily *foF2* value, X_{16}, is claimed to be abnormal if X_{16} is less than the lower bound $\tilde{X}$-IQR or greater than the upper bound $\tilde{X}+\text{IQR}$. $\tilde{X}$ and IQR are the median and inter-quartile range, respectively, of the *foF2* in the previous 15 days. The 15 day interval was selected due to a statistically determined recurrent interval for earthquakes with a magnitude higher than 5 in the Taiwan area. Correlation of anomalies with the earthquake occurrence showed the increase of anomalies within the interval 5 days before the earthquakes with $M \geq 5$.

It is of interest to compare the ionosonde data with another source of ionospheric information. The GPS TEC data is a good candidate for such a comparison. It is a completely different technique, at the same time it was demonstrated that this technique permits us to observe the pre-seismic ionospheric anomalies as well (Liu et al. 2002). Liu et al. (2004) analyzed the seismo-ionospheric anomalies monitored by the GPS TEC. A similar technique as used in Liu et al. (2003) for data processing was used for 20 strong earthquakes ($M \geq 6$) in the Taiwan area. The most sensitive appeared to be the negative ionospheric anomalies (negative deviation from an undisturbed level). The results of statistical processing are depicted in Fig. 5.1.

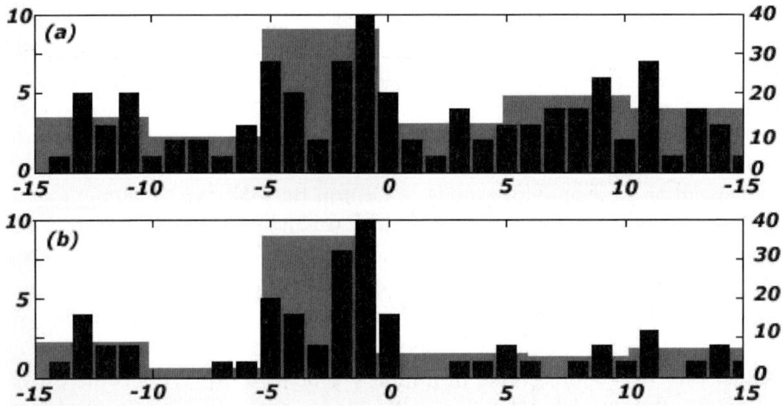

Fig. 5.1a,b The mean counting of lower anomalies occur within 15 days before and after the $M \geq 6.0$ earthquakes. (**a**) with and (**b**) without *M* 5 earthquake's and geomagnetic effects. The *dark* and *gray* histograms denote the counts of every day and 5-day period, respectively

The figure clearly demonstrates that even with noisy addins resulting from geomagnetic disturbances and smaller earthquakes (Fig. 5a) the 5-day predicting interval is clearly seen.

So we can claim that the five-day interval (which means that ionospheric precursors may be detected within the time interval 1–5 days before the earthquake) is statistically proved and established. Sometimes more extended intervals may be observed, but they still remain the case study results and will be demonstrated only in the context of complex data processing in Chapter 6.

If we go to smaller time scales, we will see that ionospheric precursors change as a function of local time. The critical frequency deviation can be positive as well as negative.

Figures 2.12 and 2.13 demonstrate the typical appearance of seismic precursors registered at different ionospheric vertical sounding stations for earthquakes with a magnitude from 5.4 up to 7.1 (Pulinets et al. 1998b). Because of the changes in plasma density associated with seismic activity in average are of the same order of magnitude as the day-to-day ionospheric variations, the question arises, how to distinguish seismic precursors from the background of the daily variations. It was noted early by Pulinets et al. (1998b) that the behavior of the seismoionospheric precursors shows some regularity in local time. The first conclusion was that, for specific intervals of local time, the deviation would be either negative or positive. But later studies (Chen et al. 1999) have shown that the situation is more complex. While the behavior in local time has some regularity, it is different for different days prior to the shock. The complex chain of physical processes contributing to its generation could explain such intricate conduct of the ionospheric precursor. Among the possible sources of variability we can mention the geological structure of the region of study and especially the structure of the tectonic fault, the seismic source mechanism for the given earthquake (slip, thrust, etc.), the sporadity or regularity of the emanations from the ground into the atmosphere, atmospheric (weather) conditions, regular daily variations of the ionosphere parameters, observation point position in relation to the future epicenter position. The last issue is illustrated in Fig. 5.2.

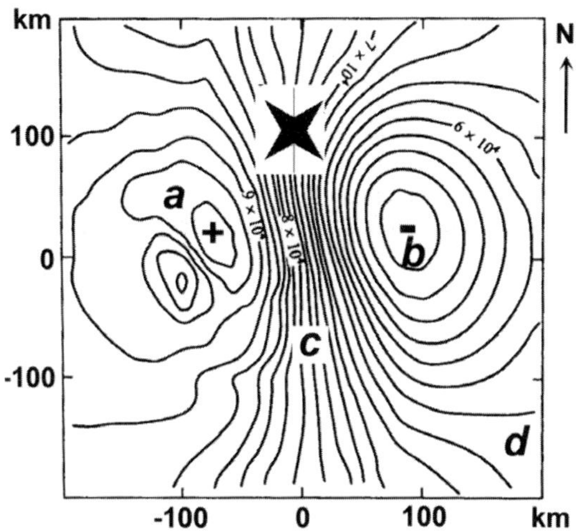

Fig. 5.2 Different relative positions of ionosonde (*a, b, c, d*) in relation to the anomalous electric field source lead to different sign and magnitude of registered precursor for the same event

Taking from our theoretical calculations the distribution of the electron concentration within the F-region at an altitude of 250 km, we can obtain completely different results for the same distribution in the ionosphere as a dependence on the ionosonde position. At position *a* – epicenter to the North-East of the ionosonde, we have a positive deviation, at position *b* – epicenter to the North-West of the ionosonde, we have a negative deviation, at position *c* – epicenter to the North of the ionosonde, we have almost zero deviation, and at position *d* – epicenter to the North-North-West of the ionosonde, we have also negative deviation but much smaller than in position *b*. This distribution was calculated for a "normal" (directed down) atmospheric field direction, but the change of the electric field direction is observed often experimentally before the seismic shocks (Nikiforova and Michnowski, 1995; Vershinin et al. 1999) which means that the signs of the dipole distribution within the ionosphere will flip, and the set of our ionosonde variation results will double for different epicenter positions in relation to the ionosonde. It is illustrated in Fig. 5.3 where the changes of sign of deviation on the ionosonde data reflect the change of the electric field direction. So we can interpret the different shape of variations observed by ionosondes for different earthquakes by the relative position of the ionosonde and epicenter and changes of the seismogenic electric field direction.

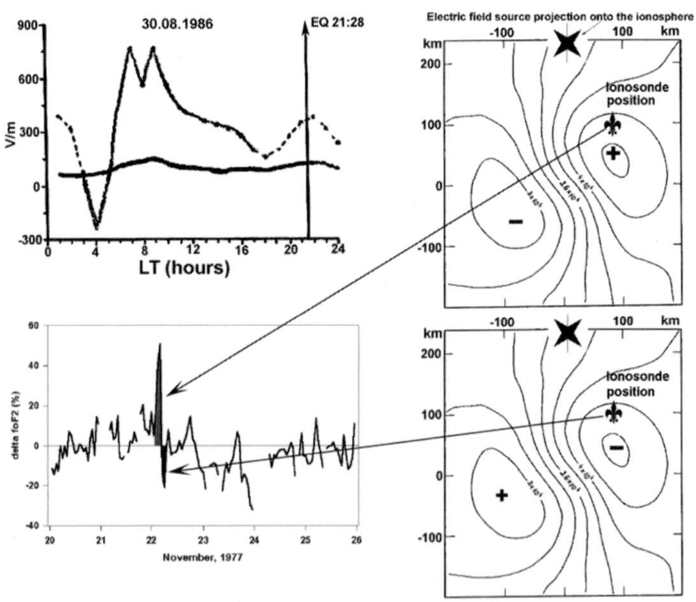

Fig. 5.3a–c Change of the electric field direction (**a**) leads to the change of electron density distribution in the ionosphere (**b**) and to change of the deviation sign registered by the ionosonde (**c**)

Such diversity of shape of precursors registered by the ground based ionosondes (actually determined to a great extent by the relative position of the ionosonde and epicenter) attracted a lot of critics. Critics affirm that observed deviations are completely casual and do not have anything in common with earthquakes. Now the question becomes more or less clear. A more direct demonstration of this fact is the result obtained in the course of statistical data processing of ionosonde data in the Taiwan area (Pulinets et al. 2002b). The authors looked for the similarity of the ionospheric precursors (the technique will be described in Chapter 6) and obtained a very strange result: for the shallow earthquakes the similarity was very low, but for earthquakes with depth h>60 km it was much higher. And only the consequential analysis of the spatial distribution of the earthquakes with a different depth (Wang and Shin 2002) allowed us to come to the correct conclusion (Pulinets et al. 2004a). One can see this in Fig. 5.4 where the spatial distribution of the earthquake epicenter in Taiwan is shown for all earthquakes (left panel) and for earthquakes with a source depth larger than 60 km. Statistically the latter group of earthquakes have a clear cluster distribution, and consequently, the relative position of the ionosonde (which is situated in the North-Western part of the island) and the epicenters is much more stable and determined which gives as a result the large similarity in the shape of the ionospheric precursors.

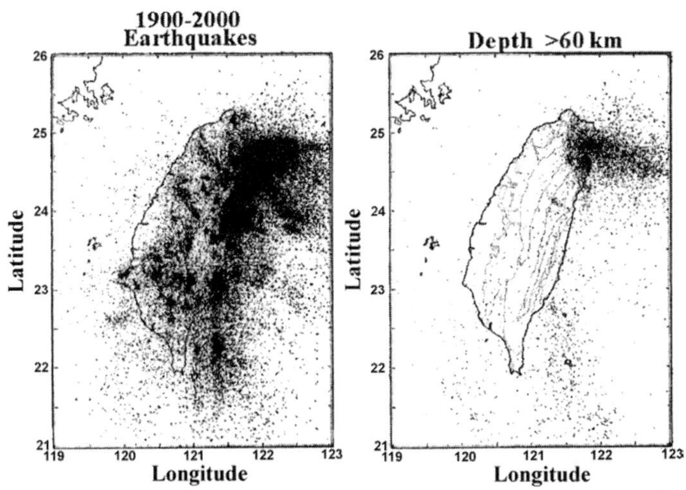

Fig. 5.4 Statistical distribution of the earthquake epicenters in the Taiwan area for all earthquakes (*left panel*) and for the earthquakes with the seismic source depth larger than 60 km (*right panel*)

Basing on our theoretical calculations we can use as the working hypothesis the following statement: *for the fixed relative positions of the earthquake source and observation point and the same source mechanism, the shape of the ionospheric precursor should be the same (or very similar) for the recurrent earthquakes occurring in the same epicenter or close to it.* Usually in the vicinity of the ionosonde (within the radius of 500 km) only one active tectonic fault is situated, therefore one can expect the reiteration of ionospheric variations observed before the earthquake in the given point.

If we accept this hypothesis one can try to look for the behavior of the precursors as a function of local time (Pulinets et al. 1998b). We analyzed 56 strong earthquakes for which records were available from vertical sounding stations located within the earthquake preparation zone. All data were reduced by the same technique. For each nine-day period (seven days before the main shock and two days after), the mean daily variations of the critical frequency were calculated and, then, the deviation Δf_0F2 from the mean value $\overline{foF2}$ was determined. For the analysis, only magnetically quiet periods were selected $(K_p<3)$. In some cases, a magnetic storm was observed on the next day after the earthquake. The maximum deviation $\Delta f_0F2_{\max}$ from the average (precursor), in percent, was represented in the plot shown in Fig. 5.5, as a function of local time. As a result, a synthesized plot was obtained which summarizes data for many earthquakes that occurred in various regions of the Earth. We understand the limitations of such an approach, nevertheless, this method of data reduction gives at least the zero order approach to the local-time dependence of seismo-ionospheric variations, which manifests itself in periodic changes of the sign of anomalous variations.

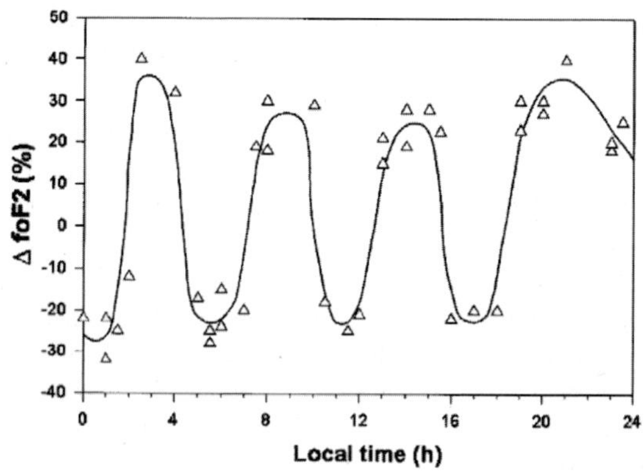

Fig. 5.5 Empirically determined dependence of the maximum critical frequency deviation $\Delta f_0F2_{\max}$ as a function of local time for 59 earthquake precursors (after Pulinets et al. 1998b)

5.2.3 Spatial Distribution of Ionospheric Precursors

The most appropriate technique to study the spatial distribution of seismo-ionospheric variations is satellite mapping. Measurements of the critical frequency *foF2* by vertical sounding from the ground or by topside sounding from the satellite give the same values (see Fig. 1.14). Therefore, we can use topside sounding for such studies, supplemented by the data of ground based ionospheric stations when they are located within the area mapped by the satellite. The spatial characteristics of the ionospheric precursors can be determined from the latitudinal and longitudinal cross-sections, as was done in Pulinets et al. (1991), and from the mapping technique applied in Pulinets and Legen'ka (2003).

Figure 2.6 presents the latitudinal distributions of the critical frequency scaled from the topside ionograms of the Alouette-1 satellite for it's passing over the impending earthquake epicenter area (curve 2) and 30° in longitude away for the case of the "Good Friday" earthquake of the 28th March 1964. From this figure one can determine the radius of the area modified by the anomalous electric field (from the maximum of deviation up to the point where both curves merge near latitude 65°). We can estimate the radius of the zone within the ionosphere near 1,200–1,300 km which gives a magnitude higher than 7 according to formula (1.8). A more accurate estimation is possible when you build the 2-D map of the ionospheric disturbance before the earthquake. The procedure of the map's construction can be explained with the help of Fig. 5.5. To exclude the solar effects, the most important of which is the daily variations of electron concentration in the ionosphere (the difference between the daytime and nighttime values of the electron concentration can reach two orders of magnitude), we use the conception of the so-called LT map. It is an artificial map using the same local time all over the globe. Such a map is an excellent tool for revealing the longitude variations within the ionosphere, i.e. to see how different the ionosphere is at different longitudes for the same geophysical conditions including local time. The orbit configuration of the high inclination satellite gives an opportunity during a short period (near 24 hours) to collect information sufficient to produce the global map for fixed local time.

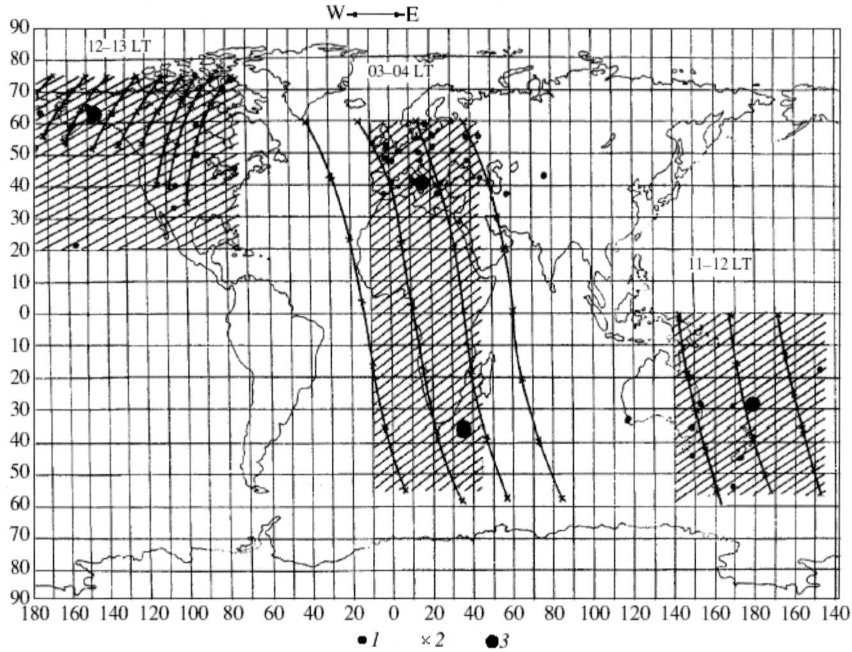

Fig. 5.6 Schematic presentation of the satellite mapping of the ionospheric irregularities associated with earthquake preparation. 1 – position of ground based ionosondes, 2 – points of the satellite sounding along the orbit, 3- earthquake epicenters position (after Pulinets and Legen'ka 2003)

Two consecutive days of satellite monitoring produces a denser grid for map construction. The topside sounding satellite measurements were made globally and the large-scale ionospheric variations before the strong earthquakes were detected at all latitudes. For high latitude variations the data of the Alouette satellite were used together with ground based ionosondes to study the effects of the Alaska "Good Friday" earthquake of 1964. For other latitudes the data of the Intercosmos-19 topside sounder were used. Maps of the distribution of ionospheric parameters were obtained using the commercial "Surfer" software (Golden Software Inc.) creating the map grid by the "kriging" technique (Oliver and Webster 1990). To make the grid denser the data of ground based ionosondes were used if they were present within the area of mapping. Figure 2.7 presents the dynamics of the development of the ionospheric irregularity over Alaska in 1964. The maximum extent of the disturbed area in longitude observed in the middle panel is of the order of 40° which gives a magnitude of ~7.7. This is less than the real magnitude nearer 9, but we should keep in mind that the satellite monitoring was not continuous and nobody can guarantee that we "caught" the maximum phase of the disturbance development. The other reason is that the earthquake was extremely strong, and one may expect some saturation of the effect within the ionosphere.

We may conclude only that for strong earthquakes the latitude and longitude spatial scale of the ionospheric effects is of the order of 20° . For all latitudes the satellite data confirm the ground based data demonstrating the positive and negative variations in different local time.

Regarding the spatial extent of the precursory effects in the ionosphere, two new features were detected by the satellites. In specific conditions (when both hemispheres are not sunlit) the conjugated effect could be observed. This means that the ionospheric anomaly over the seismically active area in one hemisphere is mapped through the geomagnetic field lines into the magnetically conjugated area of the opposite hemisphere. The example of such an effect is presented in Fig. 5.7 for the Irpinia earthquake of 1980 in Central Italy. The electron density variations in the magnetically conjugated area were accompanied by the corresponding effects in VLF noises and particle precipitation (Ruzhin and Larkina 1996). It is one more solid confirmation of the proposed physical mechanism of anomalous electric field effects demonstrating the field penetration in the magnetosphere along geomagnetic field lines.

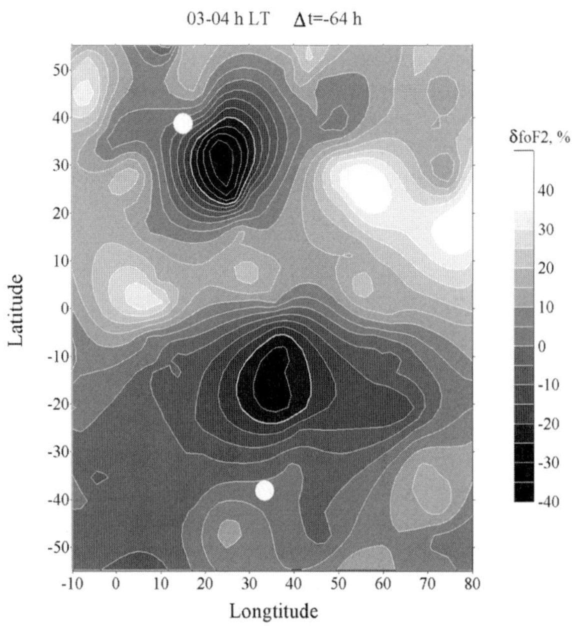

Fig. 5.7 Example of formation the conjugated modified area in the Southern hemisphere before the Irpinia earthquake 23rd November 1980, M=6.9. Data from Intercomos-19 satellite topside sounding (Pulinets and Legen'ka 2003)

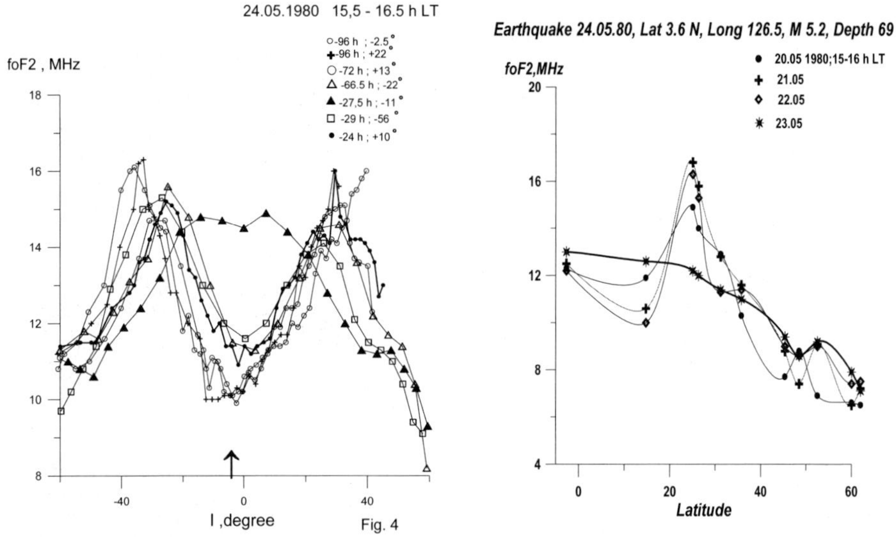

Fig. 5.8 Example of equatorial anomaly crest disappearance before a strong earthquake in the equatorial Western Pacific region. *Left panel* – the satellite data. *Triangles* show latitudinal distribution one day before the earthquake – the crests of the equatorial anomaly disappeared. The same for the ground based sounding – *right panel.* The data collected from the latitudinal chain of the ionosondes situated close to the epicenter longitude. *Asterisks* – the latitudinal profile corresponding to the satellite data – day before the earthquake

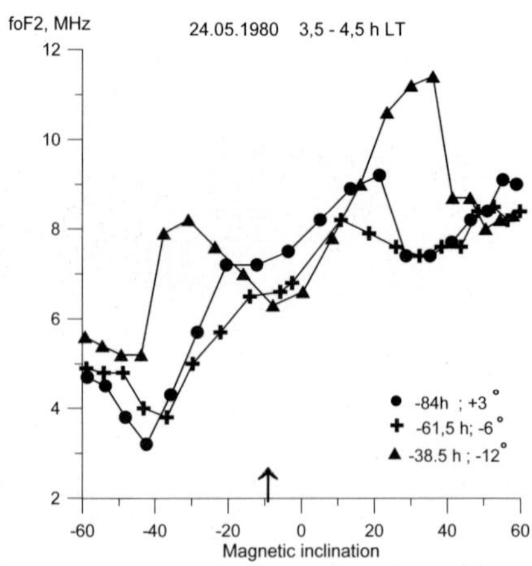

Fig. 5.9 Development of the crests of the equatorial anomaly (*triangles*) in the early morning hours before the earthquake in equatorial latitudes. Parameters of the earthquake are given in Fig. 5.8 (*right panel*)

The second feature is observed in the equatorial ionosphere and is closely related with the conjugated effects described above. In low and equatorial latitudes the geomagnetic field lines are very short which facilitates the effect of magnetic conjunction. And indeed, it was detected (Pulinets and Legen'ka 2002) that if the epicenter falls between the crests of an equatorial anomaly, the anomaly reacts as a whole structure, completely changing its shape and the level of development.

It looks like the anomaly crests disappear before the earthquake in the afternoon hours of local time (Fig. 5.8) or contrary, the development of the crests occurs in the early morning before sunrise (Fig. 5.9). This implies the existence of the zonal component of the seismically generated electric field after its penetration into the ionosphere and its interaction with the electric field responsible for the fountain effect in the equatorial anomaly. As an independent example of this phenomenon we can supply the paper of Liu et al. (2002) where the anomaly modification was reported for the last destructive earthquakes in Taiwan by the data of GPS TEC monitoring (Fig. 5.10).

Unfortunately, up to now there is no model able to calculate the electric field effects in the equatorial ionosphere penetrating from the ground level. Our model works only in middle and high latitudes. But from the experimental results we can conclude that *if the epicenter of the impending earthquake is between the crests of the*

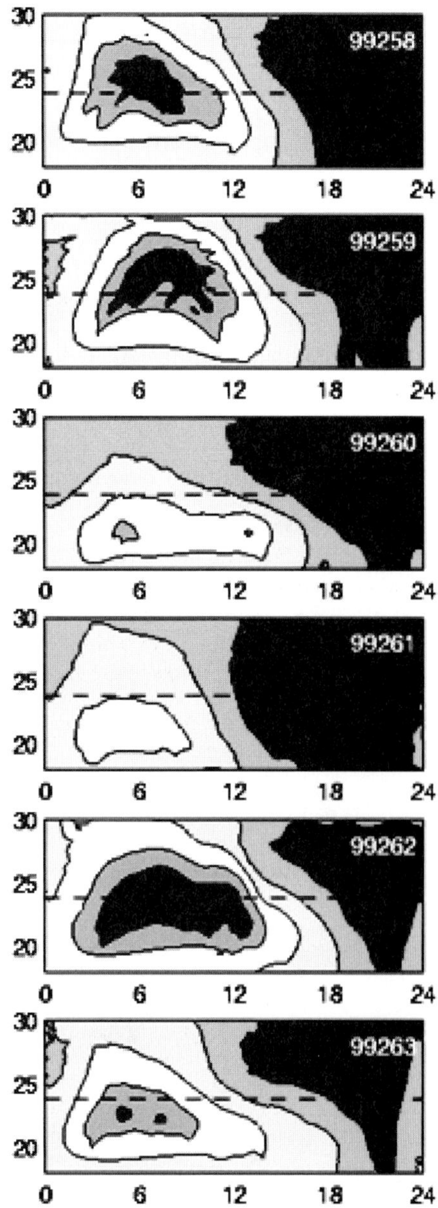

Fig. 5.10 Variations of development and position of the equatorial anomaly Northern crest before catastrophic Chi-Chi earthquake in Taiwan on the 20th Sept. 1999, *M*=8. Clearly seen is the southward displacement of the crest and its practical disappearance on days 260 and 261

equatorial anomaly, the equatorial anomaly reacts on the earthquake preparation as a whole including the crests modification from both sides of the geomagnetic equator.

5.2.4 Local Ionospheric Plasma Parameter Variations Before Earthquakes

Up to now only the critical frequency variations were described as ionospheric earthquake precursors, but they reflect the substantial changes happing within the whole bulk of the Earth's ionosphere which is reflected in variations of the local plasma parameters. To analyze the local plasma parameters variations at different altitudes the data from the AE-C local plasma probe were used for the period of January and February 1974 when the orbit of the satellite was elliptical. For the temperature and ion composition measurements we selected times of undisturbed conditions in the absence of magnetic disturbances, periods during magnetic storms, and periods a few days before the earthquakes (Pulinets et al. 2003b).

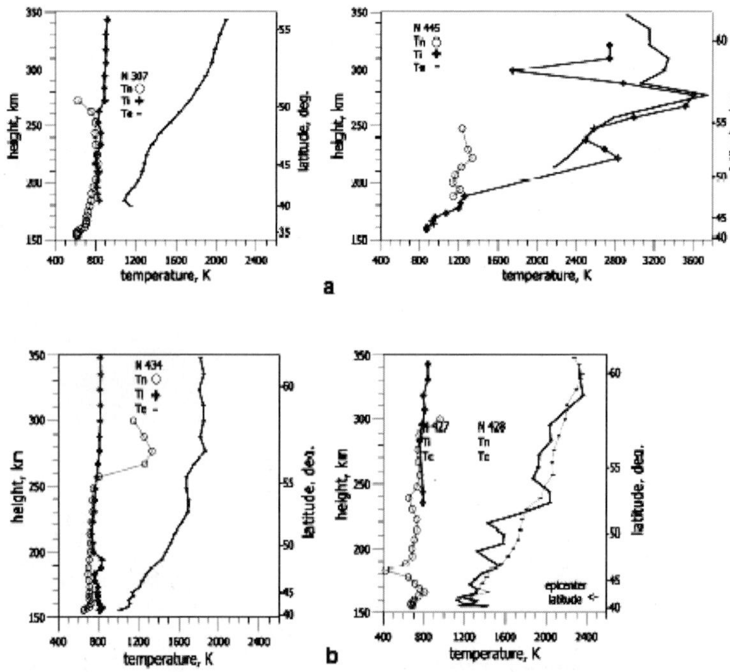

Fig. 5.11a,b a Examples of altitude temperature profiles measured on 24.01.74 before (*left panel*) and during the main phase of the ionospheric storm (*right panel*) on 25.01.74. *Tn*, *Ti* and *Te* – neutral, ion and electron temperature respectively. **b** Measurements during the quiet time (*left panel*) and several hours before the earthquake (*right panel*). Earthquake with *M*~6.4 happened on 24.01.74 at the point 42.1 N, 144.0 E (Hokkaido Isl., Japan)

In Fig. 5.11a one can see the effect of a geomagnetic storm on the local temperature parameters in comparison with undisturbed conditions, and in Fig. 5.11b the similar comparison, but there the undisturbed measurements are compared with the satellite pass over a seismically active area before an earthquake. Comparing the figures, we can note the well-known heating of the ionosphere during the main phase of the magnetic storm when augmentation of electron and ion temperatures at middle latitudes can reach 1,000 and 2,000 K, respectively. But no substantial heating was observed before the earthquake. The heating was not registered after the shock as well. Only small (up to 200 K) variations of the electron temperature T_e as well as variations of the temperature of the neutral component T_n are observed prior to earthquakes. These electron temperature variations before the earthquake exhibit a wave-like character with height – a fact that may imply an acoustic gravity wave propagation effect (Hegai et al. 1997).

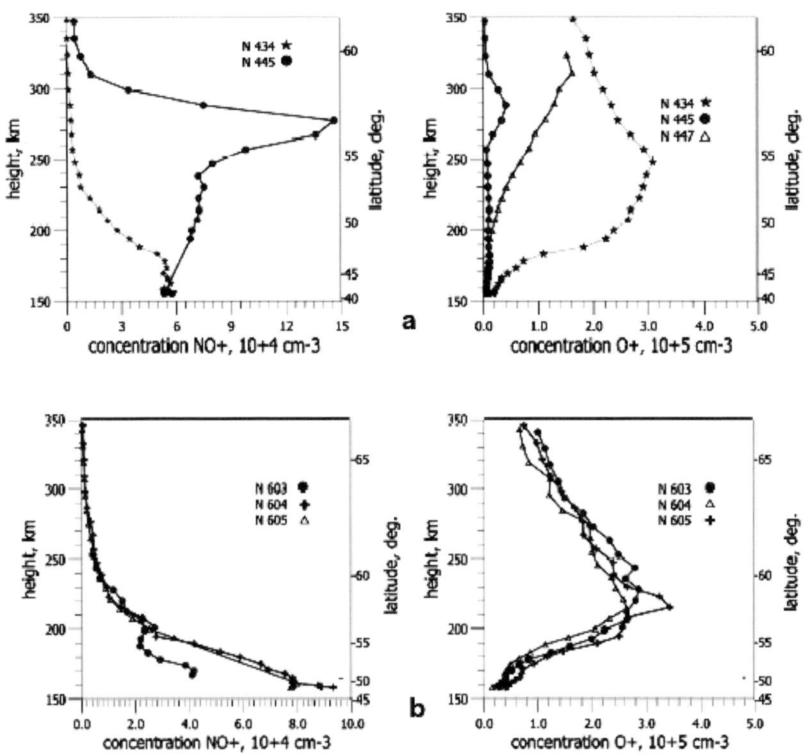

Fig. 5.12a,b **a** Variations of NO+ and O+ ions respectively during the main phase of the magnetic storm (*circles*) in comparison with quiet conditions (*stars*); **b** examples of the variations of NO+ and O+ ions respectively prior to an anticipated earthquake on the passes closest to the epicenter longitude (<10°, *points*), and far from it (>20° *crosses*)

Figures 5.12a and 5.12b represent a similar comparison of effects of the geomagnetic storm and the earthquake but for ion composition. Because even small changes of ionospheric electric field cause substantial ion drifts, the anomalous electric field appearing within the ionosphere above the region of the anticipated earthquake play an important role in observed variations. The modifications of concentration of atomic oxygen O^+ at heights of the F2-layer maximum and concentration of molecular ions NO^+ and O^+ in the bottom of the layer were marked before earthquakes. It should be noted that variations of concentration of positive ions O^+, NO^+ and O_2^+ before earthquakes have the same sign (Fig. 5.12b), contrary to variations of ion composition during a main phase of an ionospheric storm (Fig. 5.12a), when the development of negative perturbation of ions O^+ in a F2-layer is accompanied by substantial increase of concentration NO^+ and O_2^+. It is interesting to note that these “in-phase” changes of ion concentration before earthquakes are observed regardless of the sign of electron density variations (positive or negative). These peculiarities of ion composite variations may be used to discriminate the seismically induced variations from the geomagnetically disturbed conditions. During the geomagnetic storm we observe the effective increase of the mean ion mass due to the increase of the NO^+/O^+ relation as is seen from Fig. 5.12a and is known from the literature (Szuszczewicz et al. 1998). Before the earthquakes we have the opposite effect discussed in Chapter 4 – increase of the light ions concentration which is equivalent to a decrease of the mean ion mass. This effect is demonstrated in our analysis of the AE-C data and is shown in Fig. 5.13a where measurements of the mean ion mass are plotted, while the satellite passed over the anticipated earthquake epicenter one day before the earthquake (dots), and one day after the earthquake – crosses. Measuring simultaneously the electron and ion temperatures and using the formula for the ionosphere height scale (Smith and Kaiser 1967) the changes in the ionosphere height scale can be obtained:

$$H_s = \frac{T_e + T_i}{\frac{M_+}{0.85} + \left(\frac{h}{R_0} + 1\right)^2 \frac{dT_e}{dh}} \tag{5.1}$$

where M_+ – mean ion mass, and R_0 – is the Earth's radius.

Since there are no noticeable changes in neither electron, nor the ion temperature during the seismically induced perturbation of the ionosphere, we conclude that the main reason for the height scale changes is due to changes in the mean ion mass. Such changes in the height scale can alternatively be obtained by topside sounding, i.e. by deriving the height scale from the topside vertical electron profiles. Such measurements were conducted from onboard the Intercosmos-19 satellite and the examples of topside profiles for undisturbed conditions and prior to earthquakes are presented in Fig. 5.14.

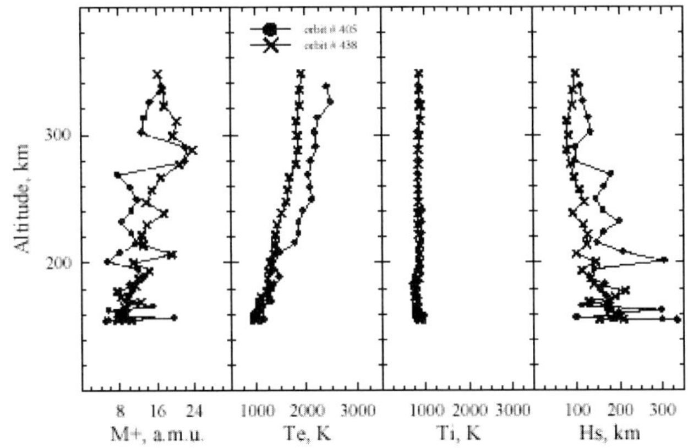

Fig. 5.13a–d From *left to right* **a** AE-C measurements of mean ion mass, **b** electron temperature, **c** ion temperature, **d** calculated height scale of the ionosphere. *Crosses* – reference level, *circles* – seismic precursory period

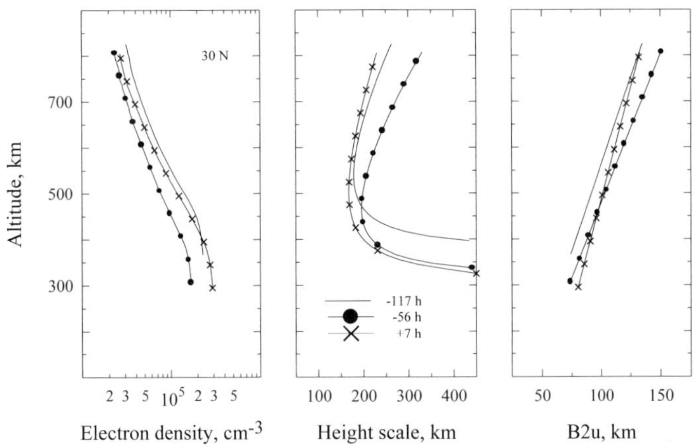

Fig. 5.14 Topside profiles (*left*), height scale (*middle*) and semithickness (*right*) for the period of the Irpinia earthquake in Central Italy, November, 1980

This figure shows in the left panel the topside profiles measured over the epicenter area 172 (clear line), 56 hours before the earthquake (dots) and 7 hours after it (crosses), in the middle panel the height scale curves calculated for the given profiles, and in the right panel the semithickness parameter B_{2u} calculated from the model (Depuev and Pulinets 2001; Nava et al. 2001). These results are consistent with estimates of mean ion mass derived from the vertical topside electron density profiles, showing effective decrease in the mean ion mass and presented in Fig. 5.15. This analysis was made for the Irpinia earthquake in Italy on the 23.11.80, 40°46'N, 15°18'E, M=6.9 and corresponds to the critical frequency distribution shown in Fig. 5.7.

5.2.5 Magnetic Disturbances and Earthquakes Precursors

All the previous discussion (except ion mass and temperature measurements) was connected with data collected during the quiet helio-geomagnetic conditions. The natural question is: how to distinguish ionospheric disturbances connected with earthquake preparing and other disturbances connected with geomagnetic storms etc. Of course, it is the favorite question of the ionospheric precursors critics.

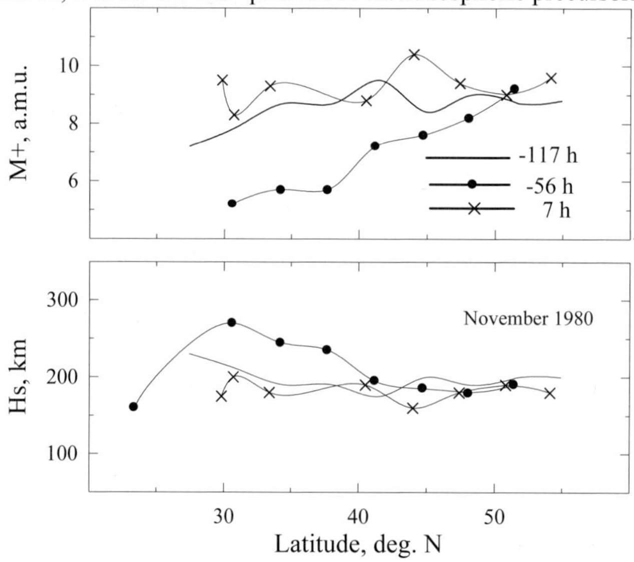

Fig. 5.15 Mean ion mass (*top panel*) and height scale (*bottom panel*) variations over the seismically active region 117 (*solid line*), 56 (*circles*) hours before the earthquake and 7 hours after it (*crosses*)

We will start by bringing together into Fig. 5.16 the Figures 2.12 and 1.20. We see that the time scale and relative amplitude of the critical frequency variations for the case of earthquakes (left panel) and for the geomagnetic storms is quite different. In Fig. 5.16a where the seismo-ionospheric variations are presented, the deviations of critical frequency are not so pronounced as during geomagnetic storms (Fig. 5.16b), and last not so long. Usually, duration of the deviation of one sign for seismic precursors does not exceed 4–6 hours. Only for very strong earthquakes (such as the Good Friday earthquake in Alaska, 1964) can the duration of anomalous deviation of the same sign can reach almost 12 hours (Pulinets and Legen'ka 2003). But for magnetic storms the long duration of the negative phase (more than 12 hours) is a normal feature. In Fig. 5.16b where data for two consecutive magnetic storms are presented, the data were collected at three midlatitude ionospheric stations. One can observe a very strong negative deviation lasting more than 24 hours, and never variations have an oscillating character (positive and negative deviations during a few hours) like one can see in Fig. 5.16a for the Milkovo station (1). Amplitude variations during a magnetic storm, as one can see from

Fig. 5.16b are also more pronounced, and could be 2 or 3 fold. At the same time estimations made in Pulinets et al. (1998b) on the basis of analysis of ionospheric effects of 54 worldwide earthquakes show ±30% maximum deviation for ionospheric precursors.

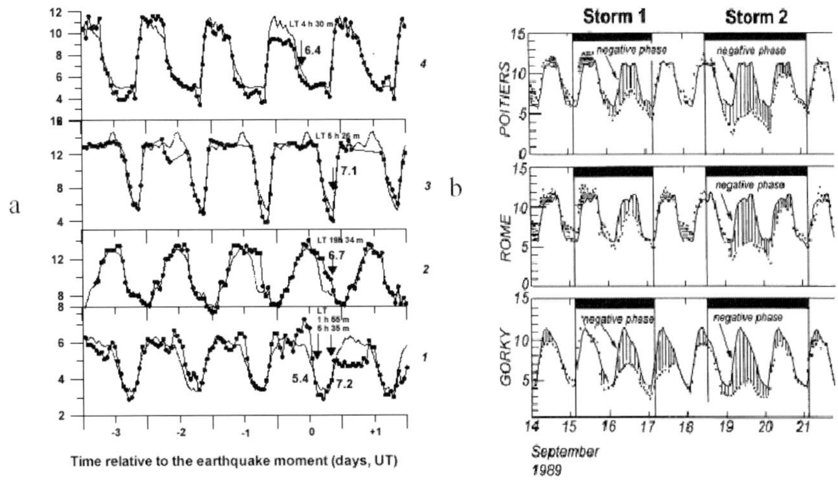

Fig. 5.16a,b a Examples of comparison of daily variations *foF2* (*points*) with their monthly median values (*stroke*), three day's before and one day after the earthquake from ground based ionosondes; **b** Comparison of daily variations *foF2* (*points*) with the diurnal quiet *foF2* (*thin solid curve*) formed by the average of quiet days in September at several European ionospheric stations during two consecutive magnetic storms: Storm 1 – Sept. 15th–16th; Storm 2 – Sept.18th–20th, 1989

The second very simple point – the storm-time variations have a global character. One can see this looking at the figures in the paper by Szuszczewicz et al. (1998). Here we bring the figure for several ionospheric stations in the Australian sector for the period of June 15th–20th 1980 (Fig. 5.17). In the upper part of the picture the geomagnetic indices D_{st}, AE, and A_p are depicted. One can see from the figure that a moderate magnetic storm took place on the 16th of June (K~4+). And the reaction of the ionosphere is global: all ionospheric stations show a growing of the critical frequency during day time, even the next day it is noticeable at the Christchurch and Campbell stations. We will note also that amplitude of the ionospheric disturbance is of the same order of magnitude at all stations. Contrary to this, only at the Norfolk station before the earthquake is the negative deviation observed due to the fact that this station is closest to the epicenter. At other stations the critical frequency is close to the median values. So, in contrast to the geomagnetic effect, which has a global scale, the seismo-ionospheric effect has a local character.

Now we can go back to the local plasma characteristics. They have an even more fundamental meaning because they are directly connected with the different physical nature of geomagnetic storm effects and the effects from preparing earthquakes. They are: the strong electron and ion heating during the geomagnetic storm and the very weak temperature changes observed before the earthquake, the increase of the mean ion mass during the main phase of the geomagnetic storm and the decrease of mean ion mass before the earthquakes. Just these features permit unambiguously to distinguish the magnetic storm effects and earthquake precursors.

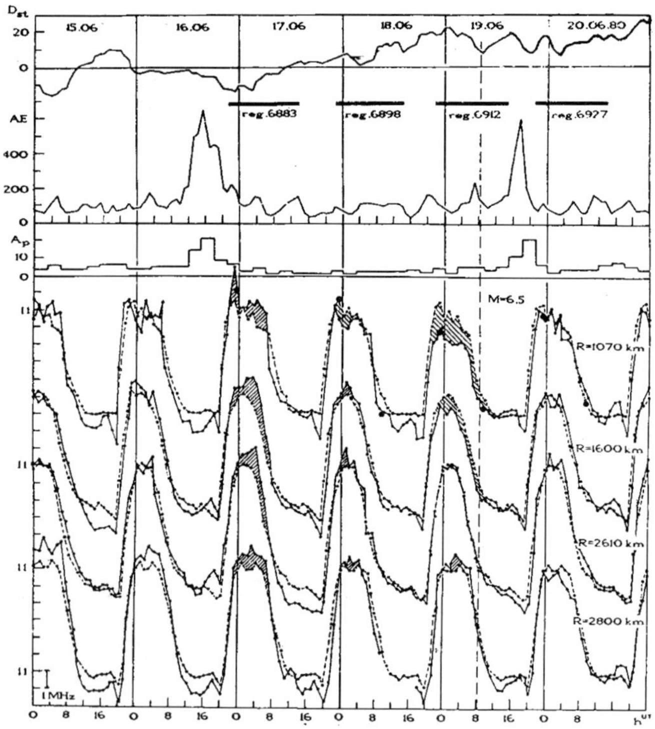

Fig. 5.17 Daily variations of *foF2* (*solid lines*) in comparison with monthly median (*dashed lines*) for the period 15th–20th of June, 1980, measured at ionospheric stations 1 – Norfolk, 2 – Christchurch, 3 – Campbell, 4 – Canberra. The earthquake moment on 09.06.80 at 08:30 UT is shown by the *vertical dashed line*. D_{st}, *AE* and A_p geomagnetic indices are shown in the *top panel* of the figure. Periods when the satellite topside sounder was conducting measurements are shown by *bold bars*

As one can see from the formula 5.1, the ionospheric scale height depends on the temperature, and on the ion mass. This means that one should observe the scale height variations during the geomagnetic storm and before the earthquakes. But the experimental data from the Intercosmos-19 satellite show that scale height changes during a magnetic storm are not so pronounced as before the earthquakes

(Depuev et al. 2001). If for the magnetic storm the average value of the scale height increase is of order of 30%, for the case of the earthquakes the scale height increase is usually higher than 50%, and may reach 100%. So, we have one more parameter to distinguish between the effects of the magnetic storm and the seismogenic electric field.

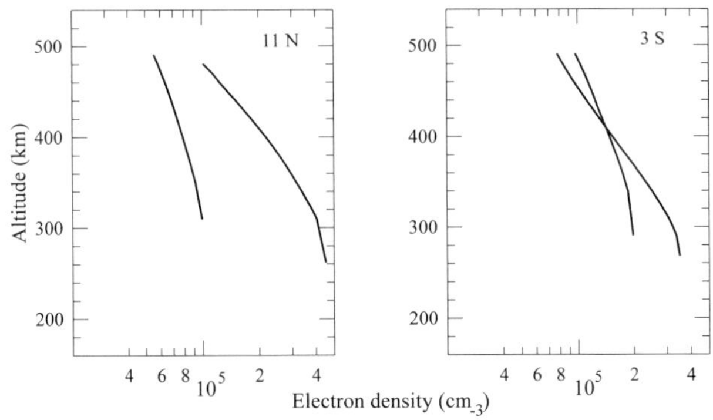

hmF2	foF2	B2u	k2u		hmF2	foF2	B2u	k2u
260	6.0	54.0	0.16	7286	270	5.3	44.7	0.17
320	2.8	(87.0)	0.29	7301	290	4.0	(69.0)	0.26

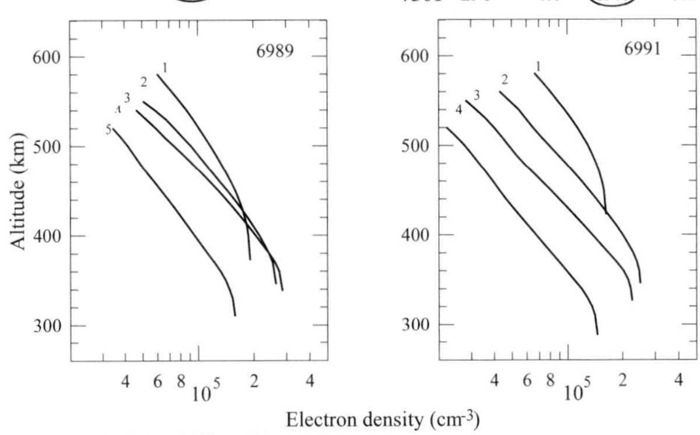

	lat	hmF2	foF2	B2u	k2u		lat	hmF2	foF2	B2u	k2u
1.	8.4	375	12.4	(75.8)	0.02	1.	9.6	425	11.5	(61.7)	0.17
2.	0.6	345	14.6	47.1	0.14	2.	1.8	345	14.2	47.5	0.12
3.	-3.3	340	15.2	43.3	0.13	3.	-2.2	325	13.5	40.6	0.12
4.	-7.2	310	14.6	44.4	0.12	4.	-10.0	295	11.4	44.6	0.13
5.	-15.1	310	11.3	46.5	0.15	5.	-14.0	290	10.8	48.4	0.11

Fig. 5.18 *Upper panel* – vertical profiles of the electron concentration reconstructed from Intercosmos-19 satellite ionograms for the case of earthquake M=7.3 at New Guinea Islands (3.2°S, 143.3°E) on 16.07.80. *Left panel* – in the point of the maximum deviation of the critical frequency, and *right panel* – in the point of the epicenter vertical projection on the ionosphere; 1 – disturbed profile, 2 – undisturbed profile. *Bottom panel* – examples of the same but for the earthquakes in New Guinea Islands (*left panel*) and Indonesia – *right panel*; 1 – disturbed profile

To have clearer idea how the scale height variations look before the earthquake we bring two more examples of the scale height changes before the earthquakes with the vertical profile model parameters (Fig. 5.18). The precursory profile scale height parameters are circled in the tables. The profiles shown in the upper panel of Fig. 5.18 correspond the critical frequency distribution shown in the middle part of Fig. 2.8. The left profile corresponds to the point of the maximum deviation of the critical frequency, and the right one – to the point of the epicenter vertical projection on the ionosphere. The correspondent increases of the scale height are 61% and 54%. The lower panel shows other examples of the vertical profiles scaled from the topside ionograms around the period of the earthquake preparation in June 1980 in New Zealand region (left panel), and in Indonesia (right panel).

As one can see from the figures, in all cases not only the scale height changes before the earthquakes but the peak height of the vertical electron distribution as well. So one more parameter to be regarded is the peak height.

5.2.6 Altitude Changes in the Ionosphere Before Earthquakes

Actually this question was touched upon in previous paragraphs during the discussion of scale height issues but we'd like to discuss more carefully the peak height variations before earthquakes. From Fig. 5.18 one can conclude that the peak height h_mF2 of the $F2$-layer can increase before the earthquakes from 50 up to 135 km. The modification of the peak height distribution before the strong earthquakes in the vicinity of the equatorial anomaly was discussed by Pulinets (1998a). This case is presented in Fig. 5.19.

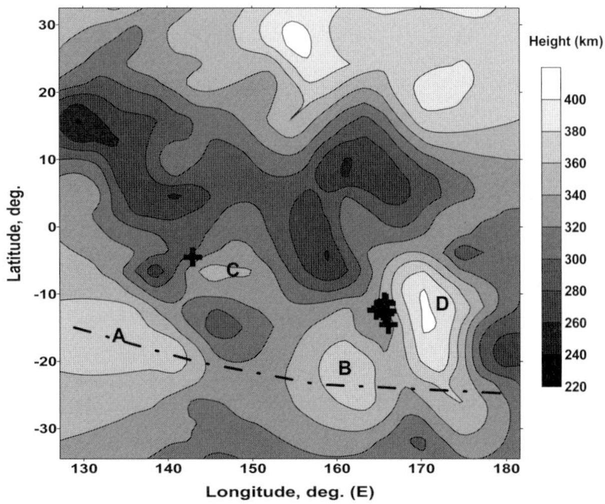

Fig. 5.19 Modification of the $F2$-layer peak height h_mF2 distribution during period of high seismic activity in the vicinity of the equatorial anomaly in the New Guinea Islands region (averaged distribution of $F2$-layer peak height h_mF2 for 13.07.80–17.07.80). The epicenter positions are shown in the figure by +. Undisturbed position of the equatorial anomaly crest is marked by *dot-stroked line*

Figure 5.19 demonstrates the map of the peak height distribution built as an integral picture for the high seismic activity period from 13 to 18 of July 1980 (the parameters of earthquakes are presented in Table 5.1), when 19 shocks of magnitude 5 or higher were registered in the New Guinea Islands region. The epicenter positions are marked by crosses. The seismic activity has an effect on the equatorial anomaly position. Its undisturbed position is shown in the figure by the dashed line. **A** and **B** regions belong to undisturbed equatorial anomaly, while **C** and **D** regions reflect modification of ionosphere peak height by seismic activity. Except the peak height modification we see again the shift of the crest of the equatorial anomaly towards the geomagnetic equator. In the case of region **D** the crest displacement is more than 10 degrees in latitude. The size of the modified area is not so large as in the case of the Alaskan or Irpinia earthquake's, which confirms the connection of the modified area size and magnitude and opens perspectives to estimate the magnitude of impending earthquakes from the ionosphere modified region parameters.

Table 5.1 Parameters of the earthquakes presented in Fig. 5.19

Date	UT	M	H	Lat	Long
10.07.1980	2.43	5.3	33	-12.758	165.905
11.07.1980	3.02	5.2	33	-13.154	166.21
11.07.1980	3.14	5.2	33	-13.363	166.381
12.07.1980	0.29	5.4	191	-5.561	149.024
12.07.1980	6.55	5.1	90	-12.884	166.443
13.07.1980	8.41	5.1	53	-12.799	166.129
13.07.1980	10.48	5	53	-13.013	165.651
16.07.1980	19.56	6.5	84	-4.456	143.521
16.07.1980	20.33	5.1	33	-14.821	166.733
17.07.1980	10.36	5.3	46	-12.512	166.332
17.07.1980	19.42	5.8	33	-12.525	165.916
17.07.1980	20.49	5.7	33	-11.828	166.44
17.07.1980	20.53	5.3	33	-12.631	165.795
17.07.1980	20.56	5.6	33	-12.887	165.222
17.07.1980	23.19	5.4	33	-11.637	166.306
17.07.1980	23.24	5.1	33	-11.826	166.24
18.07.1980	3.57	5	33	-11.724	166.115
18.07.1980	13.4	5.3	48	-12.597	166.264
18.07.1980	16	5.2	33	-12.51	166.42

A similar rise of the F-layer peak is reported based on the ground based measurements (Gokhberg et al. 1989a). But unfortunately, the authors give the apparent height which is not equivalent to the real peak height determined from the vertical profile calculations, so the real value of the peak height from the presented results is difficult to determine.

5.2.7 Periodic Variations Within the Ionosphere

Zelenova and Legen'ka (1989) and Depuev and Zelenova (1996) analyzed the ionospheric variations before the Moneron earthquake of 5 September 1971 M=7.2 at Moneron Island (46.5°N, 141.1°E). Their results are shown in Figures 2.24 and 2.22 respectively. In both figures one can clearly observe the quasi-periodic oscillation of the ionosphere parameters at all altitudes (Fig. 2.22) and for the critical frequency in Fig. 2.24. According to Zelenova and Legen'ka (1989) there were observed intensification's of the oscillations with periods 3.8, 1.25 and 1.1 hours in comparison with undisturbed days. A similar period of 1.3 hours oscillation of the critical frequency is reported by Sharadze et al. (1991) before the Spitak earthquake (7th December 1988, M=7.0, 40.9° N, 44.2° E). The authors claim the overall increase of perturbations with periods 1–4 hours one day before the earthquake. All these data fit well into the theoretical calculations of the AGW generation in the F-region of the ionosphere (Hegai et al. 1997).

The data on the periodic variations within the ionosphere are available only for several case studies and should be confirmed statistically.

5.3 Statistical Parameters of the Seismo-Ionospheric Variations in the F-Layer

The statistical confirmation was one of the weak points of the ionospheric precursors of earthquakes. Most papers published in 70–80th were case study papers and were criticized from this point of view. Chen et al. (1999) for the first time presented at the URSI General Assembly in Toronto the results of statistical studies of the ionospheric precursors registered by the ground based ionosonde in Taiwan.

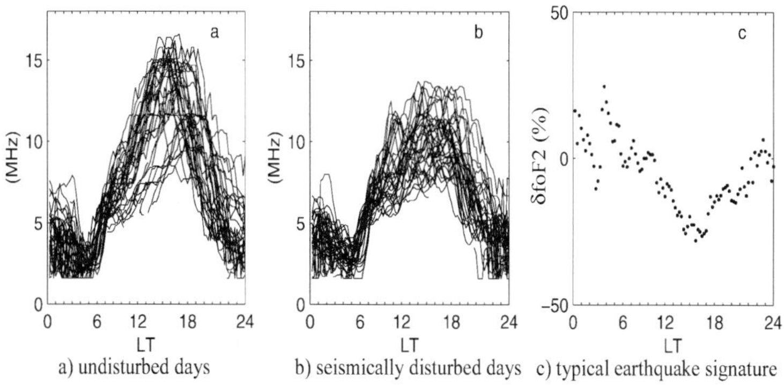

Fig. 5.20a–c a Daily variations of the critical frequency for the period of 1994–1997 at Chung-Li ionosonde put together except periods 5 days before the earthquakes $M \geq 5$. **b** The same but for the periods within 5 days before the earthquakes $M \geq 5$. **c** The difference between averaged **b** and **a**

The idea was very simple. Regardless of the fact that in the other publications the ionospheric variations were treated for a few days around the main shock time, it was decided to not select any data but process them all without any exception. In the first approach the data for 1994–1999 were used for analysis. To see if anything unusual exists in the ionospheric data at all, it was decided to stack all the daily variations outside the earthquake preparation period (i.e. 5 days before the earthquake), and contrary, put together all daily variations of the critical frequency for just this period of 5 days before the earthquake. We selected earthquakes with magnitude ≥5 within the circle 500 km around the ionosonde. The results of this processing are shown in the Fig. 5.20.

The difference between Figs. 5.20a and b is so evident that they do not need to be explained. We will explain for Fig. 5.20c. It is simply the difference between the averaged pictures b and a, and in some sense is the averaged portrait of the ionospheric precursor for the Taiwan area. One can see in the figure the oscillating variations during nighttime, and strong negative deviation for afternoon hours. But it is still a very crude guess because of averaging all the times together. The ionosphere behavior depends on a lot of parameters (season, solar cycle phase, magnetic activity, latitude, longitude, etc.). The data should be subdivided into different groups where all these factors are taken into account. The other very important parameters are the position of the future epicenter in relation to the observation point (ionosonde position), which was shown earlier), and precursor dependence on the earthquake parameters (magnitude, depth, focal mechanism). But to start with statistical estimations we will use the most pronounced negative deviation during the afternoon hours of the local time. It was decided to track this precursor statistically using the following procedure: the running 15-day median was calculated, and $\pm 2\sigma$ upper and lower bounds (σ – standard deviation). The 15-days interval was selected because of the statistically shown (for the period 1991–1993) reoccurrence period of the earthquakes with magnitude ≥5 at Taiwan of 14.2 days, to have statistically only one earthquake falling into the period of 15 days. The processing procedure was modified later by Liu et al. (2003) by introducing the interquartile range instead of 2σ for upper and lower bounds. Schematically this processing procedure is shown in Fig. 5.21.

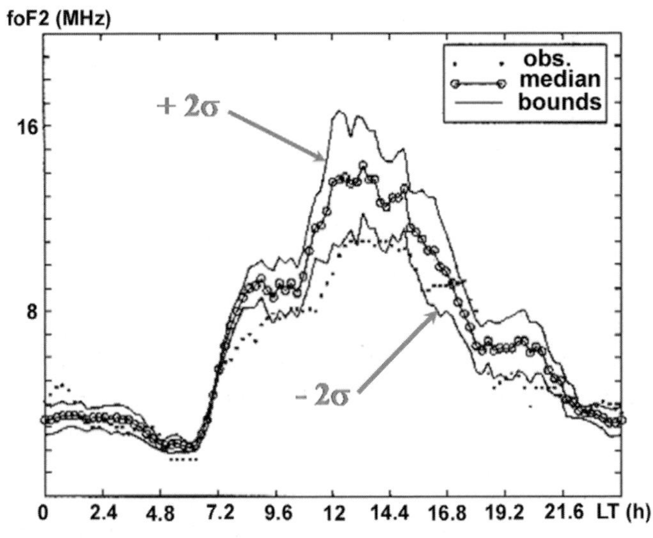

Fig. 5.21 Schematic presentation of the statistical processing procedure for the ionospheric data of the Chung-Li ionosonde to reveal the ionospheric precursors of earthquakes

The earthquake statistics for the period of 1994–1997 for the given area are presented in the Table 5.2. We should underlie that the number of days with earthquakes is smaller than the number of earthquakes because of the possibility of having several seismic shocks during one day. The authors calculated the percentage relation of cases when the deviation of the critical frequency was out from the error interval to all number of days within the intervals 1 day, 2 days, 3 days, 4 days and 5 days before the seismic shock. The authors also calculated the cumulative number of cases within the interval of 5 days before the seismic shock for the earthquakes with different intensity, namely, for $M \geq 5$, $M \geq 5.5$, and $M \geq 6$. The results are demonstrated in Table 5.3.

Table 5.2 The earthquakes statistics in the Taiwan area for the period 1994–1997

	M≥5	M≥5.5	M≥6
Number of EQ	118	46	14
Number of days with EQ	94	34	9

Table 5.3 The ionospheric precursor statistics for the earthquakes indicated in Table 5.2. P(i) probability of precursor registration for the given day *i*. $\sum^j$P(i) – cumulative probability of precursors for the *j* days before the earthquake

	M≥5		M≥5.5		M≥6	
	P(i)	$\sum^j$P(i)	P(i)	$\sum^j$P(i)	P(i)	$\sum^j$P(i)
1 day before	28%	28%	44%	44%	56%	56%
2 days before	14%	42%	21%	65%	11%	67%
3 days before	22%	64%	15%	80%	11%	78%
4 days before	9%	73%	9%	89%	11%	89%
5 days before	0.8%	73.8%	6%	95%	11%	100%

What these results mean? Let us look at the first column for the earthquakes with $M \geq 5$. The first line means that in 28% of cases for all 94 days with earthquakes with $M \geq 5$ the deviation exceeded the boundary of 2σ in 28% of cases. If we look at the second line – it corresponds to the second day before the seismic shock. The left part of the line (P(i)) means that for the all 94 days with earthquakes with $M \geq 5$ the anomalous deviation was registered in 14% of cases, but if we take the sum of 1 day before and two days before $\sum^j$P(i)=28% + 14%=42%. Moving lower up to 5th day before the earthquake we receive the cumulative probability to register the anomalous variation $\sum^j$P(i) =73%. Looking at the other columns of Table 5.3 we realize: the larger the earthquake intensity, the higher the probability to detect the anomalous variations within the ionosphere. This result is one more solid confirmation of the real existence of the ionospheric precursors of earthquakes.

Does it work in reverse? I.e., if we have the large deviation, may we say that we expect the earthquake? Unfortunately, not. We should identify the precursors taking into account all the factors we discussed in this chapter. For example, during the same period under analysis 1994–1997 there were 62 magnetic storms when the strong critical frequency deviation was observed too. So, we should apply all our knowledge to identify the precursor. The procedure of precursors identification will be discussed in Chapter 6.

If we return to the previous discussion of the period in advance of the precursors appearance, it was detected statistically within the frame of later processing (Liu et al. 2003) that ionospheric precursors appear 5 days before the seismic shock. That's why for the present processing (Table 5.3) the same period was selected too.

It should be mentioned also that the statistical processing demonstrated above we can regard only as a zero order approach. It fulfilled its main purpose: to demonstrate that the existence of the ionospheric precursors has a solid statistical background. Next steps should be in the direction of the more sophisticated processing taking into account the relative position of the ionosonde and epicenter, the possible difference in the behavior of the ionospheric variations within the interval of 5 days, etc. Some new approaches will be demonstrated in Chapter 6. As a small introduction to it will be the demonstration of the different shape of the precursors within a 5 day interval for the same dataset of 1994–1997 for Taiwan (Fig. 5.22). For example, we can note that for the 4th and 5th days before the earthquake prevail the positive, not negative deviations of the critical frequency which should be taken into account during the process of the precursors identification.

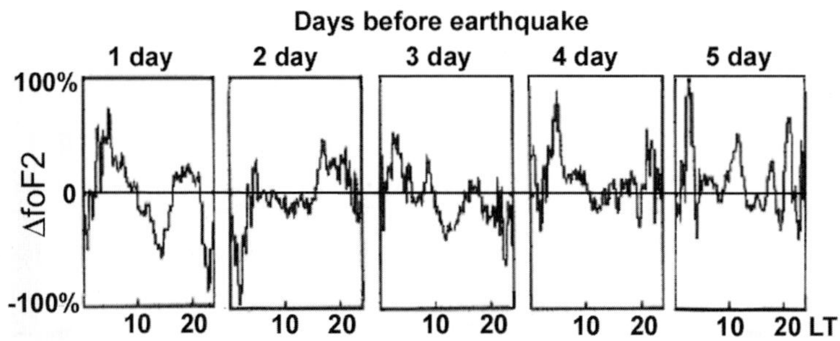

Fig. 5.22 Daily variations of the ionospheric precursor for Taiwan within the interval 5 days before the earthquake

5.4 *E*-Layer Precursors

There were a lot of publications on the *E*-layer precursory phenomena cited in Chapter 2. So as not to repeat the above-mentioned discussion we will cite only the most recent review on the sporadic *E*-layer effects associated with the earthquakes by Liperovsky et al. (2000). Many of the precursors effects described in the paper are based on the effects in the blanketing frequency f_bE_s. We regard this as not a very good characteristic to be used because it involves not only the layer characteristic itself but also the ionosonde quality (it will depend on the ionosonde emitted power and the receiver sensitivity), so will change from ionosonde to ionosonde.

The authors are also not very sure of the validity of the precursory parameters derived from the *E*-layer variability. It is enough to cite from their paper published in the same year (Liperovskaya et al. 2000): *On the basis of the six-year data sample from the station of vertical sounding, a statistical analysis has demonstrated that the probability of* E_S*-spread observation is independent of earthquake preparation processes preceding one-three days before a shock. However, the number of* E_S*-spread observations happened to increase for many earthquakes.*

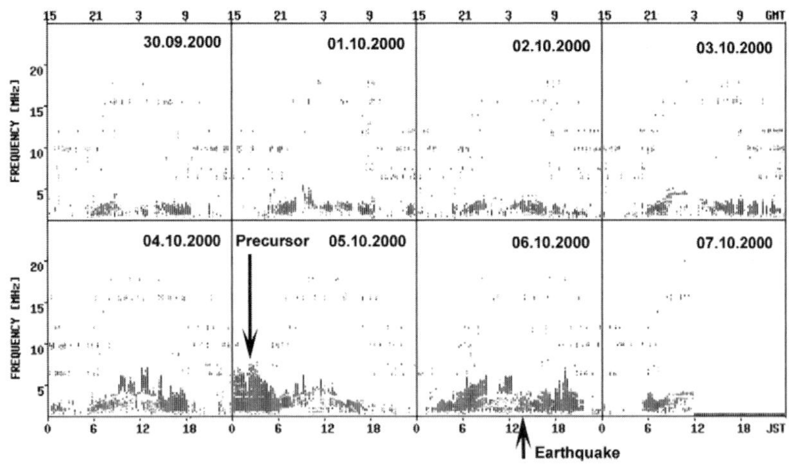

Fig. 5.23 foE_s registered by Kokubunji ionosonde (Japan) for the period 30th September – 7th October 2000 before the strong earthquake 6th October 2000, *M*=7.4, 35.46 °N, 133.13 °E. Vertical axis indicates the foE_S in MHz measured every 15 minutes and shown in the figure by *vertical bars*. *Upper horizontal axis* shows the Universal time (UT), and the *lower one* – the Japan local standard time (JST)

And only this conclusion: *increase of the sporadic E-layer activity before earthquakes* may be certain, practically without any quantitative characteristics. Receiving more and more confirmation from the experimental research and the theory the high E_s activity associated with atmospheric processes (Manson et al. 2002) makes doubtful the clear identification of the earthquake related effects due to the camouflage of these effects by atmospheric activity. Taking into account that the climatic conditions are very different for different places, the morphology of the *E*-layer behavior will be very different too. That's why, it is impossible to have universal morphology parameters for the *E*-layer globally.

Is there however any hope of using the data on the *E*-layer for short-term prediction? It seems that we can indicate at least two directions. First, it is to use the data for the *E*-layer as an additional factor for other registered precursors. We already noted that the process of the short-term prediction is based on multifactor analysis, and the increase of sporadic *E*-layer activity in addition to other precursory phenomena will be very important supporting information.

The second way – is to go from the intrinsic physical nature of the sporadic *E*-layer. There exists a well-determined sporadic *E*-layer phenomenology in the form of layer activity dependence on the local time, season, solar cycle and geomagnetic activity (together with novel information on the atmosphere-thermosphere connections). At the same time, the seismogenic anomalous electric field behavior is connected more with seismic activity, and is not so regular as E_s dependence on other geophysical parameters. So one may look for the E_s appearance, for example, in non-appropriate local time. In Fig. 5.23 one such example is depicted showing the strong sporadic E-layer development in the early morning (before sunrise) one day before the strong seismic shock in Japan (6th October 2000, M=7.4, 35.46° N, 133.13° E). As one can see from the figure, the sporadic *E*-layer never develops in the early morning hours of local time indicated in the bottom axis of the figure. And only once, on the 5th of October we can note the strong sporadic *E*-layer development with critical frequencies much higher than for the daytime conditions. This is a very strong indication for the association of the observed E_s intensification with the earthquakes following the next day.

It seems that studying the E_s morphology for the Kokubunji station it will be very easy to determine such cases of abnormal behavior of E_s and to associated them with periods of strong earthquake preparation in Japan.

It should also be noted that seismo-associated phenomena in the *E*-layer are much more local than the *F*-layer effects that's why they will be observed only by the ionospheric station close to the future epicenter. In the presented case the effects was observed only by the Kokubunji station, closest to the epicenter (the Yamagawa station closer to the epicenter did not work properly), among four ionospheric stations monitoring the ionosphere in Japan

Concluding the discussion on the precursory abilities of the E-layer we should state that the approach to the search of the ionospheric precursors in the *E*-layer should be reconsidered. Instead of looking at the effects only within a few days around the earthquake moment, the statistical processing of long data samples (several years) without any exclusion (as it was done for the Taiwan data on the critical frequency). Main attention should be directed to the search of the abnormal E-layer behavior dropping out of usual E-layer morphology.

5.5 Conclusions

Taking into account that morphological features of the ionospheric precursors of earthquakes make a foundation for the development of the techniques of short-term earthquake prediction we would once more like to bring together the main ones for classification and further use in practical application. As was mentioned earlier, some of them have a character independent of the type of measurements (like the size of the modified area in the ionosphere or precursor shift equatorward), and the other part of the precursors characteristics (like the sign of the critical frequency deviation registered by the ionosonde) depends on the relative position of the measurement point and impending earthquake epicenter. Because of the wide use of ionosondes, we will leave the features of the precursors registered by the ionospheric stations, in the list. At the same time we will not mention those features that were mentioned in the literature as a case study or were not proved statistically. If somebody is interested in more information, it is possible to find it in the cited references. We tried to collect only such characteristics, which can be checked at any time at any place by anybody. In the list we will not try to divide the precursor's characteristics into spatial, temporal, etc. because all of them are interrelated and reflect the different sides of the general physical process – seismo-ionospheric coupling.

The main characteristics of ionospheric precursors could be shortly enumerated:

- Seismically induced precursors affect the plasma density (relative to a normal non-perturbed state), and are observed between 5 days to a few hours prior to the earthquake
- The variations (relative to a normal non-perturbed state) can have a positive or negative sign
- The duration of a seismically induced deviation of a given sign is comparatively short about 4–6 hours (contrary to magnetic storm effects). Only in cases of very strong earthquakes (such as before the large 1964 Good Friday earthquake, Alaska) the duration of a seismically induced deviation can reach about 12 h
- The threshold at which seismogenic effects on the ionosphere become observable is determined by the size of the earthquake preparation zone and corresponds to a magnitude of 5
- On the average seismoionospheric variations have the same amplitude, as the day-to-day variability of an ionosphere (±30%), but can be much more pronounced, up to 100%, at particular moments of local time

- Sign and shape of seismo-ionospheric variations depend on local time. This dependence may be different for different latitudes and longitudes and will require additional research for every geophysical location
- For the case of one point measurements the sign and shape of seismo-ionospheric variations depend on the relative position of the observation point and impending earthquake epicenter
- The extent of the affected area of the ionosphere at the height of the *F*-layer maximum depends on the earthquake magnitude and can reach 40° in latitude and in longitude for very strong earthquakes with $M>7$
- The maximum of the affected area in the ionosphere does not coincide with the vertical projection of the epicenter of the future earthquake and shifted equatorward in high and middle latitudes
- The seismo-ionospheric variations have an effect on vertical distribution of the plasma leading to an increase of the height scale of the ionosphere
- The height scale changes are mainly due to changes in the ion mass (increasing concentration of light ions) or due to parallel particle fluxes. They are not due to plasma temperature changes
- The TEC seismo-ionospheric variations agree well with the measured deviations in the critical frequency
- In specific circumstances the corresponding ionospheric effects can be observed in the magnetically conjugated region
- In low and equatorial latitudes the seismo-ionospheric variations lead to equatorial anomaly modification as a whole structure. The most often observed effect is the equatorward shift of both crests of the equatorial anomaly up to their complete disappearance
- The increased sporadic *E*-layer activity is observed several days before the earthquake over the area of earthquake preparation
- Increased ionization of the *D*-region is observed several days before the earthquake over the earthquake preparation area
- Energetic particle precipitation is observed within the magnetic tube loaned onto the earthquake preparation area
- Increased concentration of the light ions is observed within the magnetic tube loaned onto the earthquake preparation area

The differences between the ionospheric variations during magnetic storm and seismo-ionospheric variations should be mentioned separately which is done in Table 5.4:

Table 5.4 Ionospheric variations during the magnetic storm (*left column*) and seismo-ionospheric variations (*right column*)

Ionospheric variations during magnetic storm	Seismo-ionospheric variations
Variations have the global character and observed worldwide	Variations are observed only over the area of earthquake preparation (with some shift equatorward)
Variations are significant in magnitude and duration and represent the continuous process lasting 40-48 hours	Variations have sporadic character and appear at the same local time every day but do not have continuous character
Disturbances in low latitudes are observed, basically, during night-time with positive and negative phases noticed, first of all in a period between sunset and sunrise	Variations are observed during daytime too, and main statistically confirmed precursor in low latitudes is negative deviation during afternoon hours
The Δh_mF2 value during a storm is 60 – 80 km with h_mF2 in quiet conditions not less than 250 km	The Δh_mF2 before earthquakes may reach 150 km
As a result of development of a storm average molecular mass at heights of *F*-layer was **increased** owing to increase of the mass relation ($[N_2] + [O_2]) / [O]$. The mean ion mass during the storm is also increased due to increase of relation $NO^+/O+$	The seismo-ionospheric variations result in **reduction** of mean ion mass at heights of *F*-layer owing to variations of magnetospheric flows of ions O^+ and increase of concentration of light ions (H^+ and He^+)
The positive and negative deviations during magnetic storm depend on the storm development phase	The positive and negative deviations of seismo-ionospheric variations depend on the local time, relative position of the observation point and epicenter, and on the sign of anomalous electric field
Large heating of electrons and ions (2000 – 3000 K) during the storm	Practically no heating before earthquakes (~ 300 K)

We would like to finish this chapter with the statement that every physical process has its own intrinsic characteristics. If you know the physical nature of the event you are able to distinguish it from other events even if they happen in the same media and at the same time as happens in the ionosphere. The multiparameter analysis of all available measurements permits you to identify the process unambiguously. With such an approach we hope that as a final stage of our research we would be able to apply our knowledge in practical applications, namely, in short-term earthquake prediction. Some ideas and first steps will be described in the following chapter.

Chapter 6 Are we Ready for Prediction? The Practical Applications

6.1 Introduction

In the course of our scientific investigation of the physical mechanism of seismo-ionospheric coupling we were trying to elaborate some practical recommendations for ionospheric precursor identification. The techniques for automatic precursors recognition were elaborated as well. It is simplest of course, if you have the means to conduct the multiparameter analysis, i.e., if you have the satellite system able to provide the ionosphere mapping, the measurements of local plasma parameters such as ion composition and mass, electron and ion temperature, small scale irregularities, particle precipitation, etc. In addition you may have the developed network of the GPS receivers, ionosondes, ground-based measurements of the geochemical precursors and electric field. We will describe in the following paragraphs the existing systems of such kinds and the systems in various stages of preparation, including the satellite constellation.

But not all nations suffering from severe earthquakes have the means to create such comprehensive systems of precursor monitoring. They may have only one ionosonde in the seismically active region. We will describe how to work only with one device.

In this regard, the chapter will be constructed in the following way: we will start with the simplest case – one observation point, and then will move towards increasing system complexity. We will demonstrate the case of comprehensive analysis of all existing information using as an example the Irpinia earthquake in Italy on the 23rd November 1980. We will describe the existing systems of ground based monitoring and the satellite systems in the implementation stage and already launched. We will try to describe the ideal (in our view) system of ionospheric precursor monitoring and will demonstrate that the ionospheric precursors satisfy all the requirements of short term prediction, i.e. are able to determine the place, the time, and intensity of the approaching earthquake, and in this regard, have a great advantage over many other types of earthquake precursors.

6.2 One Point Measurements – Precursor Mask

In chapters 3 and 4 we described the physical mechanism and main phenomenological features of the ionospheric precursors of earthquakes. The existing similarity of precursors for a given place was mentioned, especially, if the position of the seismic source is fixed in relation to the ionosondes position. The concept of the precursors mask was developed for just such a case – actually a three-dimensional wavelet permitting us to automatically identify the precursor (Pulinets et al. 2002b).

The earthquakes for the period from 1978 to 1986 with the magnitudes M>4 and epicenters located within the circle (see Fig. 6.1) with a radius of 200 km and geographic coordinates of its center φ=24.5° N and λ=122.5° E were used as a reference set of seismic events (earthquakes). Within this circle, the earthquake epicenters are located at Taiwan Island and in the surrounding sea and the distance from the ionosonde to the epicenters does not exceed 300 km.

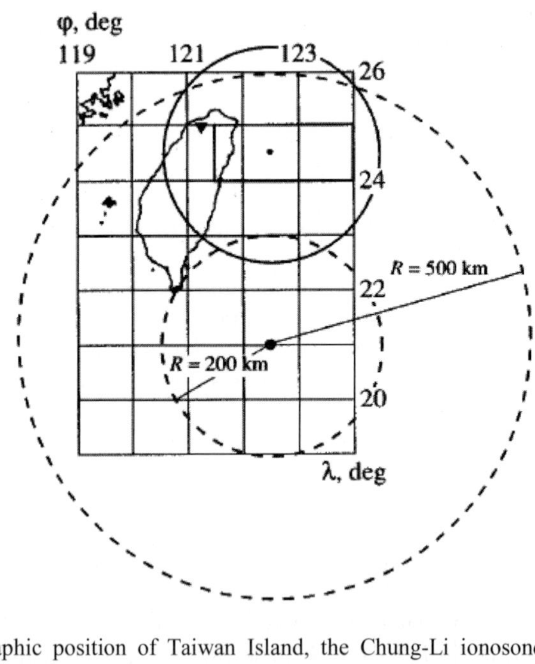

Fig. 6.1 Geographic position of Taiwan Island, the Chung-Li ionosonde (*triangle*), the earthquake epicenters considered (*within the circle drawn with a solid line*), zones of possible anomalies in the ionosphere (*within the circles drawn with a dashed line*), the zone of the earthquake epicenters to which the station is most "sensitive" (*rectangle*)

To analyze ionospheric variations related to the earthquakes considered, we used the hourly data on *foF2* at Chung-Li station (Taiwan Island, 25° N, 121.3° E) for the period from 1978 to 1986.

Since (see, for example, Davies, 1990) $N_mF2 = 1.24 \cdot 10^4 \cdot (foF2)^2$, where N_mF2 is the electron concentration in the maximum of the ionospheric F2 layer (in cm^{-3}) and *foF2* is the critical frequency of the ordinary wave (in MHz), the electron concentration behavior at the h_mF2 height may be interpreted in terms of the critical frequency measured at the ionospheric station. Therefore, we can use in the analysis only the values of the critical frequency since it is unambiguously related to N_mF2.

We analyzed ionospheric data for the selected seismoactive region obtained six days prior to the event. To get rid of noise induced by foreshocks and aftershocks, we analyzed only the earthquakes that were preceded by a period of at least 12 "quiet" days.

From existing empirical material we know that the individual precursor lasts considerably less than one day (this was mentioned in Chapter 5) and the local time of its appearance is related to the onset of an event (earthquake). In this case, it is convenient to represent the state of the ionosphere prior to the event in the form of an A_{ij} matrix whose columns contain the hourly deviations of *foF2* from its median value. The number of columns in the matrix is determined by an expected time interval between a precursor and an event. Here we assume that this interval does not exceed six days (it was statistically shown that precursor events appear 5 days before the earthquake, see Liu et al. 2004 and the discussion in Chapter 5) and, respectively, the dimension of the A_{ij} matrix is 24×6, i.e., i=1...24, j=1...6. For all events we form matrices in the same manner and obtain an $A_{ij}^{(n)}$ series, where n is the current number of an event. We now introduce the value:

$$S_n = n \frac{\sum_{i,j} \left\langle A_{ij}^{(n)} \right\rangle_n^2 n}{\left\langle \sum_{i,j} \left(A_{ij}^{(n)} \right)^2 \right\rangle_n n} \tag{6.1}$$

where $\langle \cdots \rangle_n$ means averaging over an ensemble of n events. S_n is the dispersion normalized so that S_1=1.

With the help of S_n it is convenient to characterize the degree of $A_{ij}^{(n)}$ similarity at various n. For example, if $A_{ij}^{(n)}$ values for various events do not correlate with one another, the S_n series tends to unity at $n \rightarrow \infty$. In the other extreme case, when the states of the ionosphere prior to all events are completely identical, S_n increases: $S_n \sim n$ at $n \rightarrow \infty$.

Let us assume that, if for some set of $A_{ij}^{(n)}$ matrices any corresponding series S_n reaches the values substantially higher than unity, then all $A_{ij}^{(n)}$ in this set are similar. This similarity may consist of, for example, a precursor stably appearing at the same time of day and having a stable form.

Figure 6.2 shows the behavior of S_n for two groups of deep-focus earthquakes (hypocenter depths of 60–300 km). Curves 1 and 2 correspond to a series of 23 and 30 events with maximums of S_{23}=2.965 and S_{30}=2.667, respectively. The theoretical curves for the absolute similarity of the ionosphere behavior prior to any event in a series and for the non-correlated state of the ionosphere prior to events are shown by dashed and solid lines, respectively. For the group of 26 shallow-focus earthquakes are shown by blue and red colors, respectively. The correspondent catalog data for the earthquakes presented in Figs. 6.3 and 6.4 are presented in Tables 6.1 and 6.2 (hypocenter depths of 4–50 km), the maximum value of S_{26} proved to be only 1.454. This question was discussed in Chapter 5 where the spatial distribution of the earthquake was reviewed in relation to the ionosonde position. According to our theoretical calculations the larger similarity will be in the case of the same relative position of the epicenter and the ionosonde.

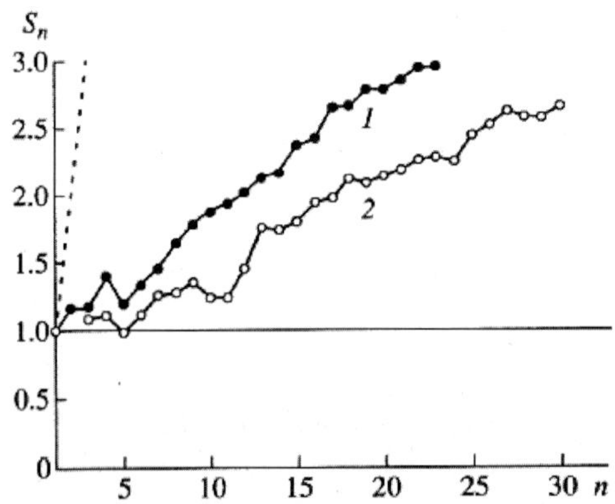

Fig. 6.2 Behavior of the S_n parameter for two groups of deep-focus earthquakes: a series of 23 (*curve 1*) and 30 (*curve 2)* events. *Dashed and solid lines*: theoretical estimations for absolute similarity and non-correlated ionosphere state prior to any earthquake in the series, respectively

Similar masks were calculated for different seismically active areas. Figure 6.4 presents the mask created for the Sofia ionospheric station for the 125 earthquakes in the Vranch Carpathian area for the period 1965–1998.

Figures 6.3 and 6.4 show the $A_{ij}^{(n)}$ matrices averaged for each of two groups of deep-focus earthquakes. Negative and positive deviations of *foF2* from the median

It is clear that for the first group of deep-focus earthquakes (Fig. 6.3), negative mean total values of *foF2* are pronounced from ~05 LT to ~15 LT prior to the shock, whereas for the second group (Fig. 6.4), positive values are observed from ~10 LT to ~20 LT, although the pattern is more mosaic-like. For the first group of deep-focus earthquakes the maximum negative and positive deviations are 1.13 and 1.58 MHz, respectively (Fig. 6.2, curve 1). For the second group of deep-focus earthquakes, these values are 0.68 and 1.1 MHz, respectively (Fig. 6.2, curve 2). The deviation values are not so much due to averaging within the group: the picture shows the tendency of the precursor's behavior in the local time and days before the earthquake, but for individual cases the deviation may be much more pronounced, of course. *The proposed technique can be used in any seismically active zone where the ionospheric station is installed, and where the distance to the epicenter is of the order 200–300 km.*

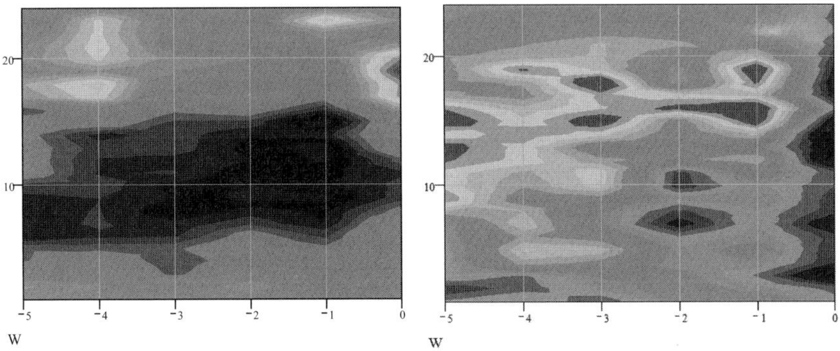

Fig. 6.3 *Left panel* – visualized presentation of the matrix, precursor "mask" for the first group of earthquakes corresponding to *curve 1* in Fig. 6.2. The abscissa is the time interval prior to the earthquake (days), and the ordinate is local time. The *color scale* reflects the mean value of the algebraic sum of the deviations of *foF2* from the median values. *Right panel* – the same as in the *left panel* but for the second group of the earthquakes corresponding to *curve 2* in Fig. 6.2

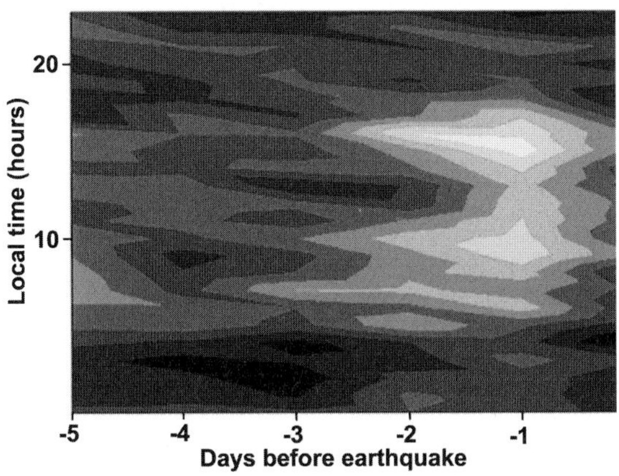

Fig. 6.4 Precursor mask for the Vranch area created from the data of Sofia ionospheric station. Intensive positive deviations are observed for the morning and afternoon hours 1 day prior to shock

Table 6.1 Parameters of earthquakes corresponding to the "mask" in Fig. 6.3. R – distance to the ionospheric station

No	Y	M	D	UT	LAT	LONG	H	M	R
1	1978	09	02	015733.40	24.90	121.99	109	6.10	68
2	1978	11	06	183805.70	24.73	123.35	97	4.70	89
3	1979	02	24	043801.40	23.91	123.40	68	4.50	112
4	1979	03	14	092649.10	24.56	122.02	81	4.00	49
5	1979	06	07	070406.40	24.83	124.03	97	4.00	159
6	1979	09	23	042407.20	24.76	122.00	96	4.30	57
7	1980	02	09	052438.10	24.79	122.28	72	4.50	38
8	1980	02	28	073621.50	23.92	123.39	67	4.40	110
9	1980	06	25	113517.70	24.84	121.91	107	4.80	70
10	1981	02	27	022733.30	24.61	121.85	71	4.70	67
11	1981	03	04	004209.90	25.06	121.86	146	4.70	89
12	1981	04	29	005329.30	23.93	121.73	60	4.20	100
13	1981	05	28	181433.60	24.86	123.61	115	5.20	119
14	1981	09	25	103446.47	24.55	122.76	110	4.00	26
15	1981	10	17	205727.69	24.67	122.55	107	4.50	19
16	1981	12	13	224003.22	24.73	122.88	139	4.50	46
17	1982	12	17	024303.66	24.59	122.55	86	6.10	11
18	1983	06	07	001143.45	24.62	122.63	103	4.90	18
19	1983	08	07	094926.79	24.58	122.68	109	4.10	20
20	1984	11	09	080610.51	24.99	122.80	151	4.60	62
21	1985	03	16	092411.22	24.35	124.07	65	4.70	160
22	1985	07	15	085259.45	25.25	123.32	193	4.70	117
23	1985	08	15	232252.20	24.85	123.49	118	4.90	107
24	1985	11	02	001330.64	24.80	122.18	84	5.20	46

Table 6.2 Parameters of earthquakes corresponding to the "mask" in Fig. 6.4 R – distance to the ionospheric station

No	Y	M	D	UT	LAT	LONG	H	M	R
1	1978	01	31	071851.90	23.58	121.71	63	5.00	129
2	1978	02	08	031033	24.82	121.98	102	4.20	63
3	1978	03	17	061621.30	24.94	122.08	100	4.50	64
4	1978	04	11	191613.90	24.75	121.93	78	4.20	63
5	1980	06	06	235518.30	24.92	124.20	92	4.80	177
6	1980	11	14	133701.20	24.57	121.95	76	5.20	56
7	1981	06	25	170415.42	24.69	122.50	113	4.00	20
8	1981	07	25	035841.01	24.90	122.31	122	4.50	47
9	1981	08	03	074548.17	25.07	123.10	163	5.30	87
10	1982	02	26	060508.07	24.68	121.99	88	4.40	55
11	1982	05	06	031548.15	24.84	122.51	99	4.80	37
12	1982	08	16	022611.25	25.11	122.11	76	4.20	78
13	1982	09	17	165405.25	24.55	123.05	73	4.50	55
14	1982	09	24	161845.47	24.57	122.48	88	5.10	7
15	1982	12	10	093503.29	25.24	124.12	139	4.40	183
16	1983	03	05	192114.92	24.56	122.64	104	4.20	15
17	1983	04	09	080824	24.54	122.31	80	4.50	19
18	1983	04	26	152640.33	24.65	122.59	116	5.70	18
19	1983	06	21	150134.50	25.55	123.93	160	4.80	185
20	1984	02	13	044857.59	25.48	122.41	268	5.50	108
21	1984	02	23	121525.61	24.64	121.93	69	4.60	59
22	1984	04	17	202407.61	25.52	122.84	242	4.80	118
23	1984	11	30	101923.27	24.40	122.64	67	5.20	18
24	1984	12	29	010701.30	24.80	122.04	85	5.40	56
25	1985	02	15	225326.24	25.13	123.45	191	4.50	119
26	1985	09	10	085729.71	24.55	122.89	105	5.10	39
27	1985	10	26	125647.90	24.48	122.83	92	4.00	33
28	1986	01	29	231604.93	25.39	122.77	95	4.60	101
29	1986	06	11	051646.71	24.26	121.89	80	4.20	66
30	1986	10	14	020846	24.04	122.38	61	4.80	52

6.3 Two or More Ionosondes Available

When one is able to select the configuration of the ionosonde position (from an existing network, or by installing new devices) it would be very useful to select two stations in the following way. From the point of view of solar-terrestrial relations they should be in the same conditions (in this case it is enough to select two stations on the same geomagnetic latitude). But from the point of view of seismo-ionospheric coupling, they should be in different conditions (it may be significantly different distance to the epicenter or they should be essentially in different areas of the two-poles structure forming in the ionosphere before the earthquake, see Fig. 5.2). Such a configuration uses the locality of the seismic effect within the ionosphere mentioned in the Conclusions section of Chapter 5 and its complex structure shown by theoretical calculations.

We experimented with such configurations of ionospheric stations in the Western Pacific area (Gaivoronaskaya and Pulinets 2002) and in Europe and obtained encouraging results. The quantitative evaluation of the ionosphere variability at heights of the F2-layer maximum was made for the pairs of ionospheric stations which were more or less in the same geophysical conditions but were at different distances from the epicenter. It was made by means of calculation the day-to-day correlation of daily variations of critical frequencies *foF2* between two stations:

$$C = \frac{\sum\limits_{i=0,k} (f_{1,i} - af_1)(f_{2,i} - af_2)}{k(\sigma_1 \sigma_2)} \tag{6.2}$$

where index 1 and 2 corresponds to the first and second ionospheric stations, f= *foF2*, k=23 and af and σ are determined by the following expressions:

$$af = \frac{\sum\limits_{i=0,k} f_i}{k+1} \tag{6.3}$$

$$\sigma^2 = \frac{\sum\limits_{i=0,k} (f_i - af)^2}{k} \tag{6.4}$$

Results for the pair Magadan – Petropavlovsk-na-Kamchatke stations are shown in Fig. 6.5 for the three-month period January-April 1974 (bottom plot). In this case the Petropavlovsk-na-Kamchatke station is inside the seismically active zone (53.0° N 158.7° E), and Magadan situated ~700 km to the West (60.0° N, 151.0° E). Earthquakes parameters for the given period are presented in Table 6.3. Looking at the picture one can note that in general, the correlation coefficient between two stations is very high (it reflects the fact of their similar geophysical conditions, and that the ionosphere behavior is in phase on both stations in quiet conditions). But sometimes one can observe the sharp drops of the cross-correlation coefficient, which can be explained in the following way. We actually have two sources of perturbations: geomagnetic disturbances and seismic activity. In the case of geomagnetic disturbances the correlation should be higher because they have a global character and should be "felt" more or less equally by both stations. In the case of seismically induced disturbances, they should be more pronounced at Petropavlovsk-na-Kamchatke, and a drop of cross-correlation coefficient should mean the ionospheric precursor registration. On the bottom axis the seismic events are marked by triangles, and the geomagnetic disturbances (with $k_p \geq 6$) – by diamonds. One tick on the horizontal axis corresponds to two days. The list of earthquakes in the Kamchatka area with a magnitude higher than 5 and a distance ≤350 km from the ionospheric station is given in Table 6.3.

Table 6.3 List of earthquakes in the vicinity of Petropavlovsk-na-Kamchatke station. R – distance to the ionospheric station

No	Date	DOY	UT	LAT	LONG	H	M	R
1	1974/01/09	9	02:50	51.3	160.3	0.0	5.8	163
2	1974/01/10	10	05:19	51.4	160.0	40.0	5.5	171
3	1974/01/22	22	13:28	55.0	162.4	4.0	5.9	333
4	1974/02/06	37	02:51	52.6	160.7	22.0	5.0	117
5	1974/02/07	38	19:04	49.7	156.7	40.0	5.5	369
6	1974/02/09	40	18:22	50.3	157.7	40.0	5.2	284
7	1974/02/17	48	08:35	50.4	157.0	19.0	5.1	298
8	1974/02/26	57	06:24	52.9	160.2	79.0	5.6	75
9	1974/03/12	71	05:06	50.7	157.2	66.0	4.3	250
10	1974/03/15	74	22:11	49.4	158.5	35.0	5.0	396
11	1974/04/13	103	20:09	55.1	162.1	21.0	5.2	327
12	1974/04/18	108	10:31	53.9	163.4	35.0	5.3	338
13	1974/04/22	112	07:18	53.7	160.8	31.0	5.3	155
14	1974/04/22	112	13:42	50.2	157.0	40.0	5.5	328
15	1974/05/01	121	21:56	53.4	160.5	38.0	5.2	126

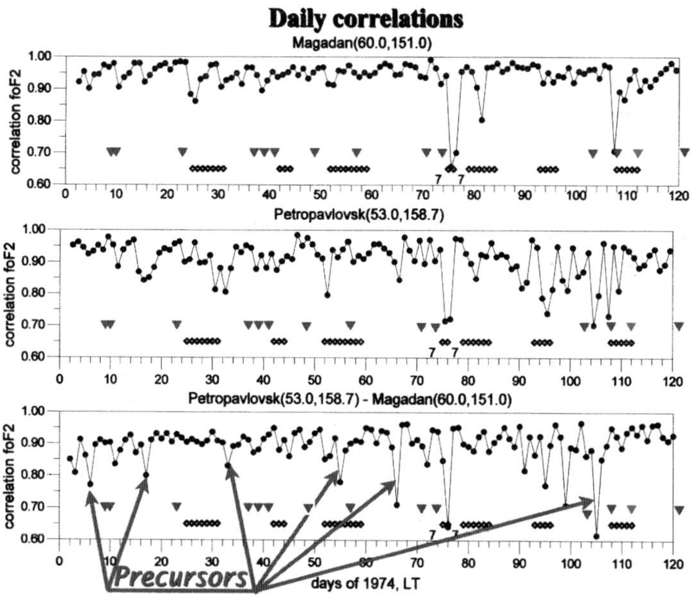

Fig. 6.5 Demonstration of the correlation technique for the ionospheric precursors identification. *Top panel* – autocorrelation coefficient of daily values of *af* for Magadan station, *middle panel* – the same for Petropavlovsk station, *bottom panel* – cross-correlation coefficient calculated according to (6.2). *Triangles* – seismic events, *diamonds* – periods of geomagnetic disturbances

One can see from the figure that most of the minima on the bottom graph appear within the interval 3–7 days before the seismic shock. The only minimum not connected with the seismic activity is on day 76 when a very strong long lasting magnetic storm was observed. Both stations tracked the minor storms in phase, but in this case their reaction was different. It may be connected with the longitudinal difference between the stations positions and different UT behavior of the storm, or the higher latitude position of the Magadan station, and the possible main trough excursion to the South during the storm main phase (Karpachev et al. 1995). Looking at the list of earthquakes, one can note that earthquakes 9 and 10 do not fit well to the criteria (9 has a magnitude lower than 5, and 10 has a distance more than 350 km). But both earthquakes are on the same longitude as the ionospheric station what make it more sensitive to the anomalous variations produced by these seismic events. At least, the pronounced minimum is observed which may be attributed to effects from one of these events.

Similar results were obtained for Europe. In Fig. 6.6 the cross-correlation coefficient is shown for two Italian ionospheric stations, Roma and Gibilmanna for the period April-May 1984. One can see the strong drop of the cross-correlation coefficient one day before the earthquake on 29 April 1984 M=6.1, 43.3° N, 12.6° E, which occurred 126 km from the Roma station. It is interesting to note that the second shock on 7 May 1984 M=6.0, 41.8° N, 13.9° E which was even closer to the Roma station (116 km) did not give a reaction which supports our idea of the importance of the relative position of the ionospheric station and epicenter. But this earthquake was detected using another technique by the longitudinal chain of European ionosondes (see Figs. 2.10–2.11). This example demonstrates the necessity not only for multiparameter data analysis, but also for multitechnique data processing to increase the ionospheric precursor detection probability.

The simple correlation technique also provides instruments for precursor detecting even in the case of only one ionosonde. It is not so evident as with cross-correlation but at least gives the possibility of detecting the suspicious days during the geomagnetically quiet periods.

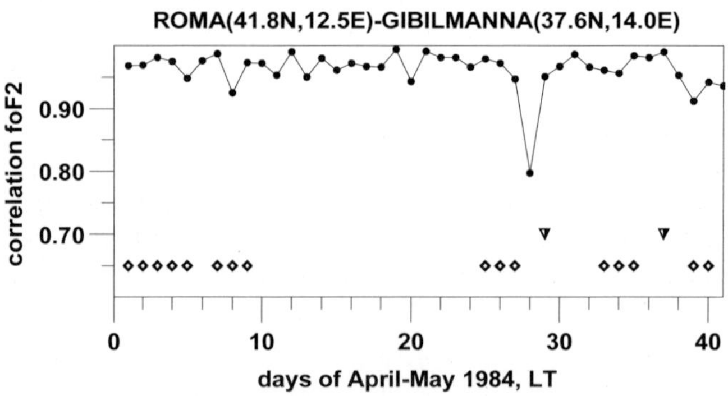

Fig. 6.6 Behavior of the cross-correlation coefficient for Roma and Gibilmanna ionospheric stations for the period April-May 1984. The earthquakes are indicated in the figure by *triangles*, the geomagnetically disturbed periods – by *diamonds*

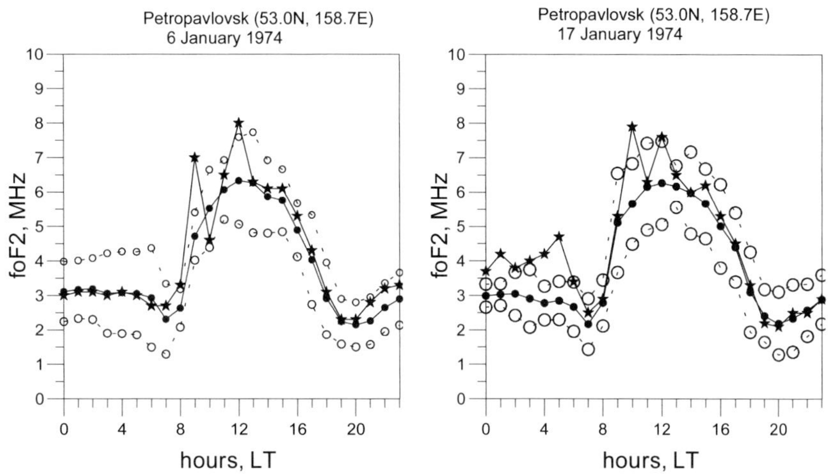

Fig. 6.7 Examples of daily variations of the critical frequency *foF2* at Petropavlovsk-na-Kamchatke for the 6th (*left panel*) and 17th (*right panel*) January 1974 when ionospheric precursors were detected by the cross-correlation technique. *Black points* – monthly median, *open circles* $\pm 2\sigma$ upper and lower bounds, *asterisk* – current values of *foF2*

Using formula 6.2, we don't calculate the cross-correlation coefficient for two stations but the correlation for the same station for two consecutive days to see day-to-day changes for the station of the ionosphere. The autocorrelation coefficient characterizes the day-to-day variability of the ionosphere. Results of such calculations for the Magadan and Petropavlovsk-na-Kamchatke ionospheric stations for the period of January-April 1974 are shown in the top and middle panels of Fig. 6.5, respectively. One can see that the main minima for Magadan are associated only with the geomagnetic disturbances, while the Petropavlovsk data shows the precursory drops, which are depicted in the bottom panel, where the cross-correlation coefficient is shown. This means that for the geomagnetically quiet periods the autocorrelation techniques works as well.

From this regard it would be interesting to check whether during precursory days when we see cross-correlation minima, the daily variations satisfy the precursory criteria 2σ (see Fig. 5.21). We depicted in Fig. 6.7 two examples of the daily behavior of the critical frequency at the Petropavlovsk station for the minima observed in Fig. 6.5 on the 6th and 17th of January. One can clearly see that in both cases the current values of the critical frequency in particular hours of local time exceed 2σ boundary, which satisfies our criterion of anomalous variations before the earthquake.

6.4 Ionospheric Mapping with Several Stations

Within the frame of European Commission of Science and Technology projects COST238, COST251 (www.cost251.rcru.rl.ac.uk/html/reports.html) and COST271 a special technique for ionosphere mapping over Europe was developed. It uses the data of European network of ionosondes, and as a result of the projects one can find now the currently updated maps of the critical frequency distribution over Europe (www.ionosphere.rcru.rl.ac.uk/maps.html). As was demonstrated in Chapter 5, the mapping technique provides the possibility to see directly the large scale ionospheric irregularities created in the ionosphere before the earthquakes. One example for Europe is demonstrated in Fig. 4.14 for the case of the earthquake of 7 May 1984 also detected by the longitudinal chain technique (Fig. 2.11).

Unfortunately, there are very few places in the world where the number of ionosondes is sufficient to produce ionospheric mapping. Even the European network does not give enough details to resolve irregularities of the order of 100 km. In this regard one can turn towards the use of GPS TEC measurements. The number of GPS receivers is growing rapidly in the World due to their wide application in geodesy and other fields of geophysics (and not only geophysics). In Chapters 2 and 5 we demonstrated the ability of the GPS receivers to register the ionospheric precursors as well. Liu et al. (2004) calculated that correlation between the variations of the ionosphere peak electron density N_mF2 and GPS TEC is 0.953. So one can use the GPS TEC data as if it is ionosonde data, especially for mapping purposes (www.kn.nz.dlr.de/daily/tec-eu). Figure 6.8 demonstrates the comparison of $\Delta foF2$ and ΔTEC variations for the period of July 1998 for Taiwan. One can see the similarity of the observed variations. So in the areas with a sufficiently dense network of CPS receivers (like California or Japan) one can try to map the ionosphere with the purpose of revealing the ionospheric precursors of earthquakes. Figure 5.10 demonstrates the ability of GPS TEC mapping (Liu et al. 2002) where one can clearly see the equatorial anomaly crest dynamics during the period of several days of the Chi-Chi earthquake preparation. This example was for low latitudes. Figure 6.9 demonstrates the example of a GPS TEC map built one day before a strong M=7.4 earthquake in California which happened on the 16th October 1999, 34.6° N, 116.3° W. One can see the positive anomaly (light colors) observed in the GPS TEC map, which according to our model is shifted to the South from the epicentral area.

Unprecedented possibilities to use the GPS TEC technique for ionospheric precursor monitoring are available to Japan where there are now near 1000 GPS receivers installed. But unfortunately (as often happens in our life) there is no interest, at least from the governmental organizations to use this information. All means in Japan are directed towards studies of the electromagnetic emissions associated with earthquakes, and ionospheric information is not taken into account. At the same time the Japanese GPS network permits us to achieve the record spatial resolution in ionospheric mapping.

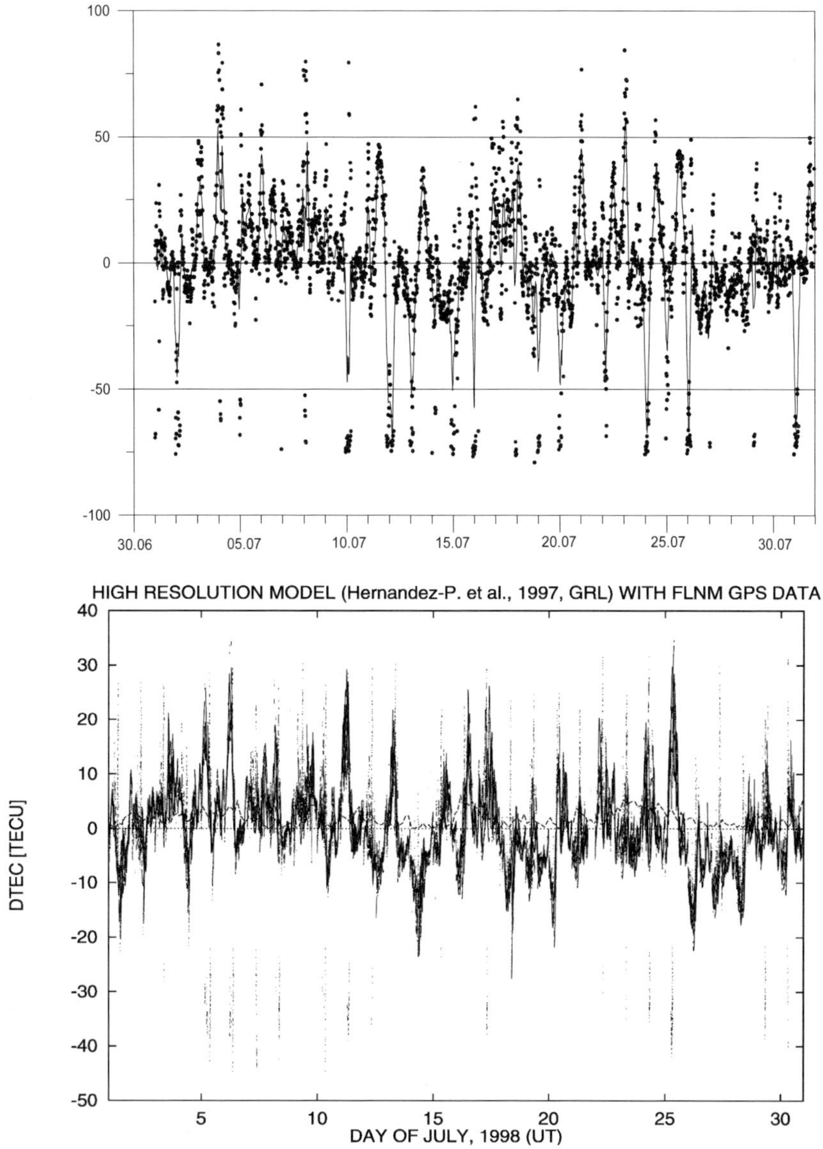

Fig. 6.8 Δ*foF2* (*top panel*) and Δ*TEC* (*bottom panel*) comparison for July 1998 for Taiwan area

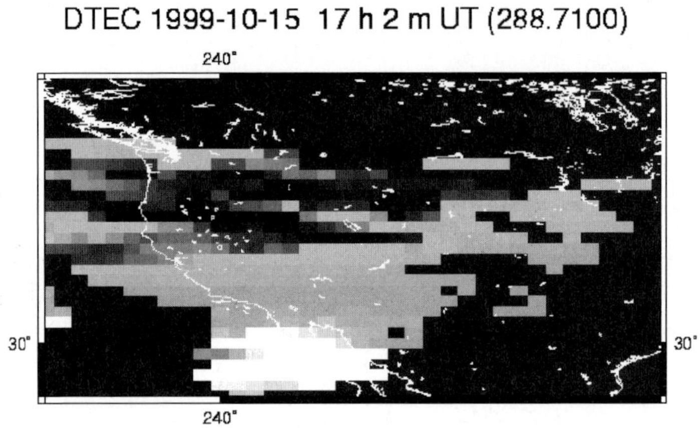

Fig. 6.9 The map of ΔTEC one day before the Hector Mine earthquake in California M=7.4, 16th October 1999, 34.6° N, 116.3° W (thanks to M. Hernandez-Pajares)

6.5 The Complex Analysis – Irpinia Earthquake 23.11.1980

Now, as the reader has some flavor of ionospheric precursor identification we will try to apply all our knowledge and to attract all available information to detect the ionospheric precursors. For this task we will select the case of the Irpinia earthquake in Italy. A destructive earthquake with M=6.9 took place in Southern Italy (under the village of Liviano), Lat ~ 40° 46'N and Long~ 15° 18'E at the depth h~16 km on November 23rd, 1980 at 18 h 34 m UT. The normal slip occurred with strike orientation 18°. It was oriented parallel to the axis of the Apennine mountains. The size of the earthquake source estimated by the aftershock activity was of the order of 14×40 km (Deschamps et al. 1983). It acquired the name Irpinia earthquake in the literature (Valensize 1993).

The closest to epicenter ionosonde was the Rome ionospheric station. From the point of view of the ionosonde data, from the first glance the situation was hopeless. Kouris et al. (2001) claim that the situation observed in November 1980 does not decline from "normal", providing the variations of the critical frequency and deviation from the monthly median for January and November 1980 for the Rome station in Fig. 3 of their paper.

We will return later to the Rome ionosonde data, but now let us leave sinful Earth and rise into space where the Intercosmos-19 satellite was on orbit with a topside sounder and other equipment onboard. This allowed us to get a latitudinal–longitudinal distribution of the ionospheric disturbance associated with the Irpinia earthquake. For plotting the maps of $\Delta foF2(\varphi, \lambda)$ the time was chosen for when

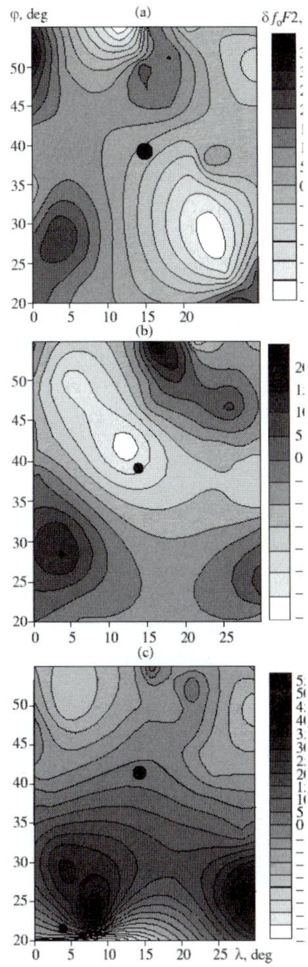

Fig. 6.10 Dynamics of the modified region in the ionosphere –64, -42 and +9 h in relation to the Irpinia earthquake seismic shock moment by the data of Intercosmos-19 topside sounding

the satellite data were available. During this period the satellite passed over the epicenter at 03–04 h LT for $\Delta t \approx -112$, –87, –64, –42, –40, and +9 h; and at 18–19 h LT for $\Delta t \approx -120$, –96, –72, and – 1.39 h. The anomalous decrease and increase of *foF*2 ("*f*–" and "*f*+") were observed during the early morning hours and evening hours, respectively. The maps of *foF*2 (φ, λ) plotted using the data of November 18th–19th were taken as the quiescent background. The maps of $\Delta foF2(\varphi, \lambda)$ for the morning hours were plotted for three different time moments: $\Delta t \approx -64$; –40...–42.5, and +6...+8 h (Fig. 6.10). One can see that the "cloud" of a decreased density gradually began to be formed already 2.5 days ($\Delta t \approx -64$ h) ahead of the main shock, but in the region displaced by ~10° in both latitude and longitude to the south-east of the epicenter. Later, at $\Delta t \approx -40...-42$ h, this "cloud" was displaced toward the epicenter and "spread" in the northwest direction. Roughly 6–8 h after the main shock, the state of the ionosphere in the epicenter region was stabilized. Recently, it was shown by Pulinets and Legen'ka (1997) and Ruzhin et al. (1998) that ionospheric effects might appear not only in the epicenter region, but also in the magnetically conjugated point in the opposite hemisphere. Using only the data of the Interkosmos-19 satellite, we succeeded in constructing a map $\Delta foF2(\varphi, \lambda)$ for the morning sector at $\Delta t \approx -64$ h in the entire region of latitudes from 60° N to 60° S, i.e., including the magnetically conjugated region (Fig. 5.7). The latitudinal cross-section of the critical frequency deviation for the passes when the conjugated effect was observed is presented in Fig. 6.11. As one can see from the figure the positive deviation was observed during the evening hours and the negative one – before the sunrise. We will use this picture later for magnitude prediction from the satellite data.

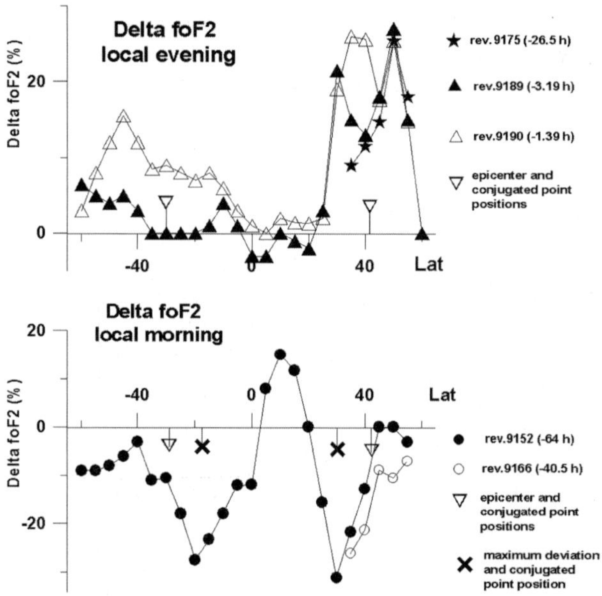

Fig. 6.11 Deviation of the critical frequency Δ*foF*2 for evening (*upper panel*) and early morning hours of local time for the passes of Intercosmos-19 satellite over Irpinia earthquake epicentral area and magnetically conjugated area

The Intercosmos-19 satellite was the complex geophysical observatory and its payload contained the devices of different destination. The multichannel VLF receiver registered the VLF emissions simultaneously with the ionosonde, and also detected the conjugated effects in anomalous VLF emissions associated with the Irpinia earthquake (Ruzhin and Larkina 1996; Ruzhin et al. 1998). The satellite orbits where both the ionospheric and VLF anomalies were registered are presented in Fig. 6.12 (Pulinets and Legen'ka 1997).

As one can see, the pre-earthquake effects within the ionosphere for the case of the Irpinia earthquake were quite significant in size (20°–30° in latitude and longitude modified region within the ionosphere) and in amplitude ±30% for the evening and morning hours. The registered anomalous VLF emissions and effects in the conjugated area manifest that all of the magnetic tube was affected by the earthquake preparation process. If we look at the altitude changes of the electron concentration, we would also detect essential changes in the ionosphere scale height (increase ~80 km, Fig. 5.15) which is equivalent to the mean ion mass drop, and peak height increase (~100 km, Fig. 5.14).

If we have such significant effects in space, why was nothing observed on the ground surface? It seems that such a situation is impossible. We start the search again from space but will look not at the ionosphere but at the ground surface. The NOAA/AVHRR satellite monitored the ground surface in different spectral bands giving the possibility to monitor the ground surface temperature (Tramutoli et al. 2001). Leaving outside the scope of this discussion the very sophisticated and confident data processing presented by the authors, we find from Fig. 6.13 that practically all the Apennine Peninsula was occupied by the temperature anomaly in November of 1980 before the Irpinia earthquake (which also give us the possibility to estimate the magnitude of the approaching earthquake).

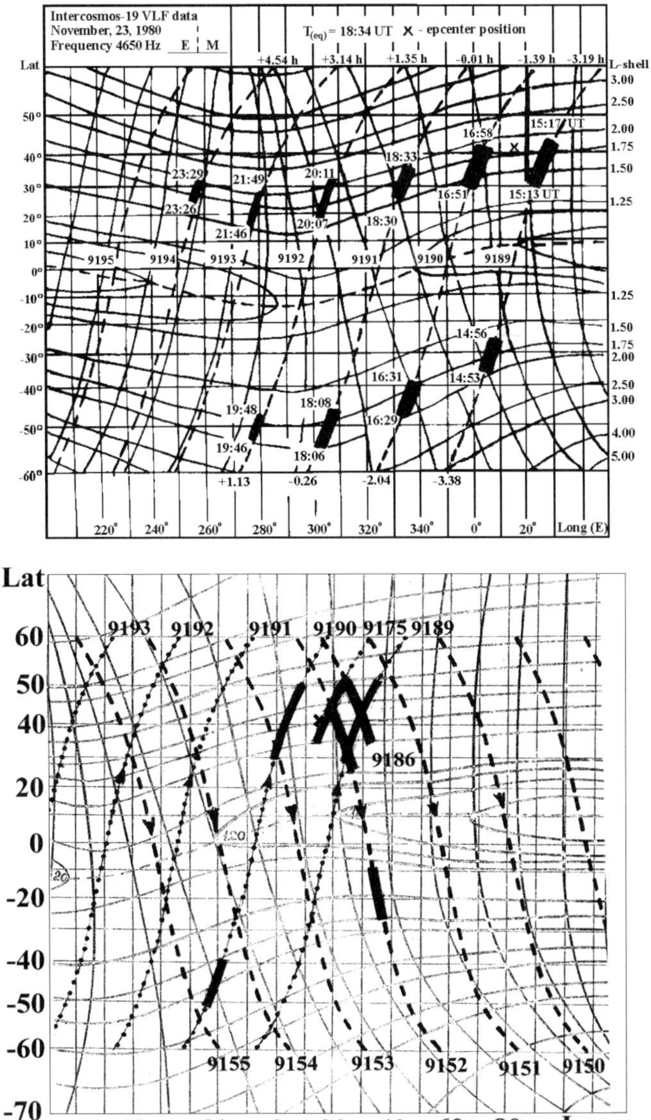

Fig. 6.12 The Intercosmos-19 satellite orbits where the anomalous VLF emissions were registered (*upper panel*) and electron concentration anomalies (*bottom panel*) *Bolds bars* indicate parts of orbits where effects were registered

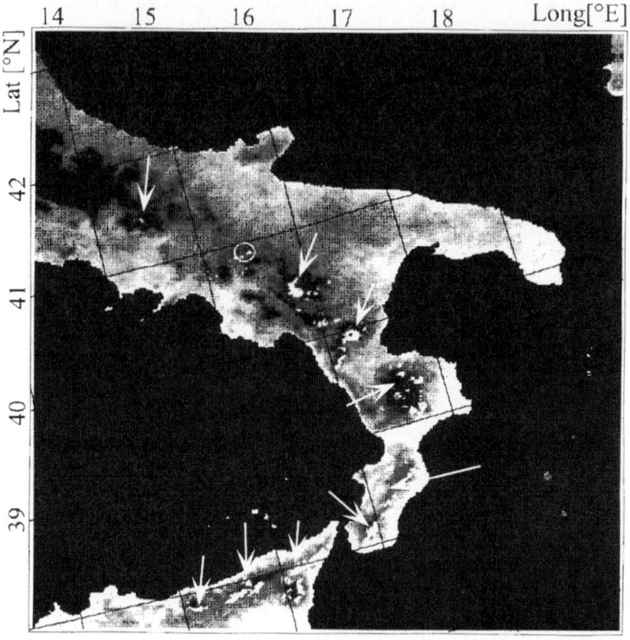

Fig. 6.13 The increased surface temperature determined from the ALICE index (bright spots indicated by *arrows*) for November 1980 by the data of the NOAA/AVHRR satellite (Tramutoli et al. 2001). Irpinia earthquake epicenter is shown by the *circle*

So we can conclude that at the ground surface also a huge area was involved in the process of the Irpinia earthquake preparation. This statement is confirmed by the radon measurements reported by Allegri et al. (1983). They presented the data of seismic activation before the Irpinia earthquake together with the data of radon anomaly measurements at significant distances from the future epicenter. The most distant earthquake from the seismic swarm registered in May-June 1980 was at 311 km from the epicenter. The radon anomalies were measured at the distances near 250–270 km from the epicenter. The possibility of the distant radon anomalies registration was stated by Fleischer (1981). Figure 6.14 demonstrates the temporal development of the radon anomalies registered on two stations (Rieti and Rome) at the distance 270 and 250 km respectively from the epicenter. One can see the sharp increase of the radon flux associated with the seismic swarm started in the middle of May 1980. Both stations demonstrate similar behavior – long duration maximum, then the drop associated with the foreshock at the end of September and the second period of the activity with gradual diminishing of the radon flux up to the background level close to the time of the Irpinia earthquake. This dynamic is in good agreement with the dilatation theory when the opening of cracks leads to increased radon emanation and the cracks close during the final stage of earthquake preparation leading to the radon flux reducing (Scholz 1990). According to the physical model discussed above the radon anomaly registered at the distance 250 km from the epicenter (which gives an area of 500 km in diameter) let us expect the existence of the precursory ionospheric effects.

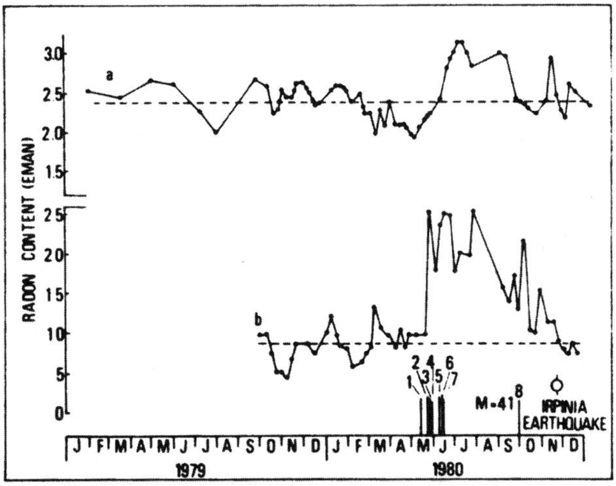

Fig. 6.14 The series of the radon content in the ground water. *a*) -Rieti station, *b*)- Rome station. The *vertical bars* in the *bottom panel* show the seismic shocks. The length of the *bars* is proportional to the magnitude

Let us return to the ionosonde data. Whether we really cannot detect anything as claimed by Kouris et al. (2001)? From Fig. 6.12 where together with rectangular geodetic coordinates, the geomagnetic ones are shown too, one can see that the magnetic field lines are inclined to the Southeast. According to the presented satellite data (Figs. 5.7 and 6.10) and our theoretical model the effect in the ionosphere should be shifted from the epicenter along the field lines, i.e. to the Southeast. What does this mean? That maybe some ionospheric station to the Southeast from the epicenter will better "feel" the process of the Irpinia earthquake preparation. Fortunately, there was the Athens ionospheric station working, which is situated just in the necessary direction. Let us compare the data from Rome and Athens. Taking into account that both stations are quite far from the maximum magnitude area shown by the satellite (Fig. 5.7), we cannot expect very pronounced variations, nevertheless both of them show a quite clear precursory effect one day before the seismic shock, i.e. 22 November. It is expressed in the fact that both stations demonstrate the extreme values for the whole month of November, especially the Athens station. In the Fig. 6.15 is shown the scatter diagram where all the data of the critical frequency for November 1980 are shown for the Athens station. One can see the minimum demonstrated on the 22nd and 23rd of November. It is important to note, that all this is going on against the background of the strong seasonal negative gradient (the electron density sharply decreases during November due to seasonal change). Nevertheless, the days 22 and 23 show the absolute minimum for the whole month. Figures 6.17 and 6.18 demonstrate this effect for other parts of the ionospheric data. In Fig. 6.17 the critical frequency variations from November 1980 for the Rome station is shown for 0600 and 1800 LT. Both parameters demonstrate the absolute minimum for the month on the 22nd of November.

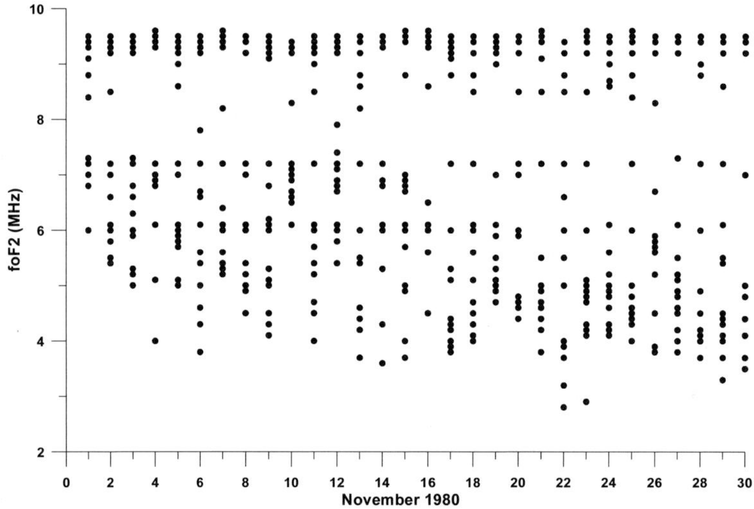

Fig. 6.15 Scattered plot for the critical frequency *foF2* for November 1980 (Athens ionospheric station)

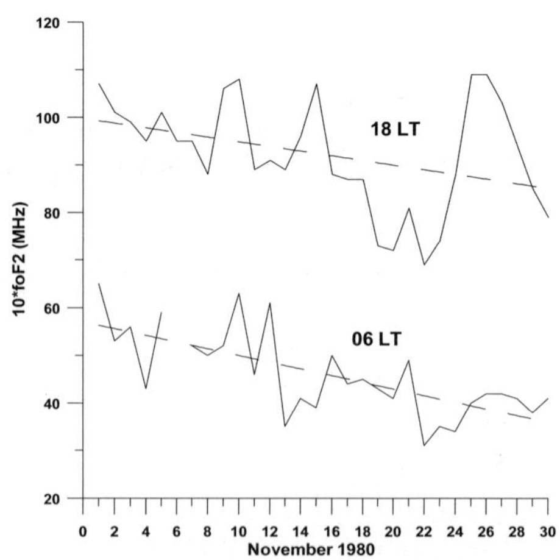

Fig. 6.16 Critical frequency for 0600 LT and 1800 LT for November 1980 (Rome ionospheric station). *Dashed lines* show the seasonal trend of the critical frequency

Gaivoronskaya and Pulinets (2002) have shown the importance of statistical parameters of the ionospheric data for ionospheric precursor identification. They are the extreme values of the critical frequency, the range of the parameter variations, the behavior of the mean daily values. All these parameters are shown in Fig. 6.17 for the Athens ionospheric station.

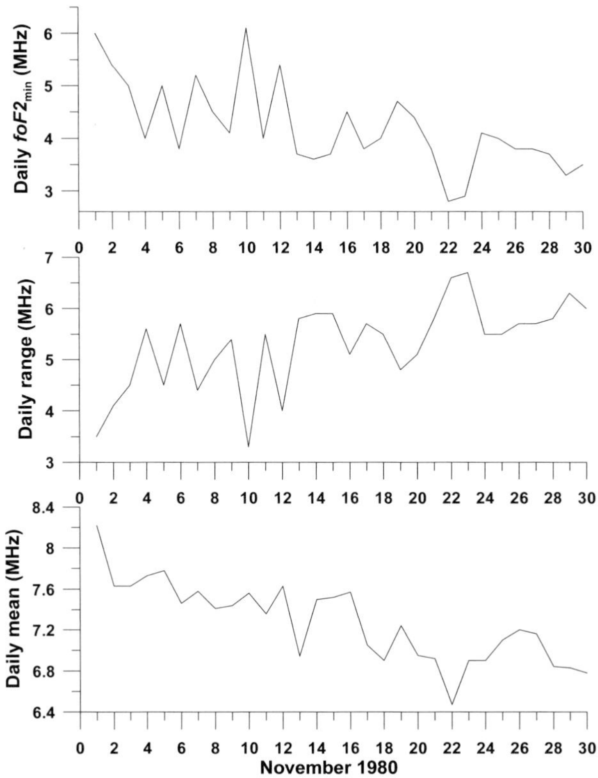

Fig. 6.17 From *top to bottom*: minimum daily values of the critical frequency, daily range of the critical frequency variation, daily mean value of the critical frequency for November 1980 (Athens ionospheric station)

From Fig. 6.17 one can see that all three parameters demonstrate the extreme values for the whole of November on days 22 and 23. This kind of analysis is usually conducted when the variation is not very pronounced as in the considered case. One can expect the more unstable critical frequency behavior on the 22nd of November. At both stations the critical frequency does not exceed 2σ (as it is expected to have, see Fig. 5.21) but is well above σ for the nighttime and evening hours as one can see from Fig. 6.18.

Concluding the Irpinia earthquake analysis we can state that most of the precursors discussed in the book were observed for the Irpinia earthquake, namely:

- Radon anomalies starting after the seismic swarm in May 1980 and continuing up to the moment of the Irpinia earthquake registered at large distances from the epicenter
- Thermal anomalies starting 5 days before the earthquake and occupying large areas (more than 500 km)

- Ionospheric anomalies registered by ground based ionosondes at Rome and Athens 2 and 1 days before the seismic shock
- Ionospheric anomalies registered by the Intercosmos-19 satellite starting three days before the shock
- Scale height anomalies manifesting the ion mass changes
- VLF noise anomalies starting 3 days before the shock
- Anomalies in the conjugated point manifesting the whole magnetospheric tube modification through the electric field mapping along the tube

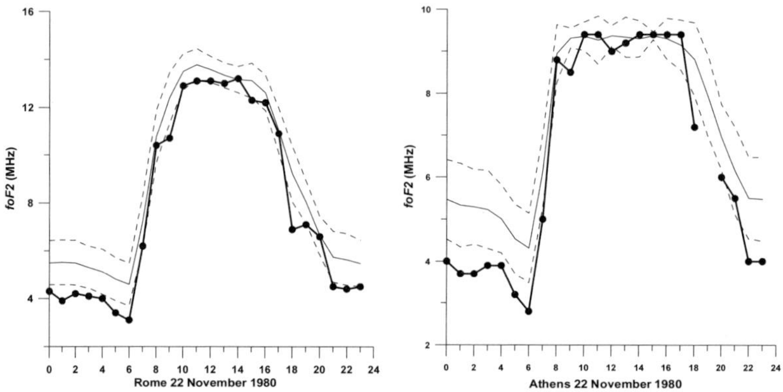

Fig. 6.18 Daily variations of the critical frequency (*bold line with points*) on 22nd November 1980 in Rome (*left panel*) and Athens (*right panel*). *Continuous line* – monthly mean, *dashed lines* $\pm\sigma$

From the obtained data on the Irpinia earthquake precursors we can write the following scenario of the events development. During several months the lower atmosphere was prepared by the radon emanation and formation of the neutral associates described in Chapter 3. Around five days before the earthquake the gas discharges started along the entire Apennine peninsula, which is registered as thermal anomalies by the satellite. The aerosols carried out with the gas increased the electric field strength which led to disruption of the neutral associates and generation of the strong electric field which was able to penetrate into the ionosphere. And starting from 3 days before the seismic shock the anomalies were registered in space by the Intercosmos-19 satellite and 2 days before the shock – by ground based ionosondes at Rome and Athens. If we know all this in 1980.

6.6 The Ground-Based and Satellite Systems of Electromagnetic and Ionospheric Precursors Monitoring

Regardless of the skepticism generated by some circles (Geller 1997) on the impossibility of earthquake prediction, more and more countries (including the highest governmental levels) realize that doing nothing is the ostrich position of dread before the real difficulties associated with real prediction system creation. The developed ground-based scientific ranges are created or are in the process of creation in such countries as Japan, Russia, Taiwan, Greece, France, China, and Mexico. It is interesting that due to large opposition in the USA, the prediction works are carried out there as a private initiative (satellite QakeFinder project). Due to tradition and historical circumstances the community working with electromagnetic emissions associated with earthquakes is much larger and has more governmental support than the ionospheric one. That's why most of the projects are mainly concerned with electromagnetic precursors of earthquakes, which are out the scope of the present book. But taking into account that electromagnetic and ionospheric effects are closely related, we will mention here most of the existing projects of the precursors studies without exception. Simultaneously we will demonstrate which advantages the ionospheric precursors have over the electromagnetic ones.

6.6.1 Ground-Based Systems

6.6.1.1 Japanese Works

During the last five years (1997–2001) in Japan were realized two projects associated with the study of earthquake precursors. RIKEN's frontier project was connected mainly with electromagnetic precursors of earthquakes, and NASDA's earthquake remote sensing frontier research project (together with electromagnetic precursors as well) included also the studies of the processes in space and radio wave propagation. Extended ground based ranges were developed in Japan by different institutions, and results of these studies one can find in many publications worldwide. In more concentrated form they were published as proceedings of the workshops (Hayakawa and Fujinawa 1994; Hayakawa 1999; Hayakawa and Molchanov 2002) and the final report of the NASDA's frontier research project (Hayakawa 2001). According to the satellite data, were analyzed with the historical data (mainly of the Russian satellites), and obtained results, which in our mind are controversial. Except confirmation of the results obtained earlier with the Intercosmos-19 satellite on the large scale of anomalies within the ionosphere associated with the earthquake preparation (Afonin et al. 2000) and that it affects the equatorial anomaly, other data including even statistical data are not convincing. This is mainly due to the fact that the authors did not divide the data by the local time (as we know from the previous discussion, the sign and amplitude of the effect within the ionosphere strongly depends on the local time). As a result they averaged the effect by statistical processing obtaining very small deviations.

The most interesting results obtained within the frame of the projects and which can be recommended for implementation, is the ULF measurements and developed techniques. The first one is connected with polarization measurements of ULF emissions (Hattori et al. 2002). The apparent increase of the magnetic field vertical component S_z was observed several weeks before the earthquake of March 26th, 1997. Polarization is also very useful to distinguish between the seismically induced signal and magnetic pulsations.

The second technique developed in Russia but installed in Japan (Ismaguilov et al. 2001) uses a triangulation procedure to detect the direction to the noises source (and probable epicenter) position. One can see the system configuration in Fig. 6.19.

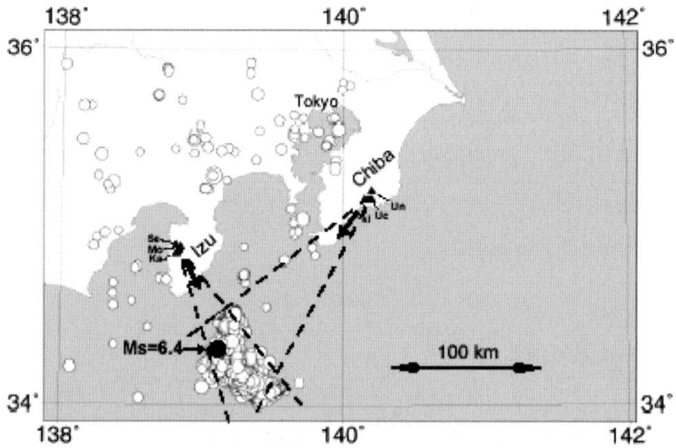

Fig. 6.19 Configuration of the magnetometer direction finding system for the earthquake epicenter location by registering ULF emission precursors (Ismaguilov et al. 2001)

The system consists of two sets of 3-component sensitive magnetometers. Every set consists of three magnetometers was installed 4–7 km one from another. And the distance between both sets was near 140 km. The system uses the vertical components indication as was mentioned above (Hattori et al. 2002) and the horizontal gradient measurements to detect the direction of the source of registered emission. So in comparison with all other techniques of electromagnetic emissions measurements, this system has an unambiguous advantage to detect the epicenter position. It can be recommended for the sites where the configuration is more or less known (this means, the position of the active fault) for the precise future epicenter determination.

As always the advantages are connected with disadvantages. The system has a limited spatial application (near 150 km), due to its high sensitivity needs a lot of care in its maintenance, and has very complex algorithms of data processing.

According to the ionospheric studies (regardless that Japan has 5 ionosondes and near 1,000 GPS receivers, and different radars and optical instruments) the ionospheric studies connected with earthquakes precursors have not been carried out in Japan except for limited attempts (Ondoh 1998, 1999, 2000).

6.6.1.2 Russian Ground Based Systems

After the USSR split to independent States the main seismic range used now in Russia is the Kamchatka Peninsula. There are two groups with a complex set of equipment for precursor study at Kamchatka. The first one (Uyeda et al. 2001; Gladyshev et al. 2001) in cooperation with Japan is directed again mainly at the registration of electromagnetic precursors. At Karimshino the experimental range in the ULF/ELF magnetic field, geoelectric potentials (telluric currents) and VLF signals from navigation radio transmitters are registered.

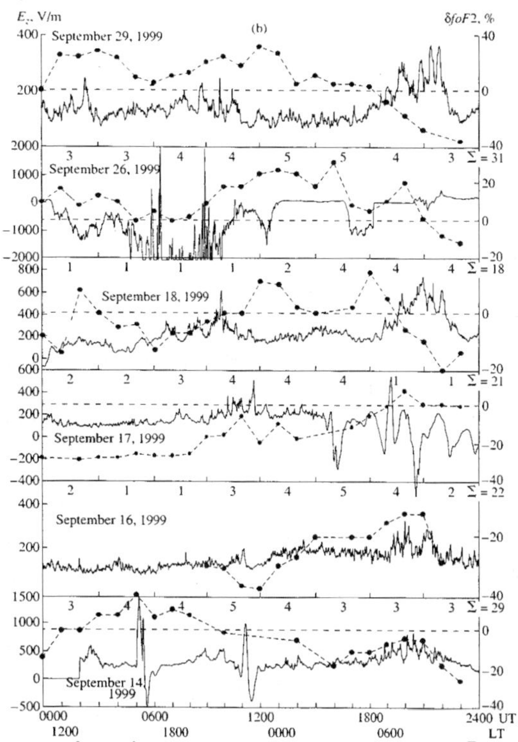

Fig. 6.20 Vertical atmospheric electric field (*bold line, left axis*) and critical frequency deviation (*dashed line, right axis*) measured simultaneously at Kamchatka for the period 14th–19th September 1999 (Mikhailov et al. 2002)

The second group develops, it seems, a more perspective direction registering many environmental parameters (Mikhailov et al. 2002, 2003). At least, this group works in the direction of the model developed in the present book. Together with meteorological parameters such as temperature, pressure, precipitation, the key for our model parameters are measured such as radon concentration, vertical electric field and ionosphere parameters. For the first time the direct comparison of the electric field variations and the ionosphere are presented (Fig. 6.20).

Their results support our conclusion on the universal effect of the electric field on the ionosphere regardless of its nature and experimentally demonstrates the existence of this effect. Of course, to make more definite conclusions, it is necessary to determine the spatial parameters of the electric field as well, installing several field mills on the distance of several hundreds kilometers, but it probably will be made in future works (as is planned within the frame of the PREVENTION project in Mexico).

6.6.1.3 Taiwanese iSTEP Project

From the point of view of governmental support and finances involved, the iSTEP Project (integrated Search for Taiwan Earthquake Precursors) is presently the largest. Started in the year 2002 it consists of five sub-projects: I – A study of earthquake precursors – seismological variations; II – A study of earthquake precursors – variations of geomagnetic and gravity fields, III – A study of earthquake precursors – radar interferometry for detection of ground surface deformation; IV – A study of earthquake precursors – ionospheric variations, V – A statistical study of electromagnetic precursors of earthquakes. For a more detailed description one can find the WEB-page of the project: http://www.ss.ncu.edu.tw/~istep. The main advantage of this project is scientists of different research directions do not spend their time trying to prove who is right as often happens between seismologists and other directions of precursors studies, but work together which is much better.

The most related to the subject of our book are the sub-projects IV and V. Actually, some results from Taiwan demonstrated in our book, formed the foundation of these projects. Taking into account the experience of Taiwanese scientists in the ionospheric precursors studies, it seems that they have long been ready to use their knowledge in practical short-term predictions.

6.6.1.4 GAVDOS ROSE Project, Greece

The interdisciplinary Research Observatory of Southernmost Europe (ROSE) project has the purpose of developing a complex geophysical observatory at Gavdos Island – the southernmost point of Greece and Europe. Its start was marked by an international symposium in Crete in September 2001. The development of the observatory occurs within a framework of European cooperation, installing the usual seismic equipment, also equipment for electromagnetic (geoelectric and magnetometric) measurements. The atmospheric and other environmental parameters will be measured at the observatory to provide the background information. The ionospheric component is provided as well, using the GPS receivers and ionospheric stations at Athens, Rome, Gibilimanna and the probable ionosonde installation at Gavdos. The state of the observatory and the results are not published yet.

6.6.1.5 PREVENTION Project, Mexico

The project PREVENTION PREVENTION (PREcursors of Volcano Eruptions and Notable Tremors Integral ObservatioN) intends to establish a complex geophysical observatory in one of the most seismically active areas of Mexico – the state of Guerrero including measurements at all levels from underground to space to provide a dataset of simultaneous multiparameter measurements for direct checks on existing physical models of seismo-ionospheric coupling. The second site of experimental survey will be organized in the vicinity of the Popocatepetl volcano to study the precursors of volcanic eruption.

The Pacific coast of Mexico is one of the most seismically active areas on our planet. The boundary between the Cocos and North American plates is the locus of numerous subduction earthquakes driven by a convergence velocity of 5–7 cm/yr. Most of the plate boundary has experienced ruptures in historic time – with the exception of the Guerrero Coast. The Guerrero region of Southern Mexico has accumulated more than 5 m of relative plate motion since the last major earthquake. Currently, the segment with the largest deficit in seismic energy release, the "Guerrero gap", is also the nearest to Mexico City (population ~20 million). Since the most recent large (M_s =7.6) earthquake in 1911, >5 m of plate convergence has generated only a few M_s~6 events near the edges of this segment. The next major interplate earthquake in Guerrero could have moment magnitude M_w=8.1–8.4 (Suarez et al. 1990; Kostoglodov et al. 1996).

It is intended to equip the observatory with measurements of DC electric and magnetic fields, ELF and VLF electromagnetic fields, optical measurements of atmospheric emissions; HF electromagnetic emissions, over-horizon propagation of VHF electromagnetic waves, radon and aerosol monitoring at surface level, use the data of the GPS receivers network for GPS TEC measurements and mapping, use the ground based ionosonde data, to use the satellite data from the dedicated satellites, such as DEMETER, Vulkan etc. (the satellite systems will be described below). The satellite thermal images of the area will be used as well. The satellite data on the ionosphere, thermosphere and atmosphere parameters such as DMSP, TIMED, CHAMP and the planned COSMIC project will be used as well (Pulinets et al. 2004b).

The project started in the second half of 2003, its status and main results can be accessed at the WEB site http://tonatiuh.igeofcu.unam.mx/~pulse/ projects/PREVENTION/.

6.6.2 Dedicated Satellite Missions for Earthquake Precursor Registration from Space

At present moment several countries in the World have already launched dedicated satellites or are in the stage of their preparation. They are: Russia, France, USA and Ukraine. Taking into account that the space projects are not possible without the support from state agencies, this means that the issues of earthquake prediction are not so hopeless as was thought at the end of the 90-th.

6.6.2.1 COMPASS Satellite, Russia

COMPASS (Complex Orbital Magneto-Plasma Autonomous Small Satellite) was developed in Russia by the State Rocket Centre "V.P. Makeyev Design Bureau" (satellite design) and the Institute of Terrestrial Magnetism, Ionosphere and Radiowave Propagation (IZMIRAN) of the Russian Academy of Sciences (scientific payload). The satellite contained the following equipment onboard:

- ULF/VLF wave complex (0.5 Hz–20 kHz);
- HF radiospectrometer (0.1–10 MHz) ;
- FM-3 K fluxgate magnetometer;
- GID-12T (GPS receiver);
- Two-frequency HF receiver.

Regardless of the successful launch on 10th December 2001, due to a failure of the telemetry system, the information from the satellite was not received. Presently a second version of the satellite is being prepared (COMPASS-2) to be launched in December 2004 (Oraevsky et al. 2003).

6.6.2.2 QuakeSat, USA

Regardless of the strong opposition to the electromagnetic precursors in the United States, the private QuakeFinder company (Palo Alto, California, http://www.quakefinder.com/) together with Stanford University and California Polytechnic State University designed the nano-satellite, which was launched 30 June 2003 from Russian Plesetsk Rocket Range by Eurockot launcher together with several other small satellites attached to the Breeze booster. The satellite was launched on a high inclination quasi-circular orbit of 840 altitude. As a construction it uses the Cubesat conception of incredibly small size 10 cm×10 cm×10 cm. The magnetometer installed onboard the satellite is intended to register the ULF and ELF emissions associated with earthquake preparation as was registered by the Russian Cosmos 1809 satellite (Chmyrev et al. 1989, 1997). Regardless of the intention of the authors of the project "to provide earthquake warning data to government officials, with an increasing level of warning on the order of weeks, days, and hours, prior to a major earthquake" it seems to be very doubtful. If one carefully looks at the papers (Chmyrev et al. 1989, 1997) there are only one or two examples of the ULF emission registration presented without any statistics. No technology for earthquake identification or epicenter finding were developed. The authors intend to combine their satellite measurements with the magnetometer network installed in California basing on the results of Fraser-Smith et al. (1990) for the ULF emission registered before the Loma-Prieta earthquake M=7.1. More detailed information one can find at http://ssdl.stanford.edu/LM-CubeSat/Team4/qs_august_15_full_report.htm.

From our point of view this mission may be regarded more as a technological one then a real application for earthquake prediction. As we mentioned above, one parameter only will never give confidence to the real precursory event.

6.6.2.3 DEMETER Satellite, France

The history of this project is very long and has its roots in the end of the 80-th when the first optimistic view was created after new results obtained onboard Intercosmos-19, Intercosmos-Bulgaria 1300, Cosmos 1809 and Aureol-3 Soviet satellites. It was decided to create dedicated satellites (the two projects appeared in Russia: Preduprezhdenie (Warning) and DEMETER. One can even find these projects in the historic plans of the Russian Space Agency. But due to a lack of financing in the middle of the 90-th, the projects (where, of course, the international cooperation was provided) migrated to partners: the Warning mission – to Ukraine, and DEMETER – to France. Probably the name of the project (Demetra – the goddess of fertility in Greek mythology) saved it from syncope, and it is ready for launch on 29 of June of 2004 (probably when the book will be published, the satellite will be already in orbit: http://smsc.cnes.fr/DEMETER/index.htm). One can find the main objectives of the projects, their main characteristics and scientific payload description in Parrot (2002). The scientific payload is composed of several sensors: three electric and three magnetic sensors (six components of the EM field to investigate from DC up to 3.5 MHz), a Langmuir probe, an ion spectrometer, and, an energetic particle analyzer. They are associated with two special items: a large onboard memory (8 Gb) in order to record data all around the Earth, and a high bit rate telemetry in the X band (18 Mb/s). There are two modes of satellite operation provided: a survey mode to record low bit rate data and a burst mode to record high bit rate data above seismic regions. In the survey mode the telemetry is of the order of 950 Mb/day, and in the burst mode, it is larger than 1 Gb/orbit.

This mission looks to be the first candidate for successful missions in terms of earthquake prediction from space. It has almost all the components necessary to register the space plasma precursors: VLF emissions, electron density (only local), ion composition and energetic particle precipitation. The only thing lacking at the satellite – is the measurements of electron concentration in the *F*-layer peak – the most sensitive parameter. It was possible to detect the critical frequency with the HF radiospectrometer scaling the HF noises from broadcast transmitters penetrating into topside ionosphere at frequencies higher than the critical frequency as is proposed in Pulinets et al (2002c), but the frequency of the receiver installed onboard the DEMETER satellite is too low (3.5 MHz) to measure the critical frequencies. Probably, this device is designated to measure only the local density with the plasma resonance technique.

6.6.2.4 Ukrainian Space Projects

At the WEB page of National Space Agency of Ukraine one can find two projects connected with earthquake precursors studies. The first one – is the "Warning" mission, which was adopted from Russia. In its primary state it provided the launch of two satellites (one of AUOS class, and the second one – a small sub-satellite). The main satellite carried mainly the payload for electromagnetic emission registration within different frequency bands and local plasma parameters study. The main purpose of the sub-satellite – to put in orbit the topside sounder TOPADS developed by the Center of Atmospheric Research, University of Lowell, Massachusetts. But due to financing limits the project is suspended.

The second mission, "Variant" *F*, is a multipurpose mission to study the Solar-Terrestrial relations. But one of the scientific goals of the missions is also to measure the electromagnetic emissions associated with the preparing earthquake. It was planned to launch the payload within the frame of Sich-1 M project in 2002, but the satellite is still not launched yet (http://www.nkau.gov.ua/nsau/catalogNEW.nsf/).

6.6.2.5 Japanese ELMOS Project

From 1994 there has been discussion in Japan on a project for electromagnetic monitoring from satellite associated with seismic activity (Kodama 1994). The proposal is a compromise between the electromagnetic emissions measurements and local plasma parameters (local plasma frequency). A key of the ELMOS mission is that the directional search of the anomaly source is by three-component measurement of electric and magnetic field:

- Fluxgate magnetometers: three-component measurement of magnetic field
- Dipole antennas: Three component measurement of ELF/ULF/VLF electric field
- Electron density monitor: plasma frequency measurement
- Lightning Imaging Sensor: optical monitoring of lightning distribution

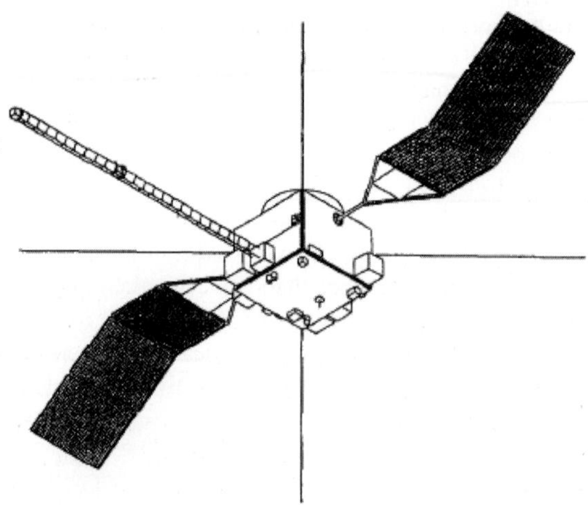

Fig. 6.21 The Japanese ELMOS satellite outlook (Kodama et al. 2000)

The satellite orbit altitude is planned to be 400–500 km with 100% duty cycle of the satellite activity and mission lifetime 2 years. The satellite outlook is presented in Fig. 6.21 (Kodama et al. 2000).

6.6.2.6 The SSTL Small Satellite for Earthquake Prediction

Surrey Space Center of the University of Surrey, UK, prepared the preliminary design of the dedicated microsatellite for ionospheric precursor research (Jason et al. 2003). The proposed platform is a 100 kg class low-cost SSTL platform, which has on-board data storage, three-axis stabilization and moderate downlink data rates to support the payloads. A 70°–85° inclination orbit between 800 and 1,000 km in altitude is baselined for the mission to enable good measurements of the topside ionosphere. Using only a single SSTL S-band ground station would enable roughly 50 min worth of payload data to be downloaded per day. The mission return may be significantly increased through use of additional low-cost ground stations. This mission solution is a balance between performance and cost. The proposed platform is based on the TopSat satellite (Wicks et al. 2001), which is jointly funded by the British National Space Centre (BNSC) and UK Ministry of Defense (MoD) and is currently under construction at SSTL. The platform has a payload mass allocation of 25 kg. A satellite payload is given in Table 6.4.

Table 6.4 SSTL Earthquake science research satellite payload

Payload accommodation	Comments
Topside sounder	Antennas: 15 m tip-to-tip > 5 MHz; 50 m tip-to-tip < 5 MHz; main antenna configuration options: three orthogonal dipoles, two V-shaped dipoles, two crossed dipoles. Based on the IS-338 flown on Intercosmos-19.
Mass-spectrometer	Mounted on platform, inlet direction along the satellite velocity vector
Magnetometer	Deployed on 2-4 m boom to minimize disturbances from the spacecraft

A swept frequency topside sounder is the main payload and 15–50 m tip-to-tip antennas would be mounted on the Earth-facing facet of the spacecraft as illustrated in Fig. 6.22. The mass spectrometer will be mounted so that the inlet is directed on the satellite velocity vector. The magnetometer would be deployed on an instrument boom, which is mounted on the space-facing facet of the spacecraft. More details on preliminary spacecraft accommodation are given in Table 6.4. Taking into account the spatial scale of the modified ionosphere region, as was discussed in Chapter 5, is of the order of 30°–40° in latitude and longitude for strong earthquakes, the seismo-ionospheric variations could be registered on two neighboring orbits which will separated by 25° in longitude from each other. So, typically up to four passes per day will be achievable for a given seismic zone. A two-year mission lifetime is targeted to provide some continuity in the collected data. The payload mass allocation is 25 kg.

There is potential for additional payloads for measurements of electric field, VLF emissions and energetic particle fluxes.

This proposal is a compromise permitting the combination of the measurements of local plasma parameters with the main source on the ionospheric precursors up today – ionospheric topside sounder.

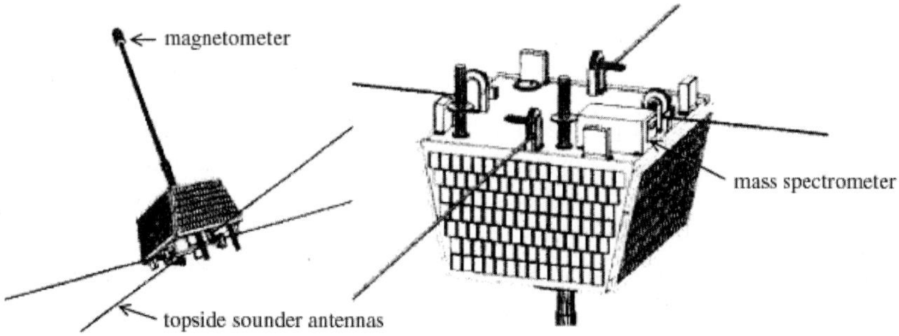

Fig. 6.22 Microsatellite concept for earthquake science research

6.6.2.7 Russian Satellite Constellation "Vulkan"

Among the satellite missions described here the "Vulkan" project is the most impressive and comprehensive one. It is not one, and even two satellites, but a satellite constellation developed for the global monitoring of natural disasters, including earthquake precursors monitoring. One can find a short description of the project in Oraevsky et al. (2002, 2003).

The system from the very beginning was directed towards the search for the ionospheric precursors of earthquakes as a main factor. From the discussion above it follows that such a system should consist of the following components:

- *Selection of seismically "sensitive" ionospheric parameters and methods of their measurements*
- *System of processing, visualization and identification of precursors*
- *Organization of an infrastructure of satellite and ground-based measurements and warning system*

What concerns the parameters to be measured onboard the satellite, basing on the retrospective analysis of the existing data and obtained results, is necessary to provide the following set of measurements of the ionospheric parameters:

- *Measurements of density in F-layer maximum (critical frequency)*
- *Measurements of height of F-layer maximum*
- *Measurements of vertical profiles of ionization (both for topside, and bottomside ionosphere)*
- *Measurements of the total electron content (TEC)*
- *Measurements of ion composition*

- *Measurements of electron temperature*
- *Measurements of ULF-ELF-VLF emissions*
- *Measurements of DC electric field*
- *Measurements of energetic particles fluxes*
- *Measurements of optical emissions*
- *To ensure a combination of satellite and ground-based measurements to have an opportunity to identify variations with local time, global and regional variations.*

Taking into account that the topside sounder and electromagnetic measurements with the receivers of high sensitivity are not electromagnetically compatible, it was decided to divide the system by two levels. The higher level with circular orbit 1,000–1,100 km will be equipped by the topside sounder and local plasma parameters measurements, and the lower one with the orbit altitude near 500 km – by the wave equipment of high sensitivity. The two levels of the system were selected also by the requirements to the measured parameters. To receive altitude distribution of electron concentration in topside ionosphere, it is necessary, that the satellite be above a maximum of electron concentration. At middle latitudes at low solar activity the maximum settles down at heights about 250–300 km. However in the area of the equatorial anomaly the peak height (depending on the degree of development of the equatorial anomaly) may reach 450–500 km. Cases are known, when the peak height within the equatorial anomaly during high solar activity periods was registered at heights about 700 km. To determine the law of electron concentration change with height (a scale height), it is necessary, that the satellite was higher at least by 100 km than the maximum of the ionization, therefore the minimal allowable height for the satellite with an ionosonde onboard is 800 km, but usually ionosondes are started on heights of 900-

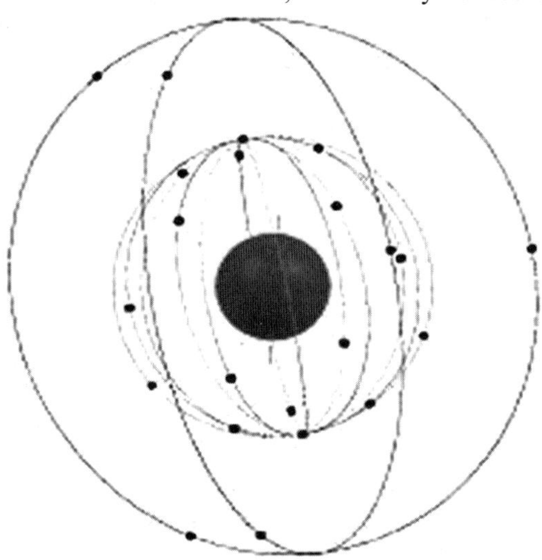

Fig. 6.23 Schematic presentation of the "Vulkan" satellite constellation

On the other hand, for optical measurements including radiometric measurements of surface temperature, and also for ELF/VLF measurements and other electromagnetic radiations connected to processes of preparation of earthquakes, it is desirable that the satellite is in a lower orbit. It is necessary to provide the orbit avoiding its essential braking in an atmosphere to increase the satellite lifetime. Such a compromise height is represented by the orbit height ~500 km. The two levels system was proposed first within the framework of the "Predvestnik" project (Chmyrev et al. 1998), which was not implemented, and then the idea was adopted by the "Vulkan" project.

The high inclination orbit selection is connected with the precursors behavior in local time. The satellite and ground-based studies (for example, precursors mask), have demonstrated that precursors may appear several days by order at the same local time, and the sign of the variation depends on local time. Besides, there are sectors of local time in which the amplitude of seismoionospheric variations is maximal, and exist such in which these variations are shown poorly (for example, the most pronounced precursor in Taiwan is registered for afternoon hours). Hence, the most preferable would be a configuration from sun-synchronous placed in sectors of local time, the most sensitive to seismoionospheric variations. Depending on launch possibilities the high inclination orbit ~83° will be quite satisfactory. For such orbits the change of local time within 5 day (the period of registration of precursors for the given earthquake) will be insignificant.

The choice of the number of satellites at each level is dependent on the necessity to provide the minimal probability of missing the target (earthquake precursor), and also to provide sufficient resolution in space for mapping the areas in the ionosphere, modified by seismic activity. As was demonstrated in Chapter 5, the longitudinal size of the modified area in the ionosphere for strong earthquakes at the peak altitude of electron concentration is from 10° up to 40° depending on the magnitude. At the same time the longitudinal interval between the neighboring orbits of the satellite makes ~25°. Thus, for the largest areas to receive even one readout not at the very edge of the modified area, it is necessary in the given longitudinal interval (or sector of local time) to have in an orbit at least two satellites. Here a compromise should be achieved between the optimal requirements and the system cost. If we take into account the ascending and descending parts of the satellite orbit, the precursor duration at least 100 min (duration of the satellite orbit) for the utmost case it would be necessary to have three satellites with orbits equidistant in longitudes. To have better resolution it was decided to have four satellites on the level.

The system of precursors recognition should work in the following mode. The information from satellites and from ground-based measurements should be collected in the processing and recognition center together with geophysical information on a condition of near-Earth space, solar and geomagnetic activity. In the first term the geophysical disturbances (such as prominent solar events, geomagnetic storms) should be traced to clarify the seismic information from interference. Then all variations according to algorithms of definition of precursors should be exposed to multi-factor analysis, which takes into account the set of observable variations, area of observation, and prehistory of earthquakes in the observable region. It is possible to construct the program of satellite activity in such a way that the information with a greater sample rate was collected above seismo-dangerous regions, in particular, with occurrence of variations satisfying the features of a precursor. In the case of an increase of the amplitude of precursors registered on all channels, it is necessary to give out a warning signal.

At the present level of our knowledge concerning precursors of earthquakes it is necessary to simultaneously work on seismic warning and to continue the fundamental studies of the effects of preparation of earthquakes, as well as their displays in various geophysical fields. All the mechanisms of formation of precursors are not completely understood, an additional volume of information is necessary to come to more certain conclusions, which can be received only as a result of expansion of a new effective system both of ground and satellite control of the precursors of catastrophic earthquakes.

It is intended to launch the first satellite of the "Vulkan" constellation in the year 2006. Then the system will be updated up to year 2010. We stay in hope that this system will be really developed. Presently, the parameters of the basic satellite for the "Vulkan" constellation are the following:

Table 6.5 Main parameters of the "Vulkan" basic platform (Salikhov et al. 2003)

Satellite parameter	Value
Payload mass	85 kg
Satellite mass	202 kg
Average daily payload consumption	90 W
Peak power consumption	210 W
Supply voltage	27 V
Attitude control accuracy along orbital coordinate system(OCS) axes	20 ang.min
Attitude control stability per each axis	0.005°/s
Accuracy of the S/C orbit position determination	± 70 m
Guarantee mission lifetime	7 years

The satellite outlook is presented in Fig. 6.24. The platform will be produced by the Research Institute for Electromechanics (NIIEM).

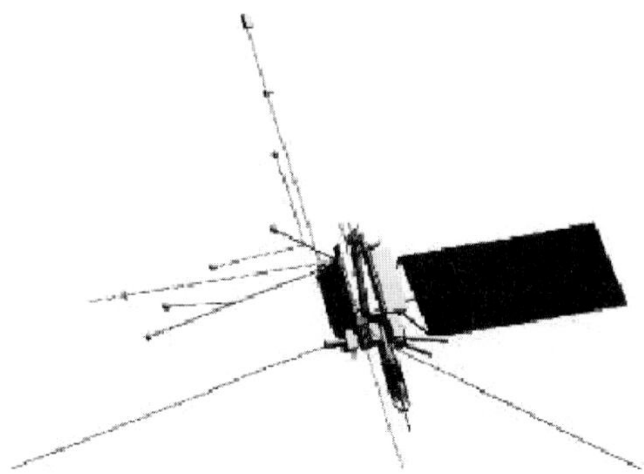

Fig. 6.24 External view of the "Vulkan" constellation satellite

6.6.2.8 *Italian "ESPERIA" Mission*

The latest international mission that has been announced is the "ESPERIA" project designed in Italy (Sgrigna et al. 2003). Its overall dimensions are is similar to the proposed SSTL mission (approximately 100 kg), Fig. 6.25. It has very promising payload consisting of:

- An Electric Field Analyser (EFA) in the frequency band DC – 10 MHz
- A Magnetic Field Analyser (MAFA) consisting of two magnetometers: fluxgate (DC--10 Hz) and search-coil (10 Hz--100 kHz)
- A Langmuir Probe & Retarding Potential Analyser (LP&RPA) to measure the local electron and ion temperatures and concentration
- A Particle Detector Analyzer (PDA) to measure the fluxes of energetic particles within the energy range 300 keV--2 GeV with high accuracy of pitch angle determination <4°

The satellite will have the geostationary circular orbit of 813 km altitude and 11.5° inclination. It means that satellite will only be able to register effects from near equatorial earthquakes. By limiting the latitudinal range of the mission, the authors of the project have complicated their task by the fast changes of the local time which are characteristic of the low orbiting satellites. In this configuration, distinguishing between the spatial, temporal and the local time variations of the precursors becomes too complicated. In contrast, such an orbital configuration gives better longitudinal resolution within the frame of one satellite mission which is impossible for higher inclination satellites. To have simultaneously in orbit DEMETER, Vulkan and ESPERIA will be a really great international constellation. The satellites will supplement one another in their orbital coverage and the space plasma parameters measured and having, as their common purpose, the earthquake precursors registration from space.

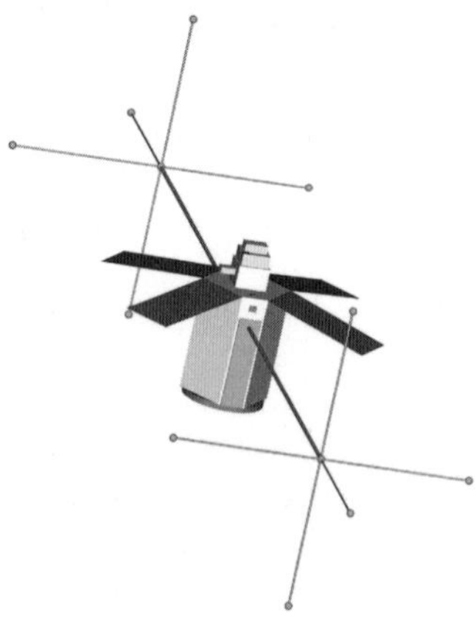

Fig. 6.25 External view of the ESPERIA mission satellite

6.7 How to use Ionospheric Precursors in Real Prediction?

We have discussed the physical model of ionospheric precursors, their main morphological features, statistical parameters, and existing techniques for identification of ionospheric precursors. Now we put the question: are we really ready provide earthquake prediction with this technique?

Official IASPEI procedure's for the evaluation of precursors, as well as for earthquake prediction evaluation exist (Wyss and Booth 1997). And before we submit officially the prediction procedures using ionospheric precursors, one should estimate their ability, advantages and deficiencies. We are still unable to discuss probabilities because nobody has as yet used ionospheric precursors for real prediction, we are in a transition period from the understanding of the physics and elaborating the practical application of the techniques, but we are able to answer three simple questions: where, when, and how strong?.

6.7.1 Where?

Among the different types of precursors, the ionospheric ones have the unique ability of being able to indicate the approximate position of the earthquake epicenter.

Theoretical calculations show that the effect from the ground (the earthquake preparation zone) is mapped onto the ionosphere along geomagnetic field lines, i.e. depends on the geomagnetic inclination. That's why the technique of epicenter location should be divided into three zones: high latitudes, middle latitudes and low latitudes including the equatorial anomaly within the ionosphere. In high latitudes the anomaly within the ionosphere will be over the epicentral area as was registered for the Alaska earthquake (Fig. 2.7). In the middle latitudes effects in the ionosphere will be observed shifted to the south from the epicenter vertical projection in the Northern hemisphere and to the north in the Southern hemisphere. In addition, not everywhere the magnetic field lines are parallel to geodetic meridians and decline. This effect is observed for the Irpinia earthquake (Fig. 5.7). So the geomagnetic field configuration should be known for the given point.

As indicated by experimental results for the equatorial areas, due to the specific electrodynamics of the equatorial ionosphere, the effect of the earthquake preparation will modify the equatorial anomaly as a whole in some longitudinal range (Pulinets and Legen'ka 2002). The effect in the equatorial ionosphere looks like the anomaly crests shift to the magnetic equator up to the complete disappearance of the anomaly (Liu et al. 2002). Nevertheless, satellite mapping sometimes is able to detect the circle-like irregularity in equatorial ionosphere too as is shown in Fig. 2.8.

To estimate the precision of the future epicenter position determination, more satellite data should be processed.

The same procedure as for the satellite mapping data can be applied to the mapping with the GPS receivers network. If the number of GPS receivers is sufficient for mapping, even better results than with the satellite can be achieved. The satellite map – is only a snapshot, but the GPS receivers are able to provide continuous monitoring of the ionosphere in real time. They can track the dynamics of the modified area and detect the moment of its maximum development which is very important for the estimation of magnitude.

In the case of the ground-based ionosonde, the epicenter position can be determined from the mask. As we mentioned above, for the fixed relative position of the ionosonde and the epicenter, the same kind of precursors mask will be registered.

This means that for the different epicenter positions different masks will be obtained. So, from the mask outlook one can conclude the direction and approximate position of the epicenter. For this procedure retrospective data analysis is necessary to obtain the masks for different sources of seismic activity. So, for the future epicenter position determination all three techniques of ionospheric monitoring are applicable: satellite topside sounding, GPS TEC monitoring and ground-based vertical sounding.

6.7.2 When?

The answer to this question from the point of view of the ionospheric precursors has only one requirement: the precursor identification. If the decision is made that the observed variation is the ionospheric precursors, then the alarm could be raised for the expectation period of 5 days (see Fig. 5.1).

Taking into account that the 5 days were obtained statistically, and there are cases of larger time intervals, maybe we can increase the expectation time up to 7 days. Probably, for every seismically active area this interval can be determined empirically from the historical data, but at least, the ionospheric precursors give a quite reasonable expectation time: not years, not months, not weeks, but only from 1 to 5 days. This is also the great advantage of the ionospheric precursors.

6.7.3 How Strong?

First of all, we should remind the reader that the ionospheric threshold exists: the ionosphere starts to "feel" the earthquake preparation from magnitudes of the order of 5. So, if the ionospheric precursor is identified, one should expect an earthquake with a magnitude of 5 or higher. To estimate the intensity of the approaching earthquake one can use the size and intensity of the anomalous area within the ionosphere from the satellite topside sounding mapping pictures (or from the maps produced by the GPS receivers network). Intensity means, how strong is the deviation from undisturbed conditions. This parameter should take into account the precursors dependence on the local time. For example, in the Taiwan area the strongest negative precursory deviations are observed for the afternoon hours. The intensity with the help of ground-based ionosondes can be estimated only if the source position is known because the results of ionosonde registration will be different for an earthquake with the same parameters but different relative positions of the ionosonde and epicenter. But for the fixed position of the ionosonde and epicenter the intensity and duration of the ionospheric anomalies can be recalculated in the earthquake magnitude.

From Fig. 6.11 where the deviation of the critical frequency for the Irpinia earthquake is shown, we can estimate the size of the modified area as ~ 1,800 km, radius ~ 900 km which gives (from 1.8): M=[log(900)]/0.43=6.9. This is exactly the magnitude of the Irpinia earthquake. Here, of course for the satellite picture one should be sure that the size of the area was determined in the maximum phase of the ionospheric irregularity development.

In procedures proposed above our task was not to elaborate the exact procedure, but only to indicate the direction one should go, to obtain this procedure.

6.8 Conclusion

Concluding the paragraph we can state that ionospheric precursors have indisputable advantages over many others, for example, seismo-electromagnetic emissions. They have a high range of detection but with the ability to determine the position of the epicenter. The EM emissions are low-range (up to 150 km) and are not able to determine the epicenter position except with the installation presented in Ismaguilov et al. (2001). In general from the EM emissions one can only make one conclusion: something happens, but where, when, and what intensity the earthquake may be – this is terra incognita for this technique.

The ionospheric precursors have a fixed time expectation interval contrary to EM emissions which are observed from several months before the seismic shock up to a few minutes without any determined temporal scale. It looks as if every reported case has its own temporal scale.

And finally, some techniques were demonstrated for future earthquake magnitude estimation from the ionospheric data. The ionospheric processes have a fixed threshold of their appearance (from magnitude ~ 5), and the intensity and area of the modified ionospheric region are the measure of the future earthquake magnitude. Of course, nobody neglects the validity of other types of precursors, but it seems, they would be the supporting factor in multiparameter analysis for the decision making after the ionospheric precursors are identified.

6.9 The Possible Ideal System for the Ionospheric Precursors Monitoring and Short-Term Earthquake Prediction

As described above, there are already several systems for precursor study in several different countries. Each of these systems has advantages and deficiencies. Let us try to imagine, using the words: "I have a dream", how one should look for the most comprehensive system for precursors monitoring.

Regardless of the probable physical mechanism we describe here, we do not try to claim that it is unique and is in a final form. The cognitive process has no end, so together with practical application, the system should have components of parameters, key for the physical models of the processes participating in the seismoionospheric coupling.

So the system starts from the geochemical precursors monitoring, including the radon concentration, other gaseous components monitoring (CO_2, CH_4, helium, hydrogen, etc). The aerosols chemical composition and sizes should be monitored also using the Lidar and spectroscopic measurements. Traditional geopotential, magnetotelluric measurements, water level, gravimetric and magnetometric measurements should be used as well.

The next level from the ground surface – is the measurement of all atmospheric parameters: temperature, pressure, relative humidity, wind, precipitation, level of turbulence etc. As one of the key parameters the vertical electric field should be measured (preferably by a network of field mills to determine the scale of the electric field disturbances). The vertical profile of the electric field and air conductivity are also very important parameters. The use of tethered balloons or airplanes are possible options.

Then ULF-ELF-VLF-HF emissions should be measured. Except for their precursory abilities their physical mechanism should be clarified. Up to now too much hypothesis exists, which means that the mechanism is still unknown. According to other radiophysical measurements, the electromagnetic emissions should be studied not only from the point of view of locally generated emissions but to study the propagation effects such as VLF navigation transmitters signals distortion, or VHF over-horizon propagation.

The next level – is the measurement of atmospheric emissions from the ground including the oxygen and nitrogen lines, as well as hydroxyl emissions. The bottom ionosphere parameters should be studied by a ground-based ionosonde, including Doppler measurements to study the movements of the ionospheric layers. It is desirable to have Lidar for measurements of the metallic ion layers in the *E*-region. In the "Predvestnik" project it was proposed to have an intermediate level, using airborne measurements. If such a possibility exists, it would be very useful because it is very difficult to monitor layers between the ground and the satellite's orbit. Balloons and airplanes – this is a good option. Study of vertical profiles of physical model parameters is important for the model's verification.

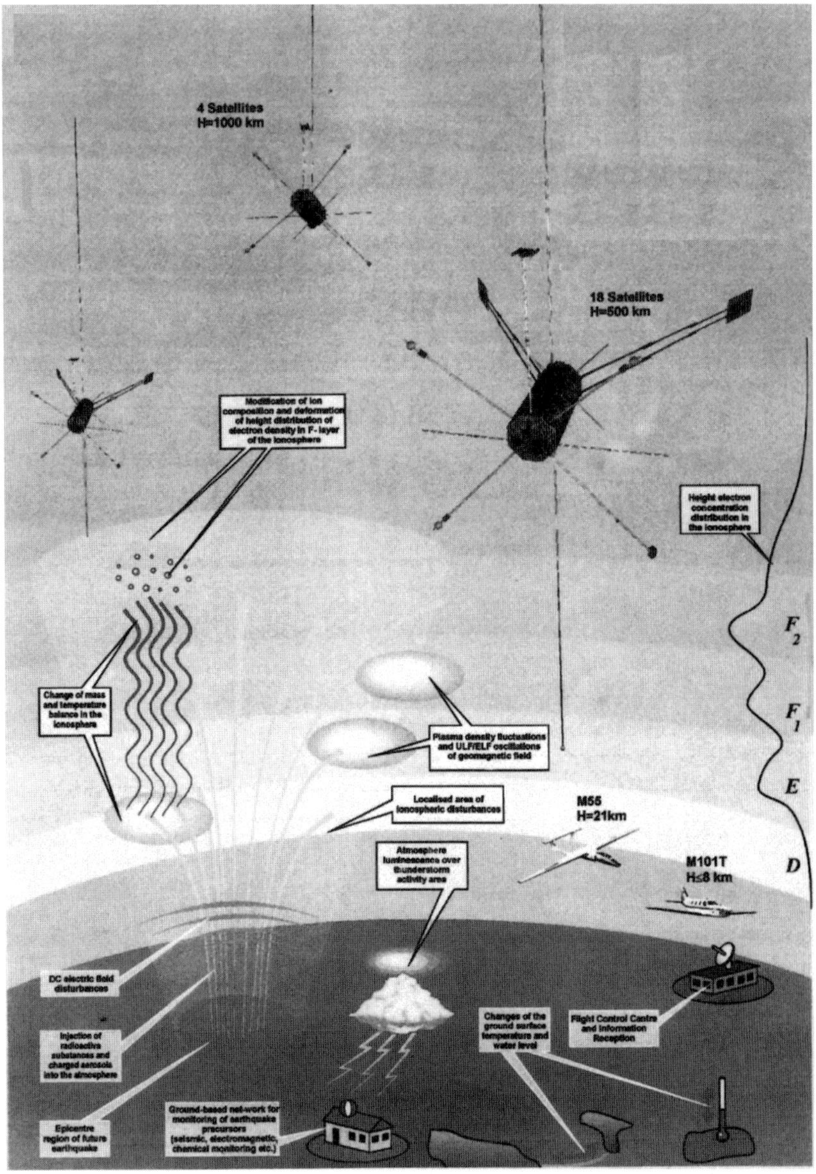

Fig. 6.26 Schematic presentation of the complex system of the earthquakes precursor monitoring (after Chmyrev et al. 1998)

The GPS receivers network is one of the most important components of the system. If the possibility exists to have the receivers within the seismically active area at least at a distance of 100 km one from another, it will present an ideal opportunity for ionospheric mapping. Together with the ground-based ionosonde it will give the possibility to have not only the total electron content but also to reconstruct the vertical distribution of the electron concentration.

And finally, the most complex element of the system – is the satellite constellation which was described earlier. It may be used not only for the given region where the ground-based monitoring system is developed but globally, including the hard-to-reach areas. It can also be used by the countries that are not able to develop such comprehensive and expensive ground-based systems.

A very important factor is data acquisition, processing, visualization and interpretation. To collect information – is only the first step. Correctly to interpret it, to identify the precursors – this is a kind of art, which has only recently developed (not in regard to traditional precursors but the ionospheric and electromagnetic ones).

The picture presented here, of course, has elements of subjectivity and author's predilection. But at least it gives to reader a flavor of this new direction in short-term earthquake prediction. We do not give a definite answer to the question, are we ready for real prediction or not? We only want to provide the vector of movement.

Our opinion is the following: we should not wait until all the things in the physical mechanism are clarified. We demonstrated that many parameters of the ionospheric precursors might already be used in for practical application. So we should use them (though probably not advertising our efforts too much before more certain results are obtained). But try to move in this direction, and don't whine that it is very difficult and hence impossible.

Chapter 7 Ultimate Results, Unexplained Phenomena, Future Tasks

The work on this book has taken quite a period of time. During this period some new ideas and results have appeared. We still have time to include them in our book, but not to rewrite the chapters already prepared (because it might be endless work); so we have decided to write this additional chapter which includes the following items.

First, we present new results just obtained and not yet approved from international conferences and journal publications.

Second, we would like to describe some results relating to the book thematic but which presently have a controversial character and are not accepted yet even by the group of scientists working in the field of electromagnetic and ionospheric precursors of earthquakes.

And the third part of the present chapter will be devoted to a question as yet not resolved, and to tasks which will determine the development of this direction of science in the nearest future.

7.1 New Results on the Pulsed HF Electromagnetic Emission Registered Before Earthquakes and its Possible Nature

7.1.1 Introduction

As often happens with new findings the reported emission was disclosed in this case by Australian radio amateur P. Hollis-Watt; listening in his car to commercial radio in the amateur band frequency 144–148 FM band. He was considerate enough to note that 24 hours later a strong (M=5.9) earthquake hit Newcastle in New South Wales, Australia. The recollection of the first event came when in October 2001 a similar anomaly was registered by him in Esperance, Western Australia, (even on the FM music band at 104.75FM) and 14 hours later the Earth trembled under Esperance (the epicenter was 200 km west at Ravensthorpe). Having had 35 years experience as a radio amateur, TV electronics engineer, ex RAAF avionics technician and part time post trade industrial electronics teacher he started to record the suspicious signals using transceivers tuned to different frequencies. His results will be described lower.

Going back to the history of HF emission registered before strong earthquakes we should mention as a first case the registration of anomalous signals on frequency 18 MHz before the Catastrophic Chilean earthquake (May 22nd, 1960), by the network of radio receivers created to register cosmic radio noise (Warwick et al. 1982). The receivers input circuits were designed in such a way that they eliminated the pulsed emission, so, they registered only an increase of the overall level of emission on the receivers' frequency 18 MHz. Later, the attention of researchers was attracted mainly to the LF, VLF and ELF frequency bands (Hayakawa et al. 1994). Nevertheless, in the VLF frequency band a pulsed character emission was observed also associated with seismic activity (Sobolev and Husamiddinov 1985). There is also a report on the emissions registered during the time of the Spitak earthquake (Voinov et al. 1992) at the frequency 75 MHz. Recently the new communications on the HF radio emissions associated with anticipated earthquakes appeared (Vallianatos and Nomikos 1998; Maeda 1999) which is evidence of the existing interest to HF EM phenomena, associated with seismic activity. Maeda (1999) also reports the pulsed emission on the frequency 22.2 MHz but only positive polarity and only around the time of the Kobe earthquake (40 min. before the shock and 1.5 hours after it).

Starting from the first publication of Warwick et al. (1982) rock fracture and the associated piezoelectric effect were proposed as a main explanation of the observed emission. This hypothesis finds support in the direct measurements of EM emission in mines in Australia (O'Keefe and Thiel 1999) but again the pulsed emission was recorded only up to the frequency of 20 kHz, and not in the HF frequency band. The comprehensive review of Parrot et al. (1993) of the HF seismo-electromagnetic phenomena provides many possible physical explanations, but the use of the term "high frequency" is very limited: it is used not in the sense accepted in plasma physics and radio wave propagation but in relation to the frequency of seismic oscillation, i.e. again we deal with the VLF signals.

The results reported in the present chapter have a specificity discriminating them from other kinds of HF emissions observed before: they have fixed shape (wavelet), they are registered not on one frequency but in a wide frequency band (132 kHz–149 MHz), and they are registered before the seismic shock (12–24 hours).

The large distances from which the emission was registered (up to 1,500 km) implies the conclusion that the source of the emission is elevated over the ground surface. We also cannot assume that the emission was reflected by the ionosphere because of the high frequency registered (up to hundreds of megahertz).

7.1.2 Experimental Setup

The measurements of the HF pulsed emission (to distinguish it from other kinds of emissions reported before we will call it the P-H emission, i.e. Pulsed High frequency) were carried out in Western Australia not far from the Burakin seismically active area. One can see in Fig. 7.1 the corresponding map showing the seismic area as a star and arrows showing the receiver sites. The seismicity spatial distribution from 1966 up to now for magnitudes higher than 2 is shown in Fig. 7.2. The dark cluster in the center of the picture is the Burakin area. The measurements were conducted not only at the sites indicated in Fig. 7.2 but directly within the Burakin area during a special expedition in the Burakin area in December 2002. During the expedition to the Burakin area the signal was recorded along the way and directly in the region.

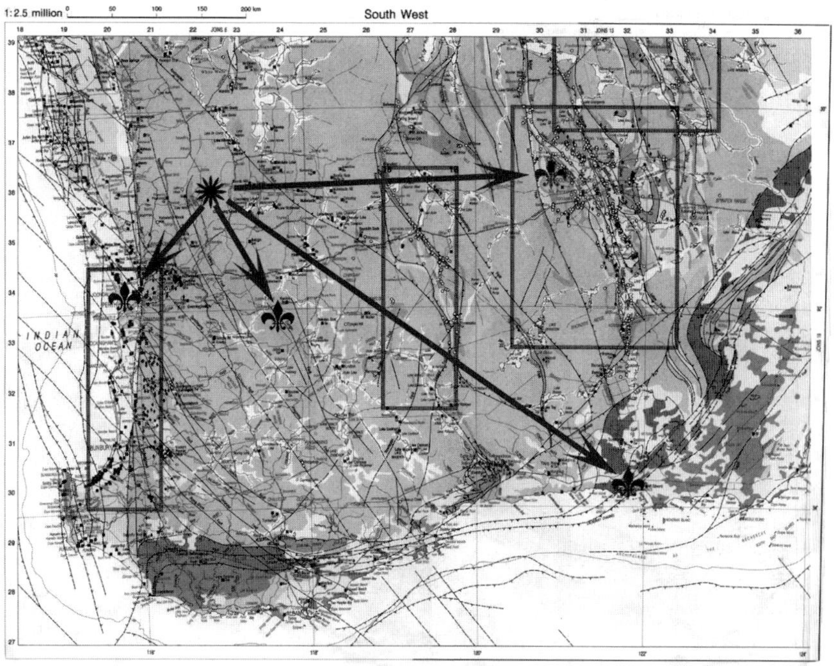

Fig. 7.1 Map of Western Australia. Burakin seismically active area is marked by *asterisk*. The *arrows* show the direction to the observation sites designated by *lilies*

The measurements were carried out on the following frequencies; 129–132 kHz, 2.3xx MHz, 4.2xx MHz, and 148.060 MHz during October 2001 – December 2002. The measurement installation diagram is shown in Fig. 7.3. For the different frequencies different professional standard amateur receivers (transceivers) were used: Yaesu VX1R (Transceiver rx 500 kHz to 999 MHz AM, FM (wide, narrow), Icom 726, all mode 100 Wssb, AM, FM CW), and the Tandy

Icom 726, all mode 100 Wssb, AM, FM CW), and the Tandy 1000 channel scanner (25 MHz-1,300 MHz). Standard antenna (for the wide band portable) was a multiband transceiver 'rubber duck' (Yaesu VX-1R as well as FT23R) also a 10 element Yagi were used, the signal was primarily from the NW direction (at Esperance), vertical polarization was ~3 dB greater than the horizontal polarization; the best signal strength was at 45 degrees azimuth (using the 10 element Yagi) for the 148.060 MHz FM narrow (455 kHz), (there was no man made interference nor when the web site NOOA for the solar activity was scanned regularly it did not coincide with the P-H pulse activity). Although when the signal pulses/min was being processed the rates were seemingly altered by the sun and the moon's position in the sky above the Burakin seismic center, i.e.: from 1,200 P-H pulses /min with sun and or moon to 200 P-H pulses/min no sun and or moon. The sensitivity of these standard receiver/transceivers is ~ 0.5 μV 12 dB SINAD. The LF signal was acquired from the ICOM 726 transceiver in the FM mode at 129–132 kHz as were the signals at 2.3xx MHz and 4.2xx MHz (this was dependant upon various man made). There were several types of data processing applied. To obtain the quantitative characteristics of the signal (pulses rate, amplitude, etc.) a special counter was designed which permitted us to detect the P-H pulses and calculate them (observing the ratio of P-H pulses/background noise). The amplitude of the pulses was estimated with the digital oscilloscope and a frequency counter. For the posterior analysis the output signal was digitized with sampling frequency 44 kHz and recorded on the hard disk. There were periods when the records were made on one selected frequency. Multifrequency recording is being implemented. A schematic diagram of the experimental installation is shown in Fig. 7.3.

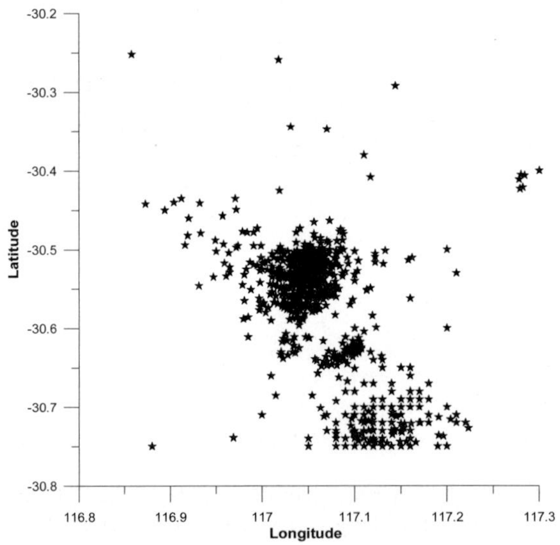

Fig. 7.2 The spatial distribution of the earthquake epicenters in the Burakin area for the period October 2001 – December 2002

7.1.3 Results of Measurements

An example of the shape of the P-H pulses is shown in Fig. 7.4a, and screenshots of low and high P-H activity in Fig. 7.4b and 7.4c respectively. The pulses rate varies from 300 per minute for a quiet period up to 2,000 12–24 hours before the earthquake. This rate depends also on the distance to the epicenter area. During the expedition it was detected that the rate and signal strength of the P-H pulses increased from LF-HF-VHF proportionally to being in the seismic area. The same tests were at 0 km with respect to Burakin, 200 km Burakin, 400 km Burakin, and 600 km Burakin, the rates of P-H pulses diminished away from the seismic area. Also whilst in a commercial aircraft flying over the area it was also observed that a larger number of P-H pulses were detected (no recordings were allowed to be made). The signal strength and the P-H rate were higher.

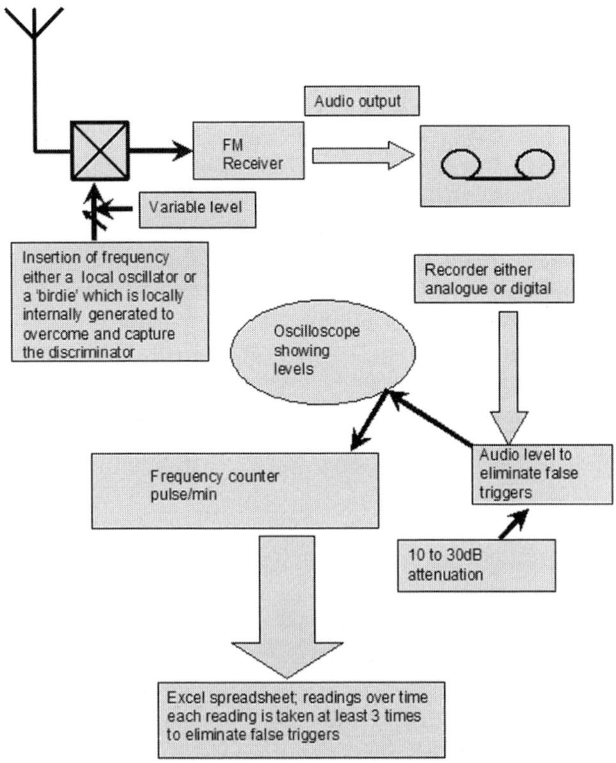

Fig. 7.3 Schematic diagram of the P-H pulses registration

One of the prominent features of the registered pulses is their stable shape regardless of changing pulse rate and intensity. As one can see from Fig. 7.4a the P-H pulse consists of two negative peaks, and one positive between them with a total pulse duration of 2–2.5 ms. This feature permits us to create the pulse wavelet for their automatic recognition and registration. Using the commercial Mathematica

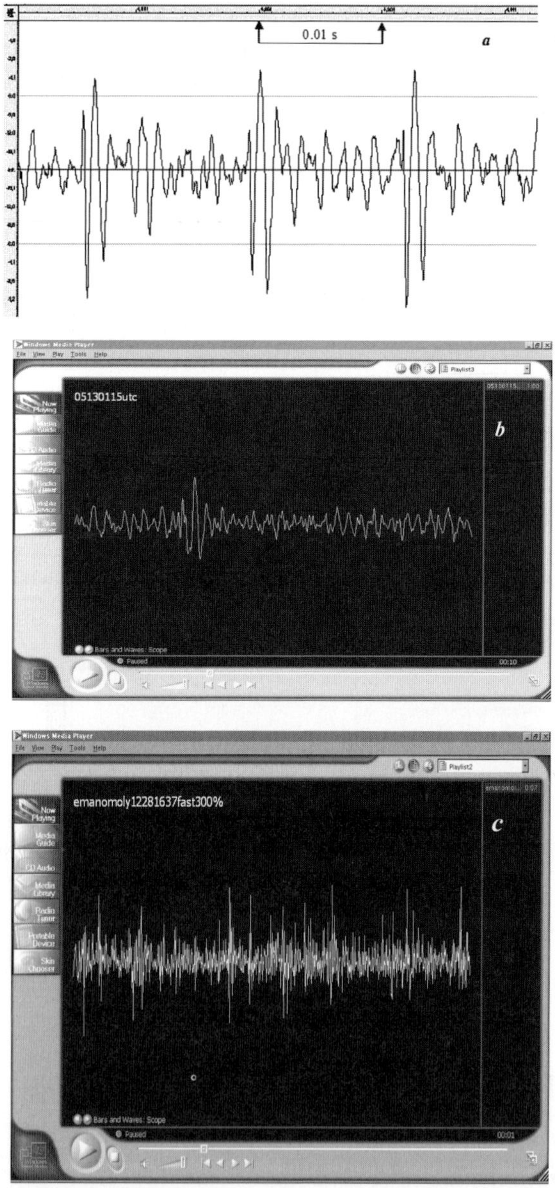

Fig. 7.4a–c **a** the shape of the registered P-H pulses. **b** Example of registration for the low activity of P-H pulses; **c** example of registration for the high activity of P-H pulses

4.1 software, with wavelet and signal analysis modules from Wolfram, the "mother" wavelet was sorted out which was determined to be a Daubechies Filter D12 (Daubechies 1988). Applying the wavelet analysis to the signal records with the determined filter will permit us to automatically identify the precursory signal, which is very important in the real time analysis of the data in probable practical application in the short-term prediction.

The software was created in the C programming language integrating with Mathematica 4.1 and its modules as well as a standard astronomy package under supervision of the computer science research laboratories at Edith Cowan University, Perth, Western Australia.

The correspondence of pulse rate daily changes and the consequent earthquake is shown in Fig. 7.5. As one can see, the technique is very sensitive even to the very small shocks like in the given example.

The distance to the epicenter of the earthquake probably associated with the registered emission was up to 1,500 km (sometimes the registrations imply the possibility of registering an even more distant earthquake in Australia).

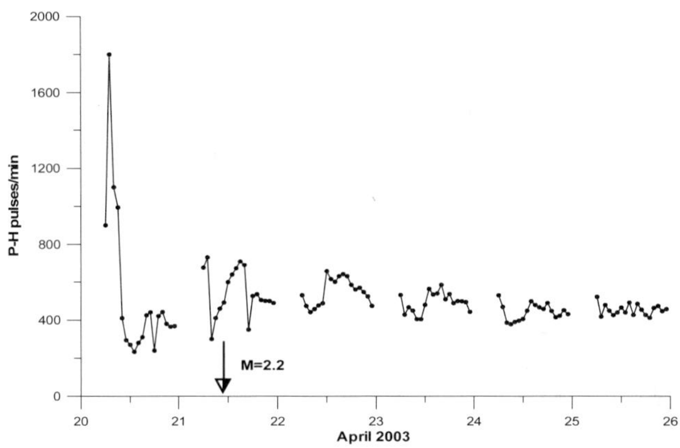

Fig. 7.5 Variation of the P-H pulses rate as dependence of the earthquake appearance in the Burakin area

7.1.4 Interpretation

The process of "ions aging" (as was described in Chapter 3) in ordinary air is finishing by the formation of composite ions – radicals $H_3O^+{\cdot}(H_2O)_m$, $NO_3^-{\cdot}HNO_3{\cdot}H_2O$, $NO_3^-{\cdot}(H_2O)_n$ etc. As a result of association of such hydrated ions – radicals, irrespective of the number of molecules of water in the neutral cluster shell, for example, such as one $NO_3^-{\cdot}(H_2O)_n{\cdot}H_3O^+$, their stability is determined both ionic and covalent coupling. Such ion clusters and free ions in turn can be linked in large dipole or quasi molecule. In this case we can regard as source of electromagnetic emission the rotationally – rotary transitions of the given dipoles.

We shall estimate the characteristics of these dipoles within the framework of the model of a rigid rotator. Let's estimate a particular case of a symmetrical whirligig with main electron and zero vibratory states. In this case the rotary constant of a molecule B_e (cm^{-1}) does not depend on a rotary quantum number J, and the radiated frequency of rotationally-rotary transition ν(cm^{-1}) of a rotary level J may be written:

$$\nu_J = 2 \cdot B_e \cdot (J+1) \tag{7.1}$$

Thus the rotary constant can be derived as:

$$B_e = \frac{\text{h}}{8 \cdot \pi^2 \cdot c \cdot \mu \cdot r_e^2} \tag{7.2}$$

where $\text{h} = 6.62 \cdot 10^{-27}$ erg·s – Plank constant; $c=3 \cdot 10^{10}$ cm/s – speed of light; μ – reduced mass of a molecule (g); r_e – equilibrium inter-nuclear spacing interval (cm).

The intensity of lines in the emission spectrum can be determined by the function:

$$I_J = N_J \cdot \frac{64 \cdot \pi^4}{3 \cdot c^3} \cdot \nu_J^4 \cdot R^2 \tag{7.3}$$

where R – the matrix element of electrical dipole moment of transition (is slightly dependent on J), and N_J – population of a J state, which can be expressed as:

$$N_j = N_0 \cdot (2 \cdot J + 1) \cdot \exp\left(-\frac{\text{h} \cdot c \cdot B_e \cdot J \cdot (J+1)}{k \cdot T} \right) \tag{7.4}$$

Here N_0 – population of a zero rotary state, k=1.38·10^{-16} erg/K – the Boltzmann constant, T – temperature of air (K). In expression (7.4) is taken into account that the relation of statistical weights of J and zero state are equal to (2·J+1).

The function N_J has a maximum on J at:

$$J_{\max} = \sqrt{\frac{k \cdot T}{2 \cdot B_e \cdot \text{h} \cdot c}} - \frac{1}{2} \tag{7.5}$$

Let's note, that the emission power (7.3) also has a maximum. However maximum of intensity does not coincide with a maximum of population, as the frequency is included in expression (7.3) in the fourth degree (in other words the quantum energy of radiation also depends on J). Substituting (7.1) and (7.4) in (7.3) and taking a derivative on J, we shall obtain the analytical expression for a maximum of electromagnetic emission intensity, also dependent only from a rotary constant molecule B_e and temperature of environment T:

$$J(r_e)=\left[\frac{18\cdot k\cdot T+h\cdot c\cdot B_e(r_e)}{216\cdot h\cdot c\cdot B_e(r_e)}+\frac{\sqrt{-6\cdot k\cdot T\cdot(500\cdot k^2\cdot T^2+44\cdot k\cdot T\cdot h\cdot c\cdot B_e(r_e)+h^2\cdot c^2\cdot B_e^2)}}{72\cdot(h\cdot c\cdot B_e(r_e))^{\frac{3}{2}}}\right]^{\frac{1}{3}}-\frac{\left[\frac{h\cdot c\cdot B_e(r_e)-10\cdot k\cdot T}{12\cdot h\cdot c\cdot B_e(r_e)}-\frac{1}{9}\right]}{\left[\frac{18\cdot k\cdot T+h\cdot c\cdot B_e(r_e)}{216\cdot h\cdot c\cdot B_e(r_e)}+\frac{\sqrt{-6\cdot k\cdot T\cdot(500\cdot k^2\cdot T^2+44\cdot k\cdot T\cdot h\cdot c\cdot B_e(r_e)+h^2\cdot c^2\cdot B_e(r_e)^2}}{72\cdot(h\cdot c\cdot B_e(r_e))^{\frac{3}{2}}}\right]^{\frac{1}{3}}}. \tag{7.6}$$

Despite the bulky look of this expression, it is quite suitable for numerical calculations. It should be pointed out, that the rotary constant, and consequently $J(r_e)$, depends significantly on the equilibrium inter-nuclear spacing interval r_e (Eq. 7.2). Thus, taking into consideration Eqs. (7.2) and (7.6) we obtain the following expression for a maximum of electromagnetic emission frequency measured in Hz:

$$f(r_e)=\frac{h}{8\cdot\pi^3\cdot\mu\cdot r_e^2}\cdot(J(r_e)+1) \tag{7.7}$$

The plots of the function (7.7) are illustrated in Figs. 7.6–7.8. In Fig. 7.6 the relation between the inter-nuclear spacing interval at ambient temperature T=270 K and the maximum emission frequency of rotationally-rotary transitions is depicted. This is valid for a molecule with a reduced mass equal to m=73 of nuclear mass units, that corresponds, for example, to quasi-molecule composed of positively and negatively charged clusters $NO_3^-\cdot(H_2O)_6$ and $H_3O^+(H_2O)_6$.

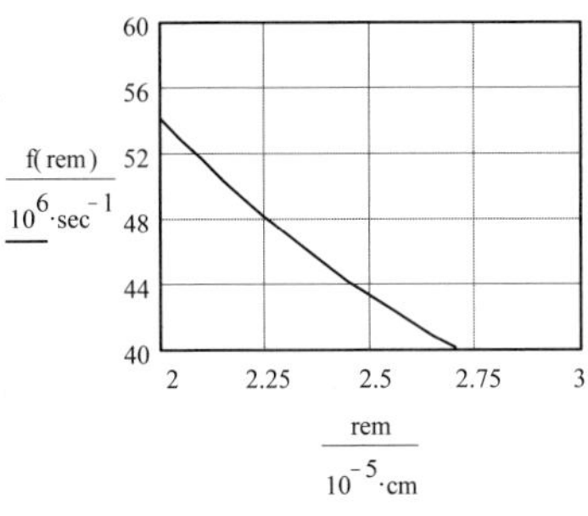

Fig. 7.6 The relation between inter-nuclear spacing interval and the maximum frequency

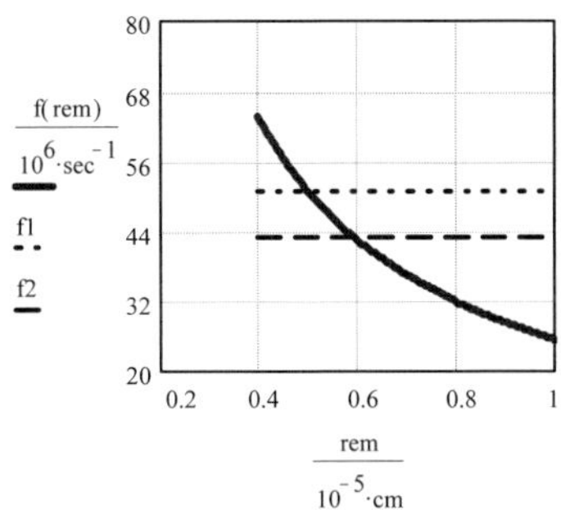

Fig. 7.7 Dependence of the maximum emitted frequency on rotational-rotational transitions of the molecule with the reduced mass equal to 1,100 a.u. on the inter-nuclear distance under the temperature T=270 K

One can see, that radiated frequencies 43 and 51 MHz which were registered in the experiments of Vallianatos and Nomikos (1998) correspond to the area of inter-nuclear spacing intervals 0.21–0.25 microns. Thus the energy of a Coulomb interaction between single charged nuclear $E_q=e^2/r_e$ does not exceed 0.007 eV, which is more than three times lower than the ambient temperature. In these conditions it is difficult to expect the stability of such a molecular complex.

Therefore it could be that emission is not from the quasi-molecule but from the dipole aerosol which has a much higher mass or that the registered emission does not correspond to the maximum part of the emission spectrum of the quasi-molecule. In the latter case the effect should be expressed more sharply under the higher frequencies of emission near 200 MHz.

In order to have the energy of interaction of positively and negatively charged clusters equal to the surrounding temperature and the frequency maximum of rotational emission within the band measured experimentally, the reduced mass of the quasi-molecule should be no less than 1,100 a.u. (Fig. 7.7), which corresponds to m and n equals approximately 60 (2–3 layers of a dense packing of water molecules). The horizontal lines in the figure correspond to experimentally measured frequencies.

The relation of energy Coulomb interaction to inter-nuclear spacing interval is shown in Fig. 7.8.

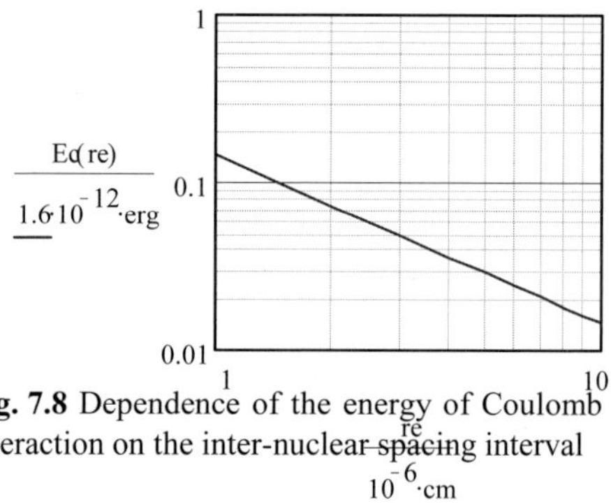

Fig. 7.8 Dependence of the energy of Coulomb interaction on the inter-nuclear spacing interval

When the detected radiation does not correspond to a maximum of quasi-molecule radiation, it is possible to estimate the applicable rotary quantum number from a condition of balance of forces of Coulomb attraction $F_q=-e^2/r_e^2$ and centrifugal retraction $F_r=M\cdot r_e\cdot\omega$, where M – is a light cluster mass, and ω – cyclic rotational speed. Thus for cyclic frequency:

$$\omega(r_e) = \sqrt{\frac{e^2}{M \cdot r_e^3}} \tag{7.8}$$

and finally we have such a relation for the rotary quantum number:

$$J(r_e) = \sqrt{\frac{4\cdot\pi^2\cdot\mu^2\cdot q^2\cdot r_e}{h^2\cdot M} + \frac{1}{4}} - \frac{1}{2} \tag{7.9}$$

Relations of energy of rotation and rotary constant from inter-nuclear spacing interval here are taken into account. Substituting Eq. (7.9) in (7.1) and taking into account Eq. (7.2), we shall obtain the dependence of the radiated frequency (in terms of cm^{-1}) on inter-nuclear spacing interval for a selected pair of clusters $H_3O^+(H_2O)_6$ and $NO_3^-(H_2O)_6$, shown in Fig. 7.9. The dashed horizontal line corresponds to the experimentally measured frequency of 51 MHz.

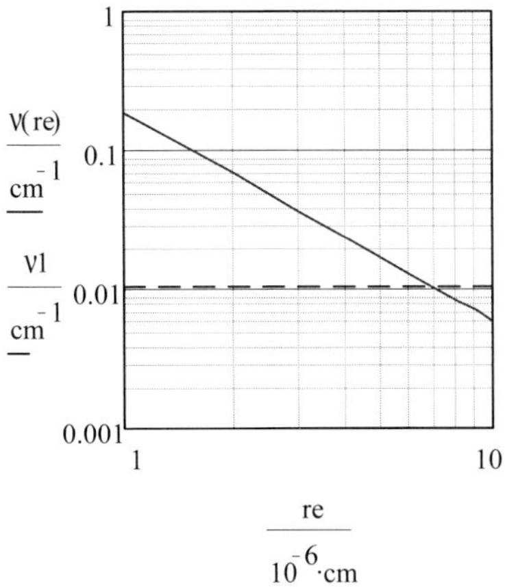

Fig. 7.9 Dependence of the radiated frequency on the inter-nuclear spacing interval for the pair of clusters $H_3O^+(H_2O)_6$ and $NO_3^-(H_2O)_6$. *Dashed line* corresponds to the frequency 51 MHz

The correspondent dependence of the rotary quantum number J on the inter-nuclear spacing interval is shown in Fig. 7.10.

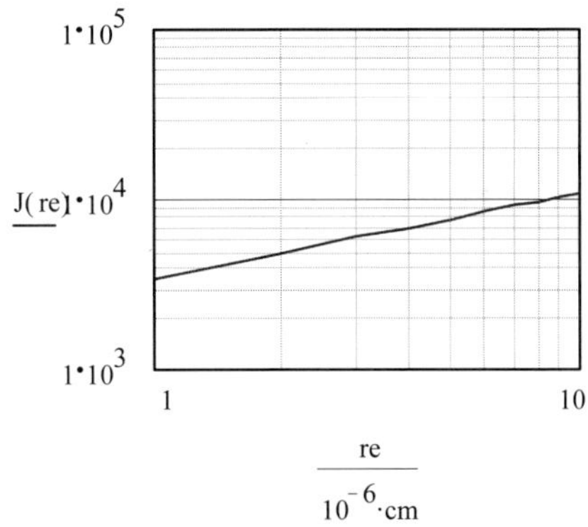

Fig. 7.10 Dependence of the rotary quantum number J from inter-nuclear spacing interval

From Figs. 7.9 and 7.10 one can determine, that the frequency 51 MHz corresponds to $r_e \approx 7 \cdot 10^{-6}$ cm, and respectively to $J \approx 9{,}000$. Thus the energy of a Coulomb interaction of charged clusters is about ambient temperature (Fig. 7.8). The distribution of quasi-molecules as dependence on the rotary quantum number for the given inter-nuclear spacing interval and full concentration of quasi-molecules $N=10^6$ cm^{-3}, agrees with Eq. (7.4), and is shown in Fig. 7.11.

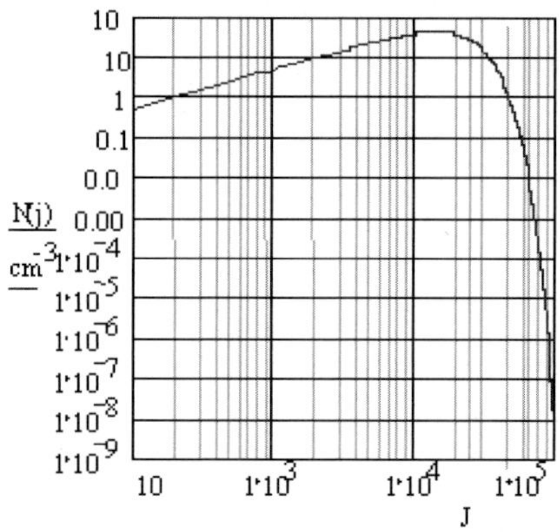

Fig. 7.11 Distribution of the quasi-molecules number as dependence on the rotary quantum number

Thus the intensity of lines in the emission spectrum of molecules (in relative units), agrees with Eq. (7.3), and will correspond to Fig. 7.12. One can see from Fig. 7.12 that the intensity maximum of emission is close to the frequency 155 MHz but the population maximum corresponds to the emission with the frequency 69 MHz. The emission intensity from the rotational levels with $J \approx 9{,}000$ is only 2% from the maximum one. It is necessary, however, to note, that at fixed inter-nuclear spacing interval the radiation from levels with J>9,000 will not be possible because of instability of such molecular complexes. Accordingly, the outlook of the curve in Fig. 7.12 will be different, but the maximum of radiation will correspond to $J \approx 9{,}000$.

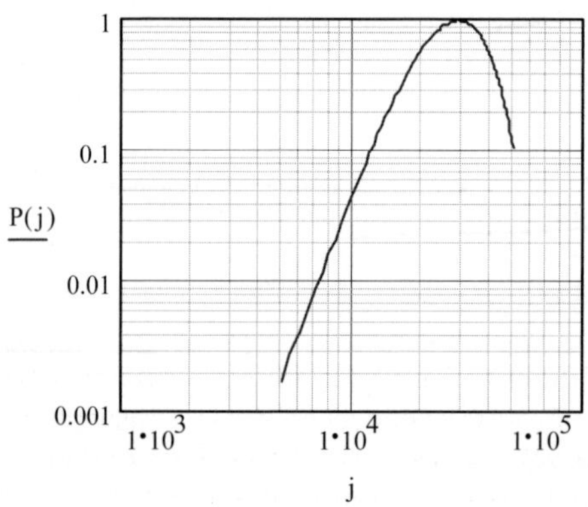

Fig. 7.12 Emission line intensity dependence in molecular emission spectrum (relative units) on the rotational quantum number

7.1.5 Discussion of the Molecular Mechanism of HF-Emission

Certainly, the experimental detection of electromagnetic emission resulting from rotationally-rotary transitions in quasi molecules, retained by Coulomb interaction, requires their noticeable concentration in air. A main reason of their disappearance should be an ion-ion recombination of elementary ions, due to which the concentration of charge carriers of one sign in dry air usually does not exceed $2{\cdot}10^3$ cm^{-3}. However, the appearance of condensation centers and the formation of heavy cluster ions in a moist air can result in essential delay of a recombination (Smirnov 2000) and, therefore, to growth of concentration of charge carriers. As was shown in Chapter 3, the effect of radon emanation can produce a final concentration of charge carriers up to 10^6 cm^{-3} and more. Such a concentration can give as a result the appreciable level of electromagnetic emission due to the mechanism described above. It may explain the HF precursors registered by Vallianatos and Nomikos (1998). Such emissions can be a result of rotationally-rotary transitions of dipole quasi molecules, consisting of large hydrated cluster ions, or dipole aerosols.

But the proposed mechanism cannot explain the pulsed character of the emission registered in Western Australia. That's why another approach was developed to explain the experimental data.

7.1.6 The Pulsed HF Emission Explanation

HF emission physical mechanism presented in previous paragraph is able to explain the observed frequency band in case of experiments of Vallianatos and Nomikos (1998) but not in case of P-H pulses. Here we want to put forward a hypothesis having in mind the following experimental facts: the registered emission does not have a continuous but pulsed character. The emission has a very wide frequency spectrum from kHz up to more than a hundred MHz. These two facts imply that the registered emission should be electric discharge-like, similar to the emission from thunderstorm flashes. The emission is connected in some way with the seismic activity because of the increase in the emission intensity and pulse rate with the quake's approach 12–24 hour before the seismic shock. Another intriguing factor is that emission is registered at large distances up to 1,500 km. Taking into account that the emission is registered in the VHF band also, the source of emission cannot be situated on the ground. The estimations show that the source altitude should be at heights from 40 to 80 km. One cannot expect thunderstorm clouds at such heights. Does something exist at all at these heights? The answer is YES. The rocket experiments demonstrate the existence on these altitudes of aerosol layers (Zadorozhny 2001), see Fig. 7.13. And here arise two questions:

1. Could the aerosol layers be formed in some way due to seismic activity?
2. Could some electromagnetic emissions be generated inside the formed layers?

Aerosols or ion clusters are always present in the lower troposphere. As was demonstrated in Chapter 3 the ion clusters are always present in the boundary layer. The diffusion cannot provide the vertical transport to the level of the order 60–90 km for the time intervals of the order of days. Moreover, the inverse temperature height intervals where the temperature lowers with the height make such transport even more problematic. But quasistatic electric fields of seismogenic origin can make such a transport of the ions essentially more effective. Using the calculations presented in Eqs. (4.17)–(4.24) we can explain the rise of the charged particles or ion clusters, or charged aerosols to the necessary altitude of 40–80 km. The altitude of the cloud formation will be determined by the balance between the vertical electric field strength and gravitational force. Starting from altitudes of 60 km, the effect of solar ultraviolet radiation becomes essential. The heavy ions or ion clusters fall into plasma from light ions and electrons and play the role of the heavy dust.

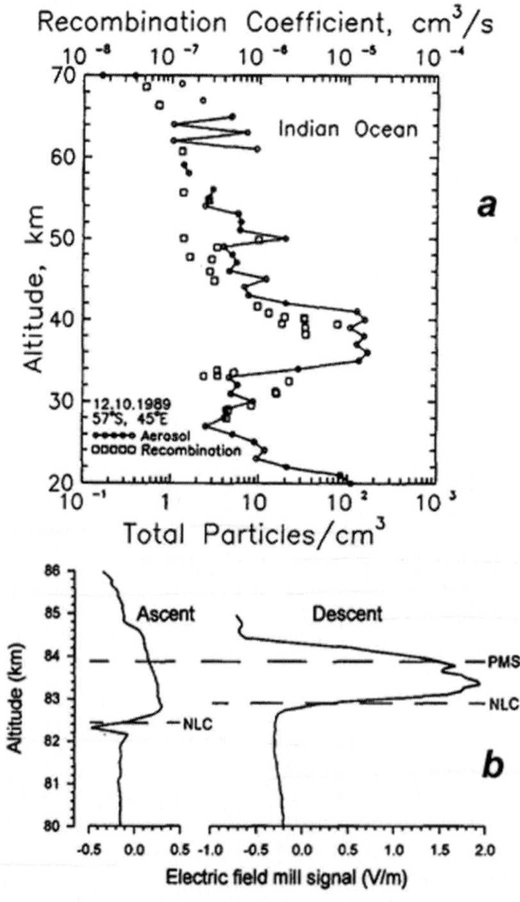

Fig. 7.13a,b **a** The dust (aerosol) content and ion-ion recombination coefficient derived from the rocket data. **b** The vertical component of the electric field measured by field mill during the NLC-92 campaign. The *dashed lines* show the heights where the peaks of NLC and PMSE were detected

So we can conclude that dusty cloud formation over seismically active zone is quite possible both from the experimental and from the theoretical points of view. The increased values of the electric field are observed within such clouds as one can see from Fig. 7.13b. But this is the macroscopic field. When we look at the structure of the cloud in the microscopic scale, the fields might be much larger.

The process of electrization in the dusty plasma was described by Kikuchi (2001). The formation of a crystal-like quadruple structure (Rosenberg and Mendis 1995, Pokhmelnykh 2003) and subsequent acceleration of the particles in the created electric field up to initiation of critical ionization phenomena can lead to spontaneous discharges within the aerosol cloud.

We can suppose that probably due to the regular quasi-crystal structure of the plasma inside the cloud is the pulse shape so similar. With the approaching earthquake the anomalous field magnitude increases, which increase the concentration of heavy ions inside the cloud, and consequently the discharge rate. Very short pulses of electric discharges inside the cloud emit a very wide band of the electromagnetic emission, which is registered by receivers tuned into different frequencies. Due to cloud elevation the emission is registered at large distances from the earthquake epicenter.

At the present moment the presented conception of the pulsed radio emission can be regarded only as speculation. But at least it permits us to explain all the morphological features of the observed emission: short duration pulses, their similarity, wide spectrum, and large distance of the emission propagation. The artistic view of the proposed conception is presented in Fig. 7.14.

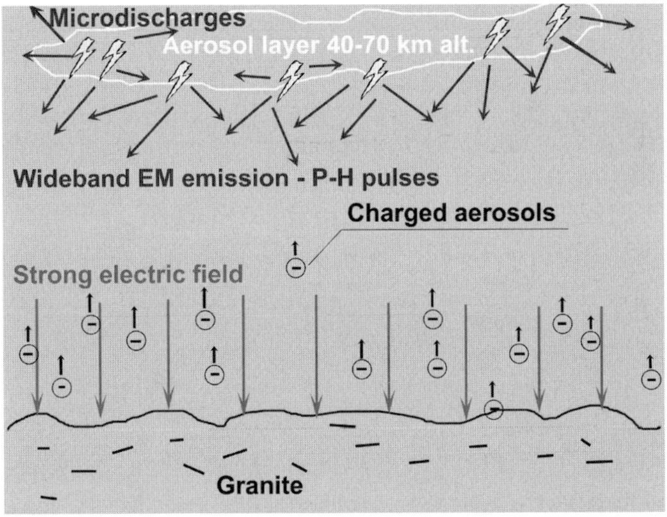

Fig. 7.14 Schematic presentation of the P-H pulses generation mechanism

7.2 On the Nature of the VLF Emission Registered on the Ground Surface Before Strong Earthquakes

Fujinawa and Takahashi (1998) presented a review where mainly the VLF emissions associated with earthquakes are discussed. The more recent review where all electromagnetic emissions associated with the Kobe 1995 earthquake is presented by Nagao et al (2002). All events concentrated around the 10th–11th January when the strong thunderstorm activity was registered, so authors are not sure of the real relation of the registered emissions to the Kobe earthquake preparation. One can find also other communications on the VLF emissions registration before strong seismic events such as Vallianatos and Nomikos (1998) and Nomikos and Vallianatos (1998).

Observed VLF emissions appear within the interval from several hours up to several days before the seismic event. Sometimes the emission has a pulsed character. It is interesting to note that if for the other kinds of registered emissions (for example ULF emissions) are proposed plenty of physical mechanisms, the VLF emission does not have the same popularity within theoretician circles. Except for the very old general idea on the rock fracturing probably being responsible for the observed anomalies (Warwick et al. 1982), there are not many other theories. Most publications have the character of a discussion and some of speculation. The most recent of Fujinawa et al. (1999) proposes the idea of an atmospheric source similar to the mechanism proposed here of P-H pulse generation.

At the same time we already mentioned the possibility of electromagnetic emissions generation as instability of atmospheric plasma created over the seismically active area. Actually we have the plasma in a strong electric field. So one can expect a different kind of current instabilities such as two-stream instability and modulation instability. Taking into account that we do not have the usual plasma with heavy ions and light electrons but the plasma of positive and negative ions with more or less the same mass, we can suppose that the main frequency excited in such plasma will be the plasma frequency determined by the plasma concentration n:

$$\omega_p = \sqrt{\frac{4\pi Ne^2}{m}} \tag{7.10}$$

where N is the concentration of plasma, e – electric charge (we suppose the single charged ions), m – ion mass.

For the cluster ion NO_3^- $(H_2O)_6$ (here n is the number of water molecules in the cluster) with 6 attached water molecules the atomic mass will be M=190 a.e. which is equivalent to m=3.15·10^{-22} g. Our estimations show that in near ground atmospheric plasma the concentration of charged particles is of the order 10^6 cm^{-3}. For such parameters the linear plasma frequency is $f_p = \omega_p/2\pi \approx 16.9$ kHz which lies just inside the VLF frequency band. Taking into account that the plasma concentration, as well as the mass of the ion cluster may change considerably, one may expect the coverage of the whole VLF band.

Of course, we have estimated only the frequency of the possible plasma oscillations. The following steps are the mechanism of plasma oscillations transformation into the electromagnetic emission and its energetic parameters estimation, which will be the subject of future work.

7.3 Biological Effects of Anomalous Electric Field

The anomalous animals behavior before strong earthquakes within the earthquake preparation area is widely described in the literature (Rikitake 1976; Mogi 1985; Bolt 1993) and it is not the aim of the present paragraph to describe these effects once more. But it is of interest how related these effects are to the physical mechanism of the ionospheric precursors formation, and especially, to the values and spatial distribution of the anomalous electric field on the ground surface.

As was stated in Chapter 1, the area of the earthquake preparation is determined by the formula (1.8) of Dobrovolsky et al. (1979). This formula was obtained based on calculations of the elastic deformations of the Earth's crust of the order 10^{-8}. As was shown later by Toutain and Baurbon (1998), the geochemical precursors obey the same law (see Fig. 1.11) which allows us to make the conclusion on the same size of the area occupied by the anomalous electric field because as a radon daughter product.

Manifestation of the biological effects of earthquakes can be also described quantitatively by studying their spatial distribution, which was done by Kozyreva (1993). A large amount of data (1,000 cases of biological precursors) was collected for the Spitak earthquake (Armenia) of the 7th Dec. 1988, M=7.0. The data of five earthquakes was processed and is shown in Table 7.1.

Table 7.1 Statistics of earthquakes where biological precursors were registered

Earthquake			Biological precursor		
Name	M	Date	No. of locations	No. of cases	R_m km
Guksayan	4.9	17.01.82	20	98	26
Paravan	5.6	13.05.86	80	> 300	96
Spitak	7.0	07.12.88	130	> 1000	220
Izu-Oshima	7.0	14.01.78	70	129	320
Spitak	7.0	07.12.88	130	> 1000	220
Gazli	7.2	20.03.84	35	175	280

Magnitude-distance dependence for these cases is presented in Fig. 7.15. The regression law was derived which looks like this:

$$\log R_m = 0.43M - 0.56 \tag{7.11}$$

Obtained regression practically identical to the theoretical law for the earthquakes preparation zone with the strains of 10^{-7} which looks like:

$$\log R_m = 0.43M - 0.40 \qquad (7.12)$$

So we can conclude that the biological precursors are revealed within the zone of earthquake preparation determined by the elastic deformation 10^{-7}. It was determined also that if the preparation zone has an elongated shape (with the largest dimension L), the biological precursors will be observed within the zone which will be from two to three times larger then this dimension in the perpendicular direction.

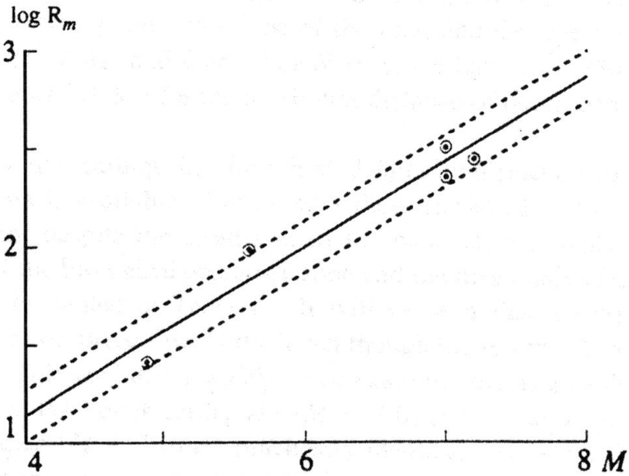

Fig. 7.15 Biological precursors magnitude-distance relation according to Table 7.1. *Dashed lines* – 70% confidence interval (after Kozyreva 1993)

If we compare the R_m with parameter a (see Fig. 4.12) of the electric field Gauss distribution on the ground surface, we can conclude that R_m may be lower than a due to the fact that the electric field decays to the edges of the earthquake preparation zone, and its magnitude is not enough to cause the evident biological behavior signs which can be identified as precursors.

But what is the electric field magnitude which causes the signs of anomalous behavior manifested by animals before the earthquake? This question was studied by Ikeya et al. (1996) in experiments with albino rats, Mongolian gerbils (sand rats), hair-footed Djungarian hamsters, guinea pigs and red sparrows. Results of their studies in the form of animals reaction as dependence of the electric field strength is shown in Table 7.2.

Table 7.2 Effects of electric field on rats and birds to attempt to

Animals	F (V/m)	$V_{//}(V)^{a)}$	$I_{//}(\mu A)^{a)}$	Responses[b)]
Rats,	2	~ 0.08	~ 0.15	Grooming (G)
(Rattus Norvegicus)	6	~ 0.3	~ 0.5	G, Nervous looking (N)
W ≈ 300g	72	~ 5.0	~ 6.0	Cramped legs (CL)
L ≈ 3.5 cm	600	~ 15	~ 52	Avoiding field (AF)
R ≈ 0.4-0.5 MΩ	1000	~ 24	~ 90	Running (R), Panic (P)
Mongolian gerbils	60	~ 1.5	~ 0.75	G, Crying?, AF?
(Meriones unguiculatus)	100	~ 2.5	~ 1.3	Standing up (SU), G, N
W ≈ 50 g, L ≈ 2.5cm	240	~ 6	~ 3	CL, AF
R ≈ 2 MΩ	400	~ 10	~ 5	R, P, Screaming(S)
Djungarian hamsters	30	~ 0.6	~ 0.3	Biting wires, AF?
(Shangarian hamster)	50	~ 1.0	~ 0.5	Running in panic?
W ≈ 20 g, L ≈ 2 cm	400	~ 8	~ 4	G, Jumping (J)
R ≈ 2 MΩ	800	~ 16	~ 8	R, P, S, Tumbling (T)
Guinea pigs	100	~ 3	~ 0.15	Nervous looking?
(Cavia porcellus)	400	~ 12	~ 0.6	Standing up
L ≈ 3cm	800	~ 25	~ 12	Grooming
R ≈ 2 MΩ	1600	~ 50	~ 25	Panic, Jump, Tumbling
Red Avadavat	100	~ 1	~ 0.5	Inflation (I), Grooming
(Amandava Amandava)	300	~ 19	2-3	Jumping, AF
R ≈ 2-3 MΩ	600	~ 50	~ 5	Flying up (FU), P, AF

a) The $V_{//}$ and $I_{//}$ were calculated as the maximum to the animal parallel to the field direction

b) Behaviors are abbreviated. The behavior that can not be judged clearly as an electric field effect is indicated by ? mark.

The most interesting is the second column where the electric field value is indicated. Again one can see the characteristic value of 1,000 V/m (which was taken for our calculations of the electric field penetration in the ionosphere) when the behavior of animals becomes a panic. With electric field decay the manifestations of the animals behavior become not so evident.

7.3.1 Conclusion

Surprisingly, the scaling parameters of the anomalous animal behavior before earthquakes are the same as for mechanical deformations, geochemical precursors, electric field precursors, and ionospheric precursors. This fact supports our primary assumption that the anomalous animal behavior is a result of the action of the anomalous electric field. It also gives arguments to use the anomalous animal behavior as a real precursors with the same levels of confidence as other precursors involved in the physical mechanism of ionospheric precursors generation.

7.4 Anomalous Neutron Fluxes Possibly Related to Earthquakes

In 1990–1991 the new phenomenon of the increase of neutron emission intensity at the time of new and full moons was discovered (Volodichev et al. 2000). The qualitatively new data obtained in 1997 shows that the neutron peaks are induced by the tidal forces acting in the Earth's crust. On the other hand, the tidal forces are known to be a possible trigger for seismic activity. An analysis of the uniform global data on earthquakes that occurred from 1964 to 1992 made by Volodichev et al. (2000) showed their two-week periodicity connected with the new and full phases of the Moon. Further consideration of the neutron bursts and their correlation with the seismic activity can reveal new types of precursors of the seismic-activity enhancement, namely, the neutron peaks governed by the solar-lunar-terrestrial relations.

The interpretation of this phenomenon was given in Volodichev et al. (2001). The tidal wave originating in the Earth's crust perpetually transforms the microstructure of the Earth's crust leading to a variation of the concentration of natural radioactive gases in the terrestrial air and to changed conditions of their leakage to the Earth's atmosphere. These variations give rise to bursts of thermal and slow neutrons in the vicinity of the Earth's crust, because the radioactive gases are sources of energetic alpha particles that induce neutron production upon the interaction with the nuclei of elements of the Earth's crust and atmosphere.

Actually, we probably deal with another type of radon emission registration through its daughter product – neutron fluxes appearing as a result of alpha-particles interaction with the Earth's crust and air molecules.

7.4.1 Experimental Setup

In the described experiments, neutrons were detected with the monitor (Kuzhevskij et al. 1996) composed of the standard cylindrical 4-at. pressure ^{3}He gas-filled counters of 220 mm length, 30 mm diameter, and 51 cm^2 effective area each. The ^{3}He filler has a large cross section for neutron capture (over 5,000 bn in the thermal range), which is much in excess of the ^{10}B cross section for capture of thermal neutrons. The neutron capture by ^{3}He occurs under the reaction

$$^3\mathrm{He} + \mathrm{n} \rightarrow \mathrm{T} + \mathrm{p} + 760\ \mathrm{keV} \qquad (7.13)$$

Most of the deposited energy is carried away by a proton that produces the necessary ionization. The counter detection efficiency for thermal and slow neutrons is about 80%. The total effective area of the site-fixed monitor varied from 2,040 cm^2 to 4,080 cm^2. Besides, a mobile 510 cm^2 area monitor was constructed, which permitted the Cadmium ratio to be determined when detecting neutrons. The site-fixed monitor was placed in a room at 20–30 m from the ground surface.

The experiment has shown that the neutrons of energies below 0.5 MeV (i.e., the slow and thermal neutrons) constitute the major fraction (above 70%) of the background neutron flux.

The concentration of the natural radioactive gases varies by four orders, depending on the measurement of site location, i.e., on the structure and elemental composition of the Earth crust. Therefore, the contribution of the second source of thermal and slow neutrons near the Earth's surface varies strongly, from fractions of one percent to tens of percent of the total neutron flux.

7.4.2 Results of Measurements

The relationships between the near-surface neutron flux variations and the tidal gravity wave were found in 1990 by the observations in the Pamirs active seismic zone at 4,200 m above sea level. A strong increase in the background neutron flux was observed during solar eclipse. The later Pamirs expedition measurements of the background neutron flux variations have shown that the neutron increases (bursts) occur actually always during phases of the Moon, new Moon, and full Moon, with the burst-defining flux being directed off the Earth's surface.

The second phase of studies was carried out recently in the Tien Shan mountains near Alma Ata at the Laboratory of Geophysical Precursors of the Seismology Institute of the Kazakh Academy of Sciences. The neutrons were detected with a LDNM scintillation counter used customarily in nuclear geophysics researches. Boron is the working medium of the counter. The counter is a cylinder of 25-mm diameter and 63-mm length. The neutron field has been observed since January 1996. Initially, the anomalously high neutron flux count rates for the March-December 1996 period were compared to the Medeo Seismic Station reports with magnitude 3 and higher, whose epicenters were located at distances of 120–900 km. A total of 119 earthquakes were compared. The anomalous neutron flux bursts were observed in 48 cases within a 12-hour interval prior to the earthquakes, which is 40% of the earthquakes total. After extending the time interval to 24 hours, more than 72% of the earthquakes belonged therein.

For the period of March 1996 to May 1998, were revealed the relationships of the earthquake magnitude to the anomalous neutron flux intensity. 146 earthquakes of magnitude 4–6 occurred within the observation period.

The obtained relationship is displayed in Fig. 7.16, and it looks like quasi-linear dependence. This agrees with the relationships found in Sobolev et al. (1998) between the count rates of the neutron monitors located at different sites of the globe and the overall planetary energy release of earthquakes.

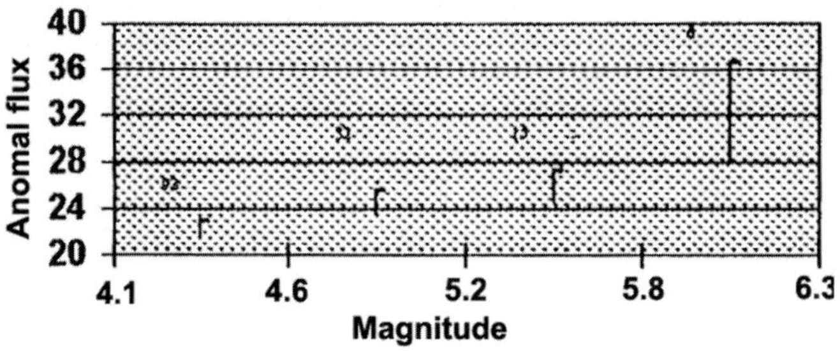

Fig. 7.16 Dependence of the anomalous neutron fluxes on the earthquake magnitude

7.5 Solar-Terrestrial Relations and Earthquakes

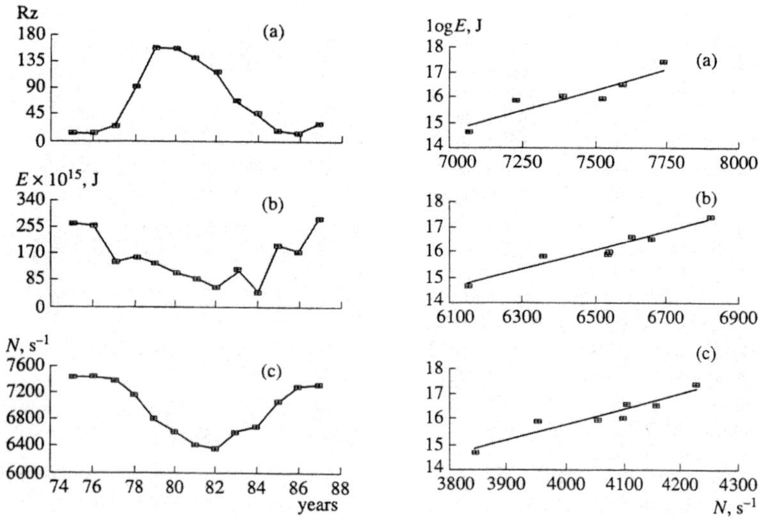

Fig. 7.17 *Left panel* **a** Wolf number, **b** global seismic activity, **c** neutron flux measured by Apatity neutron monitor. *Right panel*: correlation between the energy released by earthquakes and minimum values for Forbush decreases registered by neutron monitors: **a** Magadan, **b** Apatity, **c** Tula. (From Sobolev et al. 1998)

It is well known that the neutron fluxes being the secondary product of cosmic rays and registered by the global network of neutron monitors (http://helios.izmiran.rssi.ru/cosray/sort.pl?showtable) may serve as one of the indicators of the solar activity. The solar wind disturbances screen the cosmic ray fluxes, what leads to a decrease of the registered fluxes of neutrons.

Sobolev et al. (1998) try to link together the neutron fluxes as a measure of cosmic rays with the neutron fluxes described in the previous paragraph and possibly associated with earthquakes. In their words "Sun, Moon, interplanetary medium, magnetosphere, ionosphere, atmosphere of the Earth, and the Earth itself together with processes in its interior resulting in earthquakes form a coherent system. Active processes of terrestrial origin are constituent of the totality of phenomena in the solar system". They try to correlate the solar activity, cosmic rays and the global seismic activity for different periods of time (full solar cycle, and selected years). The example of such a correlation is presented in Fig. 7.17. At the left panel of the figure from top to bottom one can see the variation of the solar spot number for the 21 cycle of solar activity, global seismic energy released for the same period, and the neutron flux measured by the Apatity neutron monitor. The general conclusion, which could be made from this graph is that the release of the seismic energy is maximal during the minimum of the solar activity cycle. Authors try to support this conclusion with short-time records associated with strong magnetic storms. It is known that they are associated with the Forbush decrease of cosmic rays. Authors compare the seismic energy release during periods of Forbush decreases associated with solar events with the minimum values of cosmic ray fluxes during Forbush decreases. The data presented for three neutron monitors.

Acknowledging the revealed correlation we cannot agree with the second conclusion of the authors (Sobolev et al. 1998) that the fluxes registered during Forbush decreases are of the origin described by Volodichev and Kuzhevskii (2003). As claim the authors of the latter paper, their fluxes correlated with lunar tides and reflect the variations of radon emanation from the Earth's crust (neutrons are the secondary product of radon alpha-activity). In addition the direction of the fluxes arrival should be measured (which is possible with existing techniques of neutrons registration) to make such a conclusion.

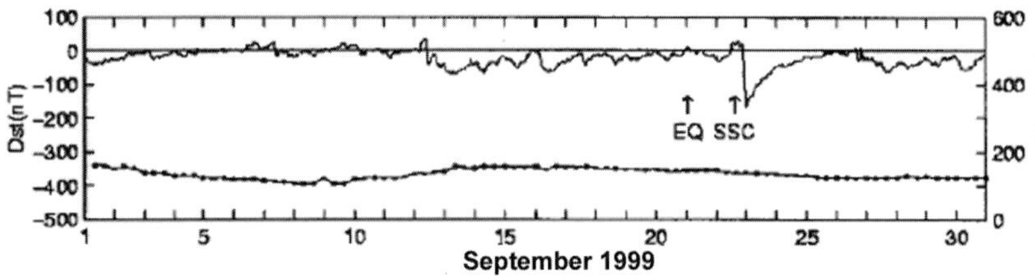

Fig. 7.18 D_{st}-index (*upper curve*) and Solar radio flux (*bottom curve*) for September 1999

The work was continued by Sobolev et al. (2001) trying to correlate the seismic activity at Kazakhstan and Kyrgyzstan with magnetic storm activity. Their conclusion is that the number of earthquakes increases after the magnetic storms with sudden commencements. But it seems that such a correlation is difficult to apply in any kind of earthquake prediction because very often for the observed cases a magnetic storm occurs the next day after the earthquake. In Fig. 7.18 one can see the period during the devastating Chi-Chi earthquake at Taiwan on 20th September 1999, M=7.3. The D_{st} index, which is a measure of the magnetic storm intensity and is submitted in the figure, and one can see that the sudden commencement (SSC) of the strong magnetic storm occurred the next day after the earthquake.

It is a controversial question and is probably connected with the geo-effectiveness of the solar events. Barsukov (1991) found some differences in the correlation between the earthquakes and sudden commencements connected with the longitude of the active area on the solar disk which is shown in Fig. 7.19.

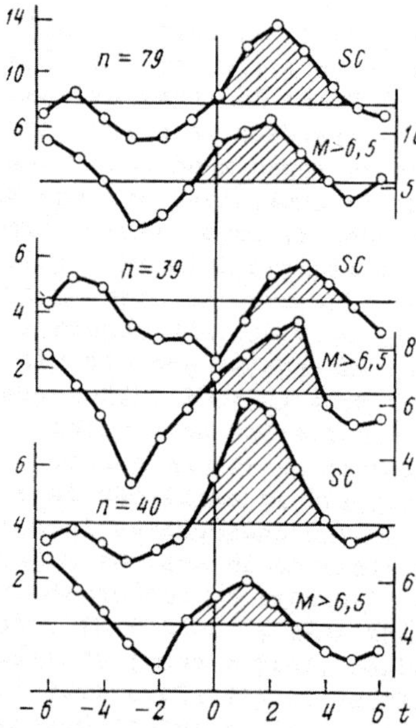

Fig. 7.19a–c Variations of magnetic storm solar commencements number *n* and number of earthquakes with $M>6.5$ in relation to solar flare moment (0 on the *x* axis). **a** active area is in the sectors 30° E 30° W on the solar disk, **b** in sector 0°–30° E, **c** in sector 0°–30° W

For the analysis were selected the strongest chromospheric flares within the interval of longitudes 30° E–30° W on the solar disk (79 events for the period June 1957 – January 1961) with a distribution of 39 events in the eastern sector 0°–30° E, and 40 events in the western one 0°–30° W. For the same period of time 191 earthquakes with the magnitude >6.5 were registered and 100 sudden commencements. The results are centered in relation to the solar flare moment, and one can see that the maxima positions differ by two days for the events registered in the eastern and the western sectors. One can also observe that the solar activity probably increases the seismic activity but there is also a significant number of earthquake's that occur before the solar flare. So the final conclusion may be that the solar activity in some way modulates the seismic activity but is not the real triggering factor.

Even more intriguing is the fact that some statistical data support the result shown in Fig. 7.18 that the earthquake occurs one day before the magnetic disturbance (Sytinskiy 1989). This was obtained also for the earthquakes with a magnitude higher than 6, and 6.5 for two datasets of strong geomagnetic perturbations (Shapley et al. 1975) for a large time span (1953–1970). The results are shown in Fig. 7.20.

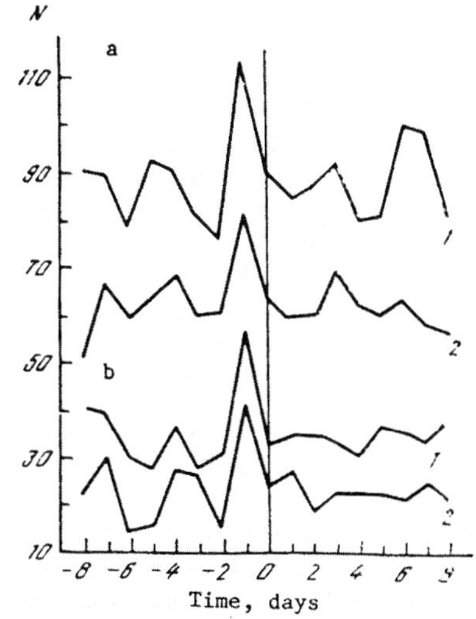

Fig. 7.20a,b a Number of days with earthquakes M>6 in relation to the day of strong geomagnetic disturbance. 1 – for the first dataset of magnetic disturbances, 2 – for the second dataset of magnetic disturbances. **b** The same for the number of days with earthquakes M>6.5

These results clearly indicate that strong earthquakes happen one day before the strong geomagnetic disturbance which is really incredible and very difficult to understand and is contradictory to results shown in Fig. 7.19. The statistical parameters for the Pearson's test χ^2 are the following: for the first dataset χ^2=25.5 (p≤0.01; v=11); for the second dataset χ^2=22.1 (p≤0.02; v=11). The existence of these contradictions and the unclear mechanism is the reason we put these results in this chapter.

To come to some definite conclusions, the same earthquake catalogs should be used, the same time intervals, the same data for the solar and magnetic disturbances, and the same periods of time and phases of the solar cycle. For example, for a much larger dataset than in Sobolev et al. (1998) Sytinskij (1989) obtains a completely different picture for the seismic energy released during the solar cycle containing three peaks (Fig. 7.21). The data were collected for the period 1902–1977. One can find more publications on the cosmic rays and earthquakes such as Morozova et al. (2000) where the authors found 22-year periodicity and another period with different correlation coefficients. But they probably do not realize that a correlation does note mean a cause-effect relationship as they claim.

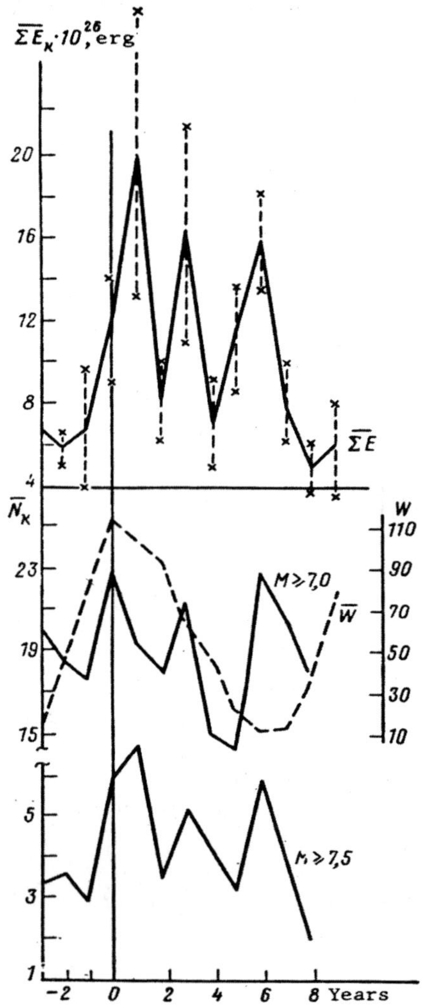

Fig. 7.21 Distribution of the average annual ΣE_k seismic energy and (*top panel*) and earthquakes number N_k for magnitudes $M \geq 7$ (*middle panel*) and $M \geq 7.5$ (*bottom panel*) versus solar spot number (*dashed line*) within the solar cycle activity. Averaged for the several solar cycles from 1902 to 1977

All these facts imply that an independent referee check should be done for the statistical data or a special international workshop organized to discuss the problems of the interrelations between the solar and seismic activity. The question remains open until some clear physical ideas appear, able to explain the observed correlations. What physical agents are responsible for the observed modulation of the seismic activity by the solar activity?

7.5.1 Exotic Cases of the Relationship Between the Magnetic Storms and Seismic Activity

We can also mention the studies where solar activity – earthquakes relation is based not on the direct comparison of days of disturbance and immediate date of the earthquake but has significant delay.

The first prediction technique is proposed by Doda (2001) and is based on the results of magneto-hydrodynamic soundings in seismically active regions (Tarasov and Tarasova 2003).

During studies of seismicity at Garm geophysical range (Tajikistan) of Institute of Physics of the Earth of the Russian Academy of Sciences the magnetohydrodynamic (MHD) generator was used for the deep magnetotelluric sounding. It has been shown, that sounding has called appreciable activization of local earthquakes. Thus it was revealed, that peak efficiency of initiating influence is observed in the top 5-kilometer layer of a crust and strongly varies within the limits of different geological structures. As a whole, the total energy released by the initiated earthquakes appeared to be five orders of magnitude larger than the pumping energy of the MHD-generator. It has allowed us to conclude, that electromagnetic influences initiated the release of the energy that has been stored in a crust in the course of natural tectonic processes, in the form of rather weak seismic events.

The studies were continued at the Bishkek (Kyrgyzia) geophysical range. The most interesting fact discovered, that the seismic activation started several days after the MHD-generator activity. The post-effect duration is 2–3 weeks after the MHD-generator impact.

Table 7.3 Results of earthquake prediction according to Doda

No	Sun active area		Date of the earthquake prediction	Real earthquake		
	Date	AE No.	D_{AE}+14 d D_{AE}+22 d	Date	Magnitude	Area
1	24.9.2001	9628	8.10.2001	8.10.2001	5.9	Mexico
		9632	8.10.2001	8.10.2001	6.5	Kamchatka
				8.10.2001	6.4	Kamchatka
2	25.9.2001	9632	9.10.2001	9.10.2001	6.5	Kuriles
		9628	17.10.2001	17.10.2001	6.0	Caribbean
3	26.9.2001	9632	10.10.2001	9.10.2001	6.5	Kuriles
		9628	18.10.2001	19.10.2001	7.5	Indonesia
4	27.9.2001	9632	11.10.2001	12.10.2001	7.3	Marian isl.
		9628	19.10.2001	19.10.2001	7.5	Indonesia

Doda (2001) proposed that we regard the electric currents induced in the Earth's crust by the strong magnetic disturbances as the same as a result of the MHD-generator impact. In this case one should expect the seismic activation several days after the strong magnetic disturbance. As Barsukov (1991) Doda also used the geoeffectiveness of the solar events using the events observed on longitudes

0°-45° W from the central solar meridian. He also empirically identified the type of the solar events that cause the correspondent seismic activation and the time delay after the solar event which is 14–21 days. With his technique it is possible to determine the approximate longitude range where one should expect the strong earthquake but not its exact position. The example of such a prediction is demonstrated in Table 7.3. The earthquake longitude sector is determined by the time of the shock wave arrival to the Earth's magnetosphere.

As stated above, the validity of such a prediction should be checked by independent experts.

The most exotic prediction technique known to authors and related to magnetic storm effects on the earthquake activity was proposed by Chinese seismologist Zhan Tie-Zheng (1998). He calculates the time interval between two magnetic storms (time is associated with the magnetic storm beginning), and then doubles this interval. The obtained date indicates the date of the anticipated earthquake. But the hint is that in some way these storms should be selected. As one can understand, all possible combinations are looked until several of them converge on the same date. For very strong earthquakes it is possible to find 4–5 pairs of magnetic storms which indicate the date of the future earthquake. Examples for two severe earthquakes (Tangshan (China) 28.07.1978 M=7.8, and Gujarat (India) 26.01.2001 M=7.8) are shown in Table 7.4.

Table 7.4 Double magnetic storm prediction results

Earthquake	Date of magnetic storm	Interval days	Date of magnetic storm	After days (2days)
Tang shan	Aug 10, 1972	724	Aug 4, 1974	723 (1)
Earthquake	Mar. 10, 1975	253	Nov 18, 1975	255 (2)
Ms 7.8, July	Nov 13, 1975	124	Mar. 26, 1976	124 (0)
28, 1978	April I, 1976	59	May 20, 1976	59 (0)
Gujarat	May 3, 1998	499	Sept 14,1999	500 (1)
Earthquake	Aug 28, 1998	437	Nov, 8, 1999	444 (7)
Ms 7.8, Jan,	March 2, 1999	348	Feb 13, 2000	348 (0)
26, 2001	Oct 23, 1999	229	June 9, 2000	231 (2)
	Oct. 6, 2000	55	Nov 30, 2000	57 (2)

In the second and for the columns are indicated the dates of predictive magnetic storms, in the fourth column – the interval (in days) between them. Doubling this interval and adding it to the date of the second magnetic storm, one can obtain the date of the future strong earthquake. In the last column one can find the real interval between the second magnetic storm and earthquake, and the prediction error. Except one case where the error is 7 days, other cases give the error no more than 2 days. As one can see too, the interval may be even more than 2 years. What is this? Playing with numbers, or does this technique has some substantiation? China was and still remains a mystery.

Concluding this paragraph we can state, that solar (and geomagnetic) activity effects on the Earth's seismicity are not proved to a sufficient extent to be used in earthquake prediction. Regardless of some indication of the existence of such a connection and even the proposal of some practical techniques, until the physics of these effects are clarified, we cannot recommend its use.

7.6 Latent Heat Flux, Thermal Effects and the Ionosphere

Recently thermal effects have been attracting more and more attention from scientists (Tronin 2000; Tramutoli et al. 2001; Tronin et al. 2002). The infrared satellite images show the surface temperature increase several days before the earthquake. The temporal evolution of the heat flux before the Irpinia earthquake one can see in Fig. 7.22.

It is interesting to note that the temporal scale of the thermal flux variations (as well as spatial one, see Fig. 6.13) is the same as for the ionosphere. The anomalies may have a steady character as demonstrated in Fig. 7.23, and can also be oscillatory with a temporal scale of days. Such an anomaly was registered before the Gujarat earthquake (India) 26.01.2001 M=7.8 (Ouzounov 2003) and is presented in Fig. 7.23.

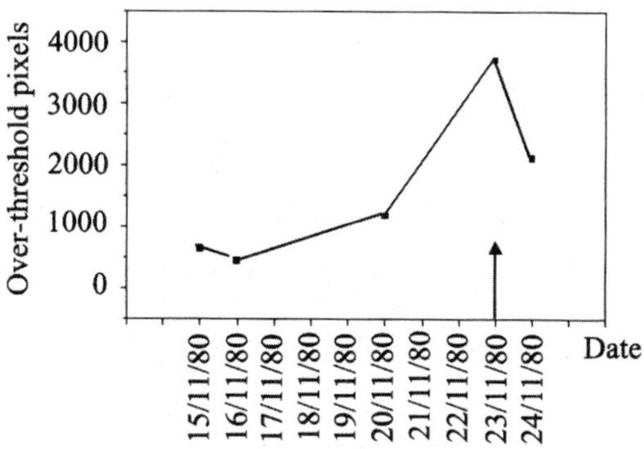

Fig. 7.22 Temporal variations of ALICE index registered by NOAA-AVHRR before the Irpinia earthquake 23.11.1980

Authors of the papers cited above believe that the temperature increase is due to the release of green-house gases before earthquakes, but we would like to present another interpretation of the observed results. The key factor for this is the latent heat flux measured by remote sensing satellites (Schulz et al. 1997). This parameter derived from the surface temperature and humidity reflects the latent heat processes, especially connected with water evaporation.

Let us imagine the stationary situation: we have the constant solar heat flux which is distributed between the heating of the ground surface, air, and the latent heat consumed for water evaporation. For the given air humidity everything is going smoothly. And suddenly, we remove a large amount of water molecules from the air violating the humidity balance. Immediately we will obtain the variations of the latent heat flux. Such a situation will be realized during the process of formation of the ion clusters (or disruption) by sharp changes of the water molecules in the air. The changes of the latent heat flux will be accompanied by the changes of the surface temperature (due to released solar heat from the latent heat), and humidity. All these changes are observed before the earthquakes (Singh et al. 2002, Dey and Singh 2004)

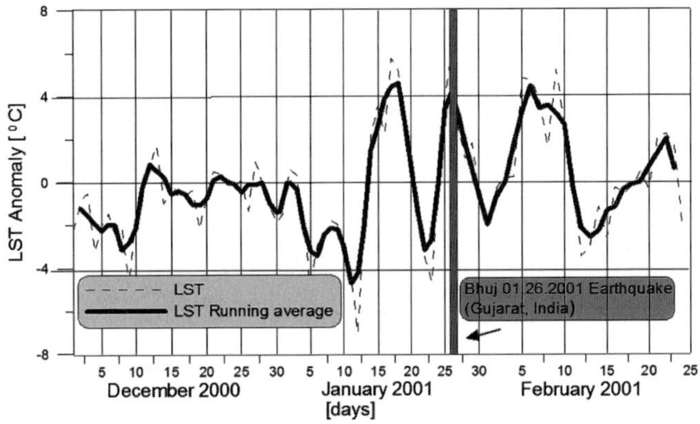

Fig. 7.23 Temperature anomaly around the time of Gujarat earthquake registered by NOAA-MODIS (After Ouzounov 2003)

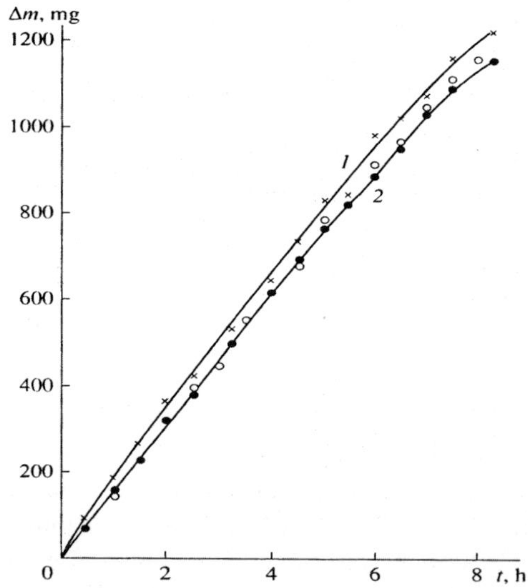

Fig. 7.24 Dependence of the water mass change under evaporation for electric field directed down (+), for electric field directed up (*open circles*), without field (*closed circles*)

The second factor that may possibly cause variations of the latent heat flux is changes of water evaporation properties under the action of the strong electric field. This effect was detected experimentally (Krasikov 2001). Figure 7.24 demonstrates the changes of the quantity of the water evaporated (located in the constant electric field) for two directions of the field, and the field off.

The electric field magnitude was 200–300 V, T=15° C, p=755–758 mm Hg. The evaporation increased under direction of the field from top to bottom.

It is interesting to note, that both effects: water molecules disappearance and electric field action on the water are present in our mechanism of seismo-ionospheric coupling. The first one is connected with the ion clusters formation and disruption, and the second one – with the anomalous electric field generation.

Therefore, one should expect the time coincidence of the observed pre-seismic ionospheric anomalies, and the registered anomalies of the latent heat flux (at least within the confidential interval one day). Such a case was found for the recent strong earthquake in Mexico 22.01.2003 M=7.8. Variations of the anomalous latent heat flux are presented in Fig. 7.25.

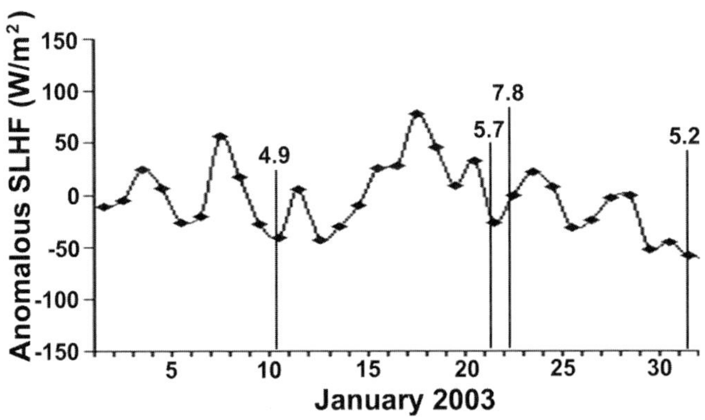

Fig. 7.25 Anomalous SLHF during registered in Colima vicinity, Mexico January 2003

One can see that the surface latent heat flux (SLHF) reached its absolute maximum in January, fore days before the strongest shock, i.e on the 18th of January. There were several foreshocks and aftershocks registered, and every of them is associated with a peak of SLHF.

The Colima GPS receiver registered GPS TEC until the main shock on the 22nd of January (according to local time it was 21st January at 2006 LT). For better resolution several days of January 2003 are presented in Fig. 7.26. To interpret the figure it is necessary to discuss the ionospheric morphology in Mexico. The ionosphere over the greater part of Mexico (excluding probably the most northern part) belongs to the outer slope of the northern crest of the equatorial anomaly. This means that the electron concentration follows the overall equatorial anomaly diurnal behavior, when the maximum of the anomaly development is reached in the afternoon. Empirically, it was determined that, at these latitudes, the daily maximum of electron concentration is reached at 1600 LT. One can see that these conditions are fulfilled on all days except the 18th of January, when the additional maximum is observed at 1010 LT. The statistical analysis shows that this peak goes beyond the limits of the upper boundary of 2σ, and can be regarded as anomalous ionosphere behavior, and according to our determination from Chapter 6 can be interpreted as the Colima earthquake precursor.

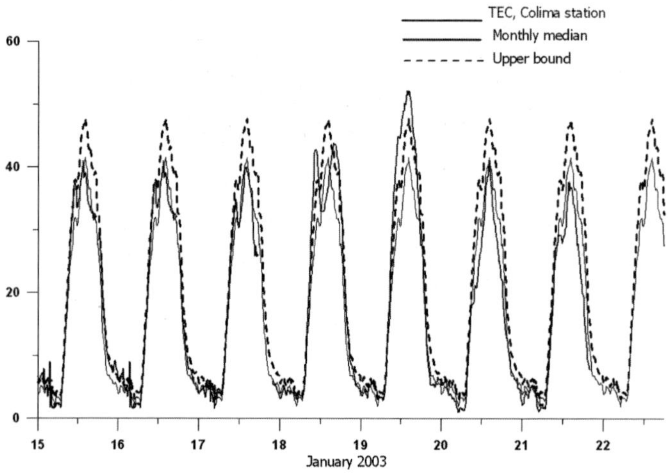

Fig. 7.26. GPS TEC registered by the Colima GPS receiver, January 2003. *Bold line* - experimental TEC measurements, *thin line* – monthly median *M, dashed line* – upper boundary $M+\sigma$

The anomalous character of the observed additional peak in daily variations of TEC on the 18th January is connected first of all with its local time position (1010 LT). Colima's position is projected at the slump of the Northern crest of the equatorial anomaly, and, consequently, is tightly related with the local time equatorial anomaly development. The maximum development of the crest at these latitudes is observed near 1500 LT and can never be associated with morning hours. The only way is to search the anomalous sources of the electron density amplification at 1010 LT, which is, in our opinion, the earthquake preparation process involving the anomalous electric field generation.

To check this, it would be interesting to track the spatial distribution of the observed anomaly. If it is in some way connected with the earthquake epicenter position, it helps to prove the observed ionospheric anomaly connection with the earthquake preparation process. The network of permanent GPS receivers of the INEGI agency of Mexico is not too dense but sufficient to be able to estimate the spatial distribution of the electron density over Mexico. Five stations were selected for this procedure (Table 7.5) and the map of TEC deviation for the time when the precursor was registered on the 18th January (Fig. 7.27) was constructed. The map clearly indicates the anomaly position over the future epicenter (Pulinets et al. 2003c). The only deficiency of the map is connected with the geographical configuration of the Mexican territory: the left bottom part of the map – is the ocean, where there are no GPS receivers. The only way to make progress in this situation is to use satellite data, which we will have (we hope) in the near future.

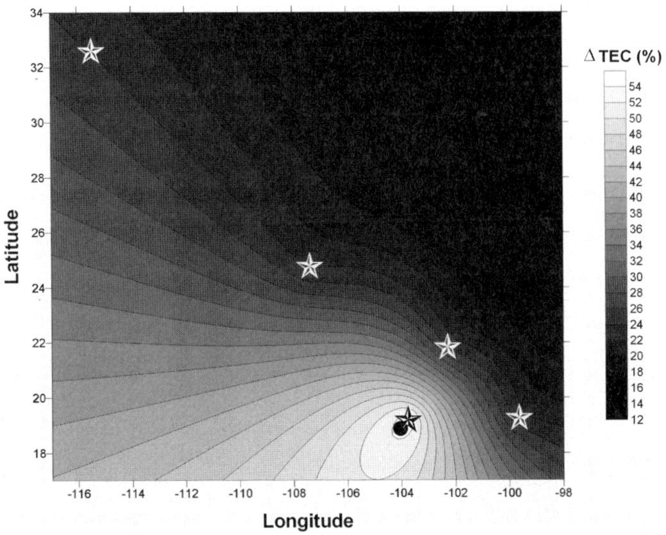

Fig. 7.27. The map of ΔTEC – deviation of vertical TEC from the monthly median for 1010 LT on 18th of January 2003 for Western Mexico 3 days before the Colima earthquake. The GPS station positions are shown by *asterisks*, the epicenter position – by ●

Table 7.5 Geodetic positions of INEGI agency GPS receivers used for the ionospheric map construction (Fig. 7.27)

Station	Latitude (N)	Longitude (W)
COL2	19.244	103.702
CUL1	24.799	107.384
INEG	21.856	102.284
MEXI	32.633	115.476
TOL2	19.293	99.643

After detecting the time coincidence of the ionospheric and SLHF flux anomalies the local meteorological data were analyzed. As the first step, the mean monthly January temperature at Manzanillo station close to the epicenter of the Colima earthquake was analyzed (Fig. 7.28). One can see from the figure that for the period of 50 years from 1954 to 2004 the temperature in January of year 2003 is the absolute maximum. So January of 2003 was analyzed in more detail using the data from two local meteorological observatories: Manzanillo (19.05 N, 104.32 W) and Colima (19.22 N, 103.7 W). The graphs of air temperature and relative humidity are presented in Figure 7.29. Both stations demonstrate the increase in the temperature range (difference between the daily maximum and minimum air temperature) starting from 12th January with correspondent diminishing of relative humidity. The humidity minimum is reached on 15th January, and then humidity increases sharply (especially at Manzanillo station) reaching the undisturbed level on 18th January.

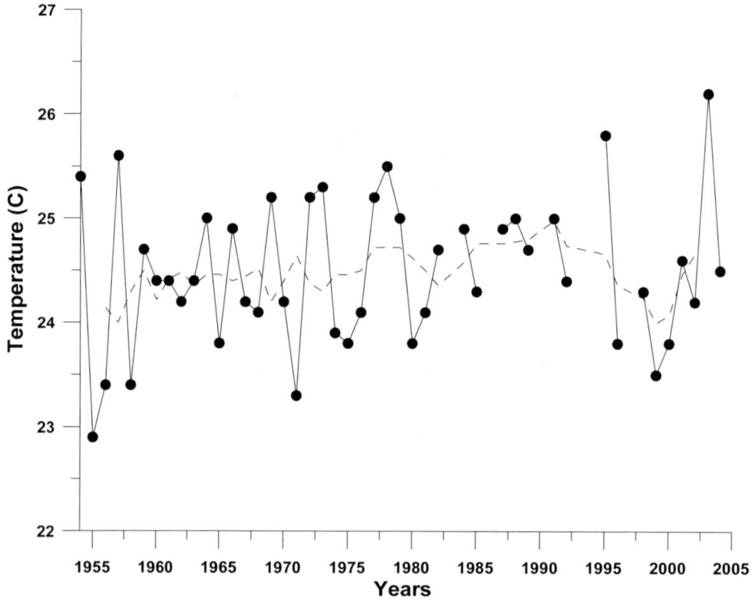

Fig. 7.28. January mean monthly temperature at Manzanillo station for the period 1954-2004

To check if the observed thermal anomaly has a local character, the map of the surface temperature distribution for the whole Mexico was built for the absolute temperature maximum observed on 14th January at Colima and Manzanillo. The map is presented in Fig. 7.30. Taking into account the complex relief of Mexico, the meteorological data were reduced to the 0 level altitude. And again, as in the case of the ionospheric anomaly, one can observe the absolute peak of the temperature close to the epicenter of the impending earthquake. The increased temperature is observed also along the activated tectonic fault.

As we can see from the data presented , the appearance of the ionospheric precursor coincides with the appearance of the main peak of the anomalous SLHF. These anomalies are preceded by the local surface temperature and relative humidity anomalies observed from 12th to 18th January. This chain of processes can be interpreted in the following way.

Starting from 12th January, the range of the temperature differences between the nighttime and daytime increases in the area of the earthquake preparation and shows a clear indication of the air humidity drop. This period can be associated with the large cluster formation accompanied by water molecule attachment and a reduction of the water content in the air. According to the existing concept, the water condensation on aerosols plays the most important role in the thermal balance of the atmosphere. Almost a half of the heat inflow into the atmosphere (~ 88 W/m^2 from the total 185 W/m^2) is the heat of water vapor condensation on aerosols (or in our opinion, water molecule attachment), (Bud'ko 1988).

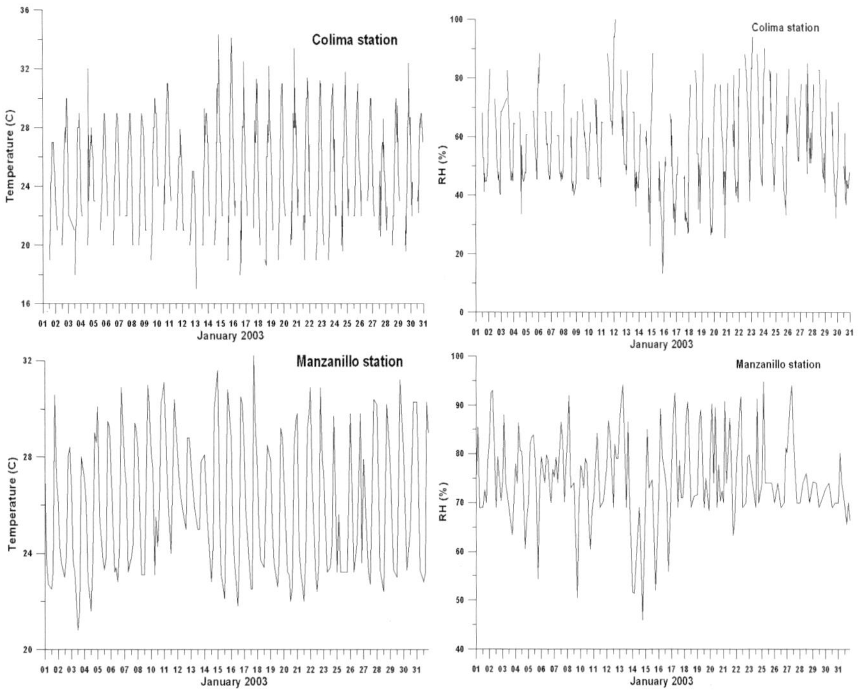

Fig. 7.29. Air temperature (left panel) and relative humidity (right panel) on January 2003 registered at Manzanillo and Colima meteorological observatories.

Under normal conditions, the condensation processes are within the normal atmosphere thermodynamics, but with the increased concentration of the condensation centers due to radon ionization, the process reaches anomalous values observed in the experimental data (Fig. 7.28).

After reaching the humidity minimum on 15-16th January, the sharp increase in humidity is observed accompanied by the anomalous SLHF growth reaching its maximum on 18th January. On the same day the ionospheric anomalies also started to be observed during the 18th and 19th January. This period can be associated with the neutral clusters disruption and the anomalous electric field generation described in Chapter 3. Reaching its maximum on 18th January, the anomalous electric field creates the anomaly in the ionosphere registered by GPS receivers.

The presented comparison is only one case study, but it gives us hope of obtaining more similar results supporting our physical mechanism and the relation of the observed thermal anomalies to the process of the anomalous electric field generation before strong earthquakes.

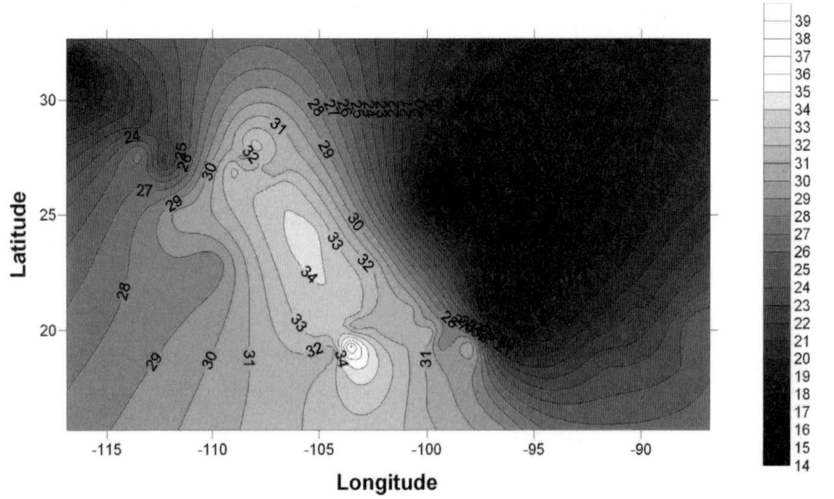

Fig. 7.30. Air temperature distribution over Mexico on 14 of January 2003 at 1400 LT (reduced to 0 m, i.e. sea level)

7.7 Future Tasks

7.7.1 Model Improving

Regardless of the fact that the model of seismo-ionospheric coupling presented in our book embraces all levels, starting from the ground surface and finishing with the magnetospheric processes, it still has some white spots. This means that further theoretical and experimental efforts are necessary to improve our model. The main problems were already discussed in the book, but here we would like to summarize them.

The first problem is not directly a problem with our model, but of our understanding of the physical processes of earthquake preparation: it is the local time dependence of the precursors. So many phenomena may contribute to this dependence that it is really very difficult to separate all of them.

Starting from underground, can we suppose a local time dependence of the Earth degassing? Probably not, only if we tie this process with atmospheric pressure, which may change regularly in local time.

The second probable reason is the weather. As was indicated in Chapter 3, strong wind can take out the new-formed ions not giving them the possibility to participate in the process of anomalous electric field generation.

The third probable reason – is the ionosphere itself. It experiences impacts from above (solar activity and cosmic rays) and from below (electric fields and movements of the neutral components). Of course the prevailing effect is the solar contribution. But there are periods in local time, when this influence weakens. The first one – is the early morning minimum, which very often deepens as a result of the seismogenic electric field effect, and the afternoon hours, when the density in the ionosphere starts to decline after passing of the main daily maximum. These afternoon hours are also characteristic for the ionospheric precursors appearance.

And the last probable reason – is the interaction of the anomalous electric field penetrated into the ionosphere with the existing electric fields, for example, the electric field responsible for the O^+ daily flux variations. Figure 7.31 (Pulinets et al. 1998b) presents a comparison of the ionospheric precursors amplitude maxima variation in local time and the O^+ plasmaspheric fluxes variation in local time according to the FLIP model (Szuszczewicz et al. 1996).

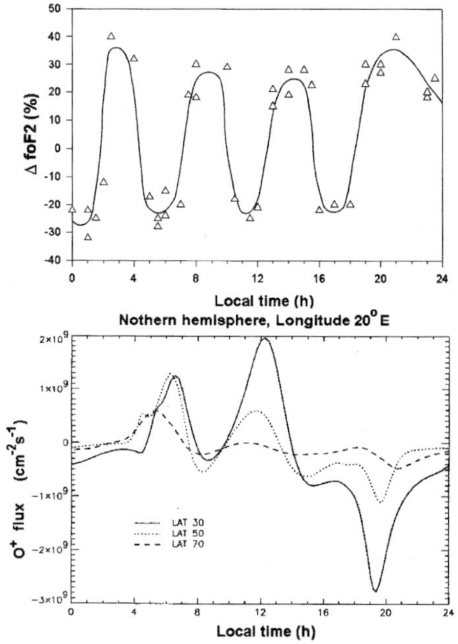

Fig. 7.31 *Upper panel* – maximum values of the critical frequency deviation as dependence on local time (composite picture for 54 earthquake precursors registered worldwide). *Bottom panel* – daily variations of the O^+ ions within the ionosphere calculated for three latitudes according to the FLIP model

Because the O^+ ion is the main constituent ion in the *F2*-region of the ionosphere, the interaction of plasmaspheric fields responsible for the ions flux magnitude with the anomalous electric field of seismic origin can modulate the flux giving variations of plasma density within the ionosphere in comparison with undisturbed conditions.

The second direction of the model's improvement is connected with the near ground plasma processes. We considered the static model not taking into account the possible plasma instabilities, which are natural to expect. The development of electrodynamics of the atmospheric dusty plasma consisting of heavy positive and negative ions is one of our future tasks for the development of our model. The possible instabilities excited in such plasma may explain the experimental results of registered electromagnetic emissions in the seismically active areas.

Next item – is the model calculation of the connection of the water molecules attachment with the latent heat flux variations. As was demonstrated in the previous paragraph, such a connection is quite possible, and may solve the existing gap in the explanation of the thermal anomalies before earthquakes.

And our last aim in the model's development – is modeling of the formation of aerosol clouds in the upper atmosphere and their electrodynamics. This may explain the anomalies of radio wave propagation in different frequency bands, and the pulsed radio emission registered over the seismically active areas.

7.7.2 Precursors Recognition and Regional Studies

It seems that we were able to demonstrate that we are very close to the practical application of our knowledge on the ionospheric precursors of earthquakes. This means that the most imminent task – is improvement of the proposed techniques of the identification of ionospheric precursors. This development should move in two directions.

The first one – is improvements of the statistical processing and pattern recognition techniques for different sources of the ionospheric information. Separate techniques should be developed for the ground-based ionosondes, other ones – for GPS TEC measurements, and special techniques for the satellite information processing.

But the pattern recognition technique should be based on the retrospective analysis of the ionospheric precursors in every given region. As we have seen from the previous discussion, the precursor's characteristics (especially, their behavior in local time) are different in different seismically active regions. So these regional peculiarities should be studied.

The last question regarding precursors recognition, which was not touched upon in the present book, is the precursors variation with the season and solar cycle phase. It is well known that there are strong variations of the ionosphere parameters as a function of season and solar cycle. It is quite possible that ionosphere sensitivity to the seismic events and the overall precursor's characteristics may also change with the season and solar cycle phase. It is our intention to reveal (or not) such dependencies.

The seasonal dependencies can be revealed not only as a result of ionospheric variations, but also as the weather conditions change with the season. Winds, rain, snow, and fog – everything may contribute to the electric field generation mechanism. These impacts should be studied both theoretically and experimentally. Probably, first such changes should be detected empirically to give some background to the theoretical modeling.

7.7.3 Interaction with Other Kinds of Precursory Phenomena

Our task was to describe the events responsible for the seismo-ionospheric coupling, but this does not mean that we propose basing all prediction on the ionospheric phenomena. The ionospheric precursors – are part of the more general physical process of earthquake preparation. This means that the prediction algorithm should be built on multiparameter analysis.

First of all the precursors should be ranked in two categories: the time of appearance before the earthquake, and their confidential merit.

All available seismic information should be used, starting from the seismic regioning, calculation of the seismic risk, and finishing with the most recent techniques based on the self-organized criticality.

Then the middle-term precursors should be tracked such as radon flux, water level etc. The cumulative principle should be used, adding every new appearing indication to the expert alarm system.

And finally, the short-term precursors should be processed, including all types of physical, geochemical, electromagnetic and biological monitoring.

7.8 Conclusion

Starting the book we had a doubt: to touch upon or not the question of earthquake prediction. First our intention was to concentrate only on the physical aspects of the problem and the model. But during the course of our work, our opinion changed. This happened due to two reasons.

Natural disasters, and earthquakes particularly are a challenge to mankind. And this challenge is directed mainly to scientists. Ordinary people expect from science and technology progress not only on effects like the ozone hole or Chernobyl, but also something which can increase their level of protection.

The second reason – is the hopeless discussion that was printed in Nature in 1996 and finished with the conclusion that earthquakes cannot be predicted. The history of science always has periods of sharp progress and stagnation but not capitulation. The earthquake preparation is not a "perpetuum mobile" or Ferma theorem, it is a real physical process. And our failures in earthquake prediction only show that we are very far from its understanding. We still do not know anything about what really happens inside the Earth at the earthquake's source. Prediction was based on some ideas on the internal structure and processes in the source based on the laboratory experiments, seismic waves analysis, and some external factors over the ground surface. The failure means only that this idea was wrong, or not correct enough to be successful in prediction. So we should look for new approaches, but not despair.

We want to propose the following. According to recent achievements in seismology, the earthquake preparation process is regarded now as chaotic. We want in this regard to make some comparisons, to explain our point of view. Let us take the typhoon or tornado, which are the classical representatives of chaotic processes in the atmosphere. We know well that August and September are the most probable months for a typhoons formation in the tropics. It is an analog of long-term prediction in seismology. Then, we go to the stage of typhoon preparation. Meteorologists also cannot tell where and when the typhoon will be formed but when it is formed, they can, with a high level of probability, predict the typhoon's propagation.

What analogy may there be for earthquakes? Up to now the earthquake preparation process is terra incognita. But it seems that there exists a stage when the process is irreversible. Something like when the typhoon has already formed. This period is not hours or minutes, but days. So we should look for the precursors that are characteristic just for this period of the earthquake's development. According to our experience, the ionospheric precursors belong just to this kind of precursors. We are working now on a statistical procedure to demonstrate this, but it seems to us that ionospheric precursors will permit us in the nearest future to break the chain of failures in earthquake prediction.

References

Afonin VV, Molchanov OA, Kodama T, Hayakawa M, Akentieva OA (1999) Statistical Study of Ionospheric Plasma Response to Seismic Activity: Search for Reliable Result from Satellite Observations, in Atmospheric and Ionospheric Electromagnetic Phenomena Associated with Earthquakes. Ed. Hayakawa M, Terra Scientific Publishing Company, Tokyo, pp. 597-618

Afonin VV, Akentieva OA, Molchanov OA, Hayakawa M (2000) Statistical study of equatorial anomaly from high apogee satellite APEX and low apogee satellite COSMOS-900, International Workshop on Seismo-Electromagnetics, Programme and Abstracts, NASDA, 19-22 Sep 2000

Afraimovich EL, Altynsev AT, Grechnev VV, Leonovich LA (2002) The response of the ionosphere to faint and bright solar flares as deduced from global GPS. Annals of Geophysics 45: 31-40

Akhmamedov KH (1993) Interferometric Measurements of the Temperature of the F2 Region of the Ionosphere during the Period of Iranian Earthquake of June 20, 1990. Geomagnetism and Aeronomy 33: 135-137

Aki K, Richards PG (1980) Quantitative Seismology. Theory and Methods. Freeman, New York

Alekseev VA, Alekseeva NG (1992) Investigation of Metal Transfer in the Biosphere During Gaseous Emission in Zones of Tectonic Activity Using Methods of Nuclear-physics. Nucl. Geophys, 6: 99-110

Alekseev VA, Alekseeva NG, Jchankuliev J (1995) On relation between Fluxes of metals in waters and radon in Turkmenistan region of seismic activity. Radiation Meas. 25: 637-639

Aleshina ME, Voronov SA, Galper AM, Koldashev SV, Maslennikov LV (1992) On the interrelation of the earthquakes sources positions and the areas of energetic particles precipitation from the radiation belt. Cosmic Res. 30: 79-83

Aleshina ME, Voronov SA, Galper AM, Koldashev SV, Maslennikov LV (1993) Precipitation of the high energy particles from the Earth's radiation belt and seismic activity. Izvestiya USSR Acad. Sci. Physics. 57: 97-99

Allegri L, Bella F, Della Monica G, Ermini A, Impora S, Sgrigna V, Biagi PF (1983) Radon and tilt anomalies detected before the Irpinia (South Italy) earthquake of November 23, 1980 at great distance from the epicenter. Geophys. Res. Let. 10: 269-272

Alouette 1. (1964) Ionospheric data ALOSYN. 16-30 April 1964. Defense Research Board. Telecommunication Establishment. Ottawa. Canada

Anderson EM (1905) The dynamics of faulting, Trans. Edinburgh Geol. Soc., 8: 387-402

Anderson EM (1942) The dynamics of faulting, 1st ed. Oliver and Boyd, Edinburgh

Antonova EA, Ivanov-Kholodny GS, Chertoprud VE (1996) Aeronomy of the E-region, “Yanus” Publ., Moscow, (in Russian)

Antselevich MG (1971) The influence of Tashkent earthquake on the Earth's magnetic field and the ionosphere. In "Tashkent earthquake 26 April 1966" FAN Publ. Tashkent 187-188

Appleton EV, Barnett MAF (1925) Local reflection of wireless waves from the upper atmosphere. Nature 25: 333-334

Araujo-Pradere EA, Fuller-Rowell TJ, Codrescu MV (2002) STORM: An empirical storm-time ionospheric correction model - 1. Model description. Radio Sci. 37: Art. No. 1070

Arijs E (1983) Positive and negative ions in the stratosphere. Ann. Geophys. 1: 149–160

Arijs E (1992) Stratospheric ion chemistry: Present understanding and outstanding problems. Planet. Space Sci. 40: 255 – 270

Arshadi M, Kebarle P (1970) Hydration of OH^- and O_2^- in the gas phase. Comparative salvation of OH^- by water and hydrogen halides. Effects of acidity. J. Phys. Chem. 74: 1483-1485

Asatryan GA, Asatryan GrA, Babayan VH (1991) The jump of intensity of ionizing component of the cosmic rays in stratosphere during period of Spitak earthquake of 7 December 1988. Izvestiya USSR Acad. Sci. Physics. 55: 500-504

Ashour-Abdalla M (1972) Amplification of whistler waves in the magnetosphere. Planet. Space Sci. 20: 639-662

Atmosphere. Hand-book (1991) Sankt-Petersburg. Gidrometeoizdat. 510 p.

Aushev VM, Ashkaliev YF, Wiens RH, Vodyannikov VV, Gordienko GI, Pogoreltsev AI, Yakovets AF (2002) Spectrum of Atmospheric Gravity Waves in the Mesosphere and Thermosphere. Geomagnetism and Aeronomy, 42: 533-541

Bak P, Tang C, and Wiesenfeld K (1988) Self-organized criticality. Phys. Rev. A 38: 364-374

Barabas' S (1991) Radon concentration in gases and water of the ore deposit and its connection with the mine blast threat. In Nuclear-geophysical studies. Sverdlovsk. Ural. Div. USSR Acad. Sci. pp. 65-70

Barsukov OM (1991) Solar Flares, Sudden Commencements, and Earthquakes. Izvestiya. Earth Physics 27: 1081-1083

Bering III EA, Few AA, Benbrook JR (1998) The Global Electric Circuit. Physics Today. 51(10): 24-30

Biagi PF, Ermini A, Kingsley SP (2001) Disturbances in LF radiosignals and the Umbria-Marche (Italy) seismic sequence in 1997-1998. Phys. Chem. Earth 26: 755-759

Bibl K (1998) Evolution of the ionosonde. Annali di geofisica 41: 667-680

Bilichenko SV, Inchin AS, Kim EF, Pokhotelov VA, Puschayev PP, Stanev GA, Streltsov AV, Chmyrev VM (1990) Ultra low frequency response of the ionosphere to the process of earthquake preparation. Doklady USSR Acad. Sci. 311: 1077–1081.

Bilitza D (2001) International Reference Ionosphere 2000. Radio Science. 36: 261-275

Biondi MA, Leu MT, Johnsen R (1972) Recombination of electrons with positive ions of the $H_3O^+(H_2O)_n$ series, - COSPAR symposium on D- and E-region ion chemistry. Aeronomy Rept. 1972, No. 48, pp. 266–269

Birfeld YG (1974) Seismo-ionospheric coupling. Application to ionospheric earthquake prediction, in "Search for Earthquake Precursors at Prognostic Networks" Nauka Publ. Moscow, pp. 200-202

Blaunstein N (2000) Large-Scale Stratification of the Ionosphere During Earthquake Preparation. Phys. Chem. Earth (A) 25: 789-791

Boborykin AM (1972) The study of the possibility of earthquake energy transfer into the ionosphere by acoustic-gravitational oscillations generated near the earthquake epicenter. Doklady Bielorus. Acad. Sci. 16: 355-357

Bolt BA (1964) Seismic air waves from the great 1964 Alaskan earthquake. Nature, 202: 1095-1096

Bolt BA (1993) Earthquakes. W. H. Freeman and company, New York

Bonchkovsky VF (1954) Changing of the atmospheric electric potential gradient as one of possible precursor of earthquakes. Proceedings of the Geophysical Institution No.25(152): 192 - 206 (in Russian)

Boška J, Šaul P (2001) Observation of Gravity Waves of Meteorological Origin in the F-Region. Phys. Chem. Earth (C). 26: 425-428

Bošková J, Šmilauer J, Jiříček F, Tříska P (1993) Is the ion composition of the outer ionosphere related to seismic activity? J. Atmos. Terr. Phys. 55: 1689-1695

Bošková J, Šmilauer J, Tříska P, Kudela K (1994) Anomalous behaviour of plasma parameters as observed by the Intercosmos 24 satellite prior to the Iranian earthquake of 20 June 1990. Studia Geoph. et Geod. 38: 213-220

Bowhill SA (1969) Interaction between the stratosphere and ionosphere, Annals of IQSY (International Quiet Sun Year) 5: 83–95

Bowman DD, Ouillon G, Sammis CG, Sornette A, and Sornette D (1998) An observation test of the critical earthquake concept, J. Geophys. Res., 103: 24,359-24,372

Boyarchuk KA (1997) Kinetics of elementary ions in the lower atmosphere acted upon by ionizing radiation. Izvestiya Atmos. Ocean. Phys. 33(2): 236 – 240

Boyarchuk KA (1999) Estimation of the concentration of complex negative ions resulting from radioactive contamination of the troposphere. Tech. Phys. 44(3): 292 – 294

Boyarchuk KA, Svirko YP (1996) Estimate of the concentration of elementary ions in radioactive contamination of the troposphere. Tech. Phys. Lett. 22(7): 575 – 577.

Boyarchuk KA, Lomonosov AM, Pulinets SA (1997a) Electrode Effect as an Earthquake Precursor, BRAS Physics/Supplement Physics of Vibrations 61(3): 175-179

Boyarchuk KA, Lomonosov AM, Pulinets SA, Hegai VV (1997b) Impact of Radioactive Contamination on Electric characteristics of the atmosphere. New Remote Monitoring Technique. BRAS Physics/Supplement Physics of Vibrations. 61(4): 260-266

Boyarchuk KA, Karelin AV, Lomonosov AM (2000) Peculiarities of the atmospheric air ionization under radioactive pollution. J.T.Ph. 70: 139 - 142

Bradley PA, Cander LR (2002) Proposed terminology for the classification and parameters for the quantification of variability in ionosphere morphology. Annals of Geophysics 45: 97-104

Brasseur G, Chatel A (1983) Modeling of stratospheric ions: a first attempt. An. Geophys. 1: 173-185

Breit G, Tuve MA (1926) A radio method of estimation the height of the conducting layer. Nature 116: 357-359

Briant CL, Burton JJ (1976) A Molecular Model for the Nuclearion of Water on Ions. J.Atm. Sci.. 33: 1357-1361

Brice N (1964) Fundamentals of very low frequency emission generation mechanisms. J. Geophys. Res. 69: 4515-4522

Brumbaugh DS (1999) Earthquakes. Science and society. Prentice Hall, Upper Saddle River, New Jersey

Brunelli BE, Namgaladze AA (1988) Physics of the ionosphere. Nauka, Moscow (in Russian)

Bud'ko MI (1988) The Earth's Atmosphere. In: The Encyclopedia of Physics, V. 1, Sovetskaya Entsyklopedia.: 133-147

Budnik AP, Vakulovsky AS, Dobrovolskaya IV (1994) Plasma track structure effect on the kinetic processes in lasers with pumping by fission fragments. Lett. J.T.Ph. 20: 67–72

Bufe C, Nanevicz J (1976) Atmospheric electric field observations, animal behavior, and earthquakes, - Proc. EHRP Conf. I., 23 – 24 Sept., 1976, California, pp. 95 – 106

Bulashevich YP, Utkin VI, Yurkov AK, Nikolaev VV (1996) Radon concentration changes in connection with mine blasts in deep mines. Doklady RAS 346: 245-248

Bullen KE, and Bolt BA (1985) An Introduction to the Theory of Seismology. Cambridge University Press, Cambridge

Castleman AW, Holland Jr. P M, Keesee RG (1982) Ion Association Processes and Ion Clustering: Elucidating Transitions from The Gaseous to The Condensed Phase. Radiat. Phys. Chem, 20: 57–74

Cerfas KE (1971) Atmospheric electricity phenomena preceding earthquakes. FAN Tashkent, pp. 184-187

Chamberlain JW (1995) Physics of the Aurora and Airglow. Classics in Geophysics. American Geophysical Union, Washington, 704 p.

Chen YI, Chuo JY, Liu JY, Pulinets SA (1999) Statistical study of ionospheric precursors of strong earthquakes at Taiwan area. XXVI URSI General Assembly, Toronto, 13-21 Aug. 1999, Abstracts, p.745

Chernyavski EA (1936) Atmosphere-electrical and electrotelluric phenomena during earthquakes. Soc. Sci.&Tech. No. 12: 26–35

Chernyavski EA (1955) The atmosphere-electric precursors of earthquakes in Meteorology and hydrology in Uzbekistan, Tashkent: 317-327 (in Russian)

Chimonas G, Hines GO (1970) Atmospheric gravity waves launched by auroral currents. Planet. Space Sci. 18: 565-582

Chmyrev VM, Isaev NV, Bilichenko SV (1986) Electric fields and hydromagnetic waves in the ionosphere over earthquake source. Geomagnetism and Aeronomy 26: 1020-1024

Chmyrev VM, Isaev NV, Bilichenko SV, Stanev G (1989) Observation by space-borne detectors of electric fields and hydromagnetic waves in the ionosphere over an earthquake centre. Phys. Earth. Planet. Interr. 57: 110-114

Chmyrev VM, Isaev NV, Serebryakova ON, Sorokin VM, Sobolev YP (1997) Small-Scale Plasma Inhomogeneities and Correlated ELF Emissions in the Ionosphere over an Earthquake Region. J. Atmos. Solar-Terr. Phys. 59: 967-974

Chmyrev VM, Poletaev BI, Pulinets SA (1998) Satellite constellation "Predvestnik". APEC-98 exibition, Kuala-Lumpur, Malaysia, November 1998

Chmyrev VM, Sorokin VM, Pokhotelov OA (1999) Theory of Small Scale Plasma Density Inhomogeneities and ULF/ELF Magnetic Field Oscillations Excited in the Ionosphere Prior to Earthquakes, in Atmospheric and Ionospheric Electromagnetic Phenomena Associated with Earthquakes. Ed. Hayakawa M, Terra Scientific Publishing Company, Tokyo, pp. 759-776

Christian HJ, Blakeslee RJ, Boccippio DJ, Boeck WL, Buechler DE, Driscoll KT, Goodman SJ, Hall JM, Koshak WJ, Mach DM, Stewart MF (2003) Global frequency and distribution of lightning as observed from space by the Optical Transient Detector. J. Geophys. Res. 108(D1) Art. No. 4005

Chuo YJ, Chen YI, Liu JY, Pulinets SA (2001) Ionospheric *foF2* variations prior to strong earthquakes in Taiwan area. Advances in Space Research. 27: 1305-1310

Chuo YJ, Liu JY, Pulinets SA, Chen YI (2002) The ionospheric perturbations prior to the Chi-Chi and Chia-Yi earthquakes. Journal of Geodynamics. 33: 509-517

Chuo YJ, Liu JY, Kamogawa M, Chen YI (2002) The anomalies in the *foEs* prior to M ≥6.0 Taiwan earthquakes. In "Seismo-Electromagnetics: Lithosphere-Atmosphere-Ionosphere Coupling", Eds. Hayakawa M and Molchanov OA, TERRAPUB, Tokyo, pp. 309-312

Cleary EN, Few AA, Bering III EAA (1997) Statistics and trends of global atmospheric electricity measurements. J Geophys Res.

Clilverd MA, Rodger CJ, Thomson NR (1999) Investigating seismoionospheric effects on a long subionospheric path. J. Geophys.Res. 104: 28171-28179

Cole RK Jr, Pierce ET (1965) Electrification in the Earth's Atmosphere for Altitudes Between 0 and 100 Kilometers. J. Geophys.Res. 70: 2735-2749

Curie J (1960) Radioactivity

Danilov AD (2000) New ideas of the D-region modeling. Adv. Space Res. 25: 5-14

Danilov AD, Lastovička J (2001) Effects of geomagnetic storms on the ionosphere and atmosphere. Int. J. Geomagn. Aeron. 2(3): 209-224

Datchenko EA, Ulomov VI, Chernyshova CP (1972) Electron density anomalies as the possible precursor of Tashkent earthquake. Dokl. Uzbek. Acad. Sci. No. 12: 30-32

Davies K (1990) Ionospheric radio. Peter Peregrinus, London

Davies JB, Archambeau CB (1998) Modeling of atmospheric and ionospheric disturbances from shallow seismic sources. Phys. Earth Planet. Interr. 105: 183-199

Davies K, Baker DM (1965) Ionospheric effects observed around the time of the Alaskan earthquake of March 28 1964. J. Geophys. Res. 70: 2251-2253

Denisenko PF, Nastasyina NV, Sotsky VV (1998) Application of the regularization method for electron density height profile reconstruction by the data of vertical topside sounding. Geomagnetism and Aeronomy. 38(1): 172-179

Depuev V, Zelenova T (1996) Electron density profile changes in a per-earthquake period. Adv. Space Res. 18: (6)115-(6)118

Depuev VH, Pulinets SA (2001) Epstein function global topside profile modeling on the basis of Intercosmos-19 topside sounding data (Quiet conditions), IRI Task Force Activity-2000. Proceedings, Trieste, 2001, p.35-40

Depuev VH, Pulinets SA, Radicella SM (2001) Epstein function topside profile modeling on the basis of Intercosmos-19 topside sounding data (Disturbed conditions), IRI Task Force Activity-2000. Proceedings, Trieste, 2001, p.41-46

Derr JS (1973) Earthquake lights: a review of observations and present theories. Bull. Seism. Soc. Amer. 63: 2177-2187

Deschamps A, King GCP (1980) The Campania Lucania (southern Italy) earthquake of November .3,1980, Earth and Planetary Sci. Let., 62: 296-304

Dey S, Singh R (2003) Surface Latent Heat flux as an earthquake precursor. Natural Hazards and Earth System Sci. 3: 749-755

Didebulidze GG (1997) Amplification/damping processes of atmospheric acoustic-gravity waves in horizontal winds with linear shear. Physics Letters A 235: 65-70

Dieminger W, Hartmann GR, Leitinger R eds. (1996) The Upper Atmosphere. Data Analysis and Interpretation. Springer-Verlag

Dobrovolsky IR (1984) Preparation Mechanics of a Tectonic Earthquake. IFZ Akad. Nauk USSR, Moscow (in Russian)

Dobrovolsky IR, Zubkov SI, and Myachkin VI (1979) Estimation of the size of earthquake preparation zones, Pageoph., 117: 1025-1044

Dobrovolsky IR, Gershenzon NI, Gokhberg MB (1989) Theory of electrokinetic effects occurring at the final stage in the preparation of a tectonic earthquake. Phys. Earth Planet. Inter. 57: 144-156

Doda L (2001) Geoseismic echo of solar storms. Air fleet. No. 3: 56-57

DohnWL, Posmetier ES (1964) Ground-coupled air waves from the great Alaskan earthquake. J.Geophys.Res. 69: 5357-5361

Dungey JW (1959) Effect of a neutral field turbulence in an ionized gas. J. Geophys. Res. 64: 2188-2194

Dungey JW (1963) Loss of Van Allen electrons due to whistlers. Planet. Space Sci. 11: 591-602

Dysthe B (1971) Some studies of triggered whistler J. Geophys. Res. 76: 6915-6931

Fatkullin MN (1982) Physics of the ionosphere. Totals of Science and technology, Series "Goemagnetism and high layers of the atmosphere", VINITI Publ. (in Russian)

Fehsenfeld FC, Ferguson EE (1969) Origin of water cluster ions in the D-region. J.Geophys. Res. 74: 2217-2222

Fishkova LM, Gokhberg MB, Pilipenko VA (1985) Relationship between night airglow and seismic activity. Ann. Geophys. 3: 679-694

Fleischer RL (1980) Radon Flux From The Earth: Methods of Measurement by the Nuclear track Technique. J. Geophys. Res. 85: 7553 - 7556

Fleischer RI (1981) Dislocation model for radon response to distance earthquakes. Geophys. Res. Let. 8: 477-480

Florida CD (1969) The Development of a Series of Ionospheric Satellites. Proc. IEEE 57: 867-875

Forbes JM, Palo SE, Zhang X (2000) Variability of the ionosphere. J. Atm. Sol. Ter. Phys. 62: 685-693

Fraser-Smith AC, Gernardi A, Mcgill PR, Ladd ME, Helliwell RA, Villard Jr OG (1990) Low-Frequency Magnetic Field Measurements near the Epicenter of the Ms 7.1 Loma Prieta Earthquake, Geophys. Res. Lett. 17: 1465-1468, 1990

Freund F (2000) Time-resolved study of charge generation and propagation in igneous rocks. J. Goephys. Res. 105: 11001-11019

Friedrich M, Pilgram R, Torkar KM (2001) A novel concept for empirical D-region modelling. Adv. Space Sci. 27: 5-12

Fujinawa Y, Takahashi K (1998) Electromagnetic radiation associated with major earthquakes. Phys. Earth Planet. Interr. 105: 249-259

Fujinawa , Takahashi K, Matsumoto T, Kawakami N (1999) Sources of Earthquke-Related VLF Electromagnetic Signals. In: Atmospheric and Ionospheric Electromagnetic Phenomena Associated with Earthquakes. Ed. Hayakawa M, Terra Scientific Publishing Company, Tokyo, pp. 597-618

Fukmoto Y, Hayakawa M, Yasuda H (2002) Reception of over-horizon FM signals associated with earthquakes. In Atmospheric and Ionospheric electromagnetic phenomena associated with earthquakes. Eds. M.Hayakawa and O.A.Molchanov, TERRAPUB, Tokyo. pp. 263-266

Füllekrug M (2003) ULF/ELF Interferometry. XXIII General Assembly of the International Union of Geodesy and Geophysics, Abstracts, Week B, GAII.04.07A/A8-002, p. B.193

Fux IM, Shubova RS (1995) VLF signal anomalies as response on the processes within near-earth atmosphere. Geomagn. Aeronom. 34: 130-135

Gaivoronskaya TV (1991) The seismic activity effects on the ionosphere. The Review. Preprint IZMIRAN No.36(983) Moscow, 25 p. (in Russian)

Gaivoronskaya TV, Pulinets SA (2002) Analysis of F2-layer variability in the areas of seismic activity. Preprint IZMIRAN No 2(1145) Moscow 20 p. (in Russian)

Galper AM, Grachev VM, Dmitrienko VV, Kirillov-Ugryumov VG, Polukhina NG, Tzarkov RN, Ulin SE (1983) Saptial-temporal correlation of the earthquakes and variations of high energy flux in the inner radiation belt. Cosmic Res. 21: 707-717

Galper AM, Dmitrienko VV, Nikitina NV, Grachev VM, Ulin SE (1989) Connection of the fluxes of charged particles of high energy in radiation belt with the Earth's seismicity. Cosmic Research 27: 789-792

Galper AM, Koldashov SV, Voronov SA (1995) High energy particle flux variations as earthquake predictors. Adv. Space Res. 15: (11)131- (11)134

Galperin YI, Gladyshev VA, Jorjio NV, Larkina VI, Mogilevsky MM (1992) Energetic particle precipitation from the magnetosphere above the epicenter of approaching earthquake. Cosmic Research 30: 89-106

Garavaglia M, Dal Moro G, Zadro M (2000) Radon and Tilt Measurements in a seismic Area: Tempreature Effects. Phys. Chem. Earth, 25: 233-237

Garmash SV, Lin'kov EM, Petrova LN, Shved GM (1989) Generation of Atmospheric Oscillations by Seismic Gravitational Vibrations of the Earth. . Izvesitya Atm. Ocean Phys. 25: 1290-1299

Gavrilov NM, Fukao S (1999) A comparison of Seasonal Variations of Gravity Wave Intensity Observed by the MU Radar with a Theoretical Model. J. Atm. Sci. 56: 3485-3494

Geller RJ (1997) Earthquke prediction: a critical review. Geophys. J. Int. 131: 425-450

Gendrin R (1975) Waves and wave-particle interaction in the magnetosphere; a review. Space Sci. Rev. 18: 145-200

Gladyshev V, Baransky L, Schekotov A, Fedorov E, Pokhotelov O, Andreevsky S, Rozhnoi A, Khabazin Y, Belyaev G, Gorbatikov A, Gordeev E, Chebrov V, Sinitsyn V, Lutikov A, Yunga S, Kosarev G, Surkov V, Molchanov O, Hayakawa M, Uyeda S, Nagao T, Hattori K, Noda Y (2001) Study of electromagnetic emissions associated with seismic activity in Kamchatka region. Natural Hazards and Earth System Sci. 1: 127-136

Gohberg MB, Morgunov VA, Yoshino T, Tomizawa I (1982) Experimental measurement of electromagnetic emissions possibly related to earthquakes in Japan. J. Geophys. Res. 87: 7824-7828

Gokhberg MB, Pilipenko VA, Pokhotelov OA (1983a) Satellite observations of electromagnetic radiation above the epicentral region of an imminent earthquake. Dokl. of USSR Acad. Sci. 268(1): 53-55

Gokhberg MB, Pilipenko VA, Pokhotelov OA (1983b) Seismic Precursors in the ionosphere. Izvestiya Earth Physics 19: 762-765

Gokhberg MB, Gershenzon NI, Gufeld IL, Kustov AV, Liperovsky VA, Khusametdinov SS (1984) Possible Effects of the Action of Electric Fields of Seismic Origin on the Ionosphere. Geomagnetism and Aeronomy 24:183-186

Gokhberg MB, Gufel'd IL, Marenko VF, Ponomarev EA, Rozhnoy AA, Yampolsky VS (1987) Studies of Perturbations of Natural and Artificial Electromagnetic Fields by Sources of Seismic Origin. Izvestiya, Earth Physics. 23: 102-108

Gokhberg MB, Kustov AV, Liperovskiy VA, Liperovskaya RK, Kharin EP, Shalimov SL (1989a) Perturbations of the Ionospheric F-Layer Prior to Strong Earthquakes. Izvestiya, Earth Physics 24: 254-260

Gokhberg MB, Gufeld IL, Rozhnoy AA, Marenko VF, Yampolsky VS, Ponomarev EA (1989b) Study of seismic influence on the ionosphere by super long wave probing of the Earth-ionosphere waveguide. Phys. Earth Planet. Inter. 57: 64-67

Gokhberg MB, Morgounov VA, Pokhotelov OA (1995) Earthquake Prediction. Seismo-electromagnetic phenomena. Gordon and Breach Science Publishers, Amsterdam

Gokhberg MB, Nekrasov AK, Shalimov SL (1996) On the influence of unstable release of green-effect gases in seismically active regions on the ionosphere. Izvestiya Earth Physics, No. 8, pp. 52-57

Gong SS, Yang GT, Wang JM, Liu BM, Cheng XW, Xu JY, Wan WX (2002) Occurrence and characteristics of sporadic sodium layer observed by lidar at a mid-latitude location. J. Atm. Sol. Ter. Phys. 64: 1957-1966

Gonzalez WD, Joselyn JA, Kamide Y, Kroehl HW, Rostoker G, Tsurutani BT, Vasyliunas VM (1994) What is a geomagnetic storm? J. Geophys Res. 99: 5771-5792

Griffith AA (1920) The phenomena of rupture and flow in solids, Phil. Trans. R. Soc. London Ser. A 221:169-198

Grigoriev IS, Melikhov EZ Eds. (1991) Physical quantities. Handbook Energoatomizdat Moscow 1232 p.

Grimalsky VV, Hayakawa M, Ivchenko VN, Rapoport YG, Zadorozhnii VI (2003) Penetration of an electrostatic field from the lithosphere into the ionosphere and its effect on the D-region before earthquakes. J. Atm. Sol. Terr. Phys. 65: 391-407

Gudzenko LI, Yakovlenko SI (1978) The plasma lasers. Atomizdat. Moscow. 112 p.

Gufeld IL, Rozhnoy AA, Tyumentsev SN, Sherstyuk SV, Yampolsky VS (1992) Radio Wave Field Disturbances Prior to Rudbar and Rachinsk Earthquakes. Izvestiya, Earth Physics. 28: 267-270

Gufeld IL, Gusev AG, Pokhotelov OA (1994) Is the prediction of earthquake date possible by VLF radio wave monitoring method? In: Hayakawa M, Fujinawa Y (eds.) Electromagnetic Phenomena Related to Earthquake Prediction. Terra Sci. Publ. pp. 381-390

Gulyaeva TL (1995) EISCAT catalog specification of the ionospheric quiet and disturbed conditions. Adv Space Res. 16 (1): 139-142

Hao J (1988) The anomalous of atmospheric electric field at the ground level and earthquakes. Acta seismol. sin. 10: 207 – 212

Hao J, Tang T and Li D (2000) Progress in the research of atmospheric electric field anomaly as an index for short-impending prediction of earthquakes. J. Earthquake Pred. Res. 8: 241-255

Hata M, Takumi I., Yabashi S. (1998) A Model of Earthquake Seen by Electromagnetic Observation,- Proc. of European Geophysical Society XXIII General Assembly, Nice, France 20 -24 April 1998. Annales Geophysicae, Supplement IV, 16: 1188

Hattori K, Akinaga Y, Hayakawa M, Yumoto K, Nagao T, Uyeda S (2002) ULF magnetic anomaly preceding the 1997 Kagoshima Earthquakes. In: Seismo-Electromagnetics: Lithosphere-Atmosphere-Ionosphere Coupling, Eds. M. Hayakawa and O. A. Molchanov, TERRAPUB, Tokyo, 2002, pp. 19-28

Hayakawa M (ed) (1999) Atmospheric and Ionospheric Electromagnetic Phenomena Associated with Earthquakes. Terra Scientific Publishing Company, Tokyo

Hayakawa M (ed) (2001) NASDA's Earthquake Remote Sensing Frontier Research. Seismo-Electromagnetic Phenomena in the Lithosphere, Atmosphere and Ionosphere. Final Report. The University of Electro-Communications, Chofu, Tokyo, Japan

Hayakawa M, Fujinawa Y (eds.) (1994) Electromagnetic Phenomena Related to Earthquake Prediction. Terra Sci. Publ.

Hayakawa M, Molchanov OA, Kodama T, Afonin VV, Akentieva OA, (2000) Plasma Density Variations Observed on a Satellite Possibly Related to Seismicity. Adv. Space Res. 26: 1277-1280

Hayakawa M, Molchanov OA (eds.) 2002 Seismo-Electromagnetics: Lithosphere-Atmosphere-Ionosphere Coupling, TERRAPUB, Tokyo

Hegai VV, Kim VP, Illich-Svitych PV (1990) The formation of a cavity in the night-time midlatitude ionospheric E-region above a thundercloud. Planet. Space Sci. 38: 703-707

Hegai VV, Kim VP, Nikiforova LI (1997) A Possible Generation Mechanism of Acoustic-Gravity Waves in the Ionosphere before Strong Earthquakes. J. Earthquake Predict. Res. 6: 584-589

Heinicke J, Italiano F, Lapenna V, Martinelli G, and Nuccio PM (2000) Coseismic Geochemical Variations in some Gas Emissions of Umbria Region (Central Italy). Phys. Chem. Earth (A) 25: 289-293

Hernandez-Pajares M., Juan JM, Sanz J. (1997) High resolution TEC monitoring method using permanent ground GPS receivers. Geophys. Res. Lett. 24: 1643-1646

Hines CO (1974) The upper atmosphere in motion: A selection of papers with annotation. Geophys. Monogr. 18. Amer. Geophys. Union, Washington D.C.

Hirshberg JR, Currie G, Breiner S (1967) Long period geomagnetic fluctuations after the 1964 Alaskan earthquake. Earth Planet.Sci. Lett., 3: 426-428

Holub RF and Bready BT (1981) The Effect of Stress on Radon Emanation from Rock. J. Geophys. Res. 86: 1776-1784

Holzer RE (1972) Atmospheric Electrical Effects of Nuclear Explosions. J. Geophys. Res. . 77: 5845-5855

Hoppel WA (1962) Electrode effect. Comparison of theory and measurement. In: Planetary Electrodynamics, Vol.2, pp. 167 – 181

Hoppel WA (1967) Theory of electrode effect. J. Atmos. Terr. Phys. 29: 709-721

Hoppel WA, Anderson RV, Willett JC (1986) Atmospheric Electricity in the Planetary Boundary Layer, - in Studies in Geophysics. In: The Eart's Electrical Envirinment. National Academy Press, Washington. D.C. pp. 149 – 165

Horanyi M, Goertz CK (1990) Coagulation of Dust Particles in a Plasma. Astrophys. J. 361: 155-161

http://sidc.oma.be/index.php3

http://spidr.ngdc.noaa.gov/spidr/

http://swdcdb.kugi.kyoto-u.ac.jp/

http://thunder.nsstc.nasa.gov/data/OTDsummaries/ (Optical Transient Detector)

Huang X, Reinisch BW (1996) Vertical electron density profiles from the Digisonde network. Adv. Space Res. 18: 121-129

Ikeya M, Furuta H, Kajiwara N, Anzai H (1996) Ground Electric Field Effects on Rats and Sparrows: Seismic Anomalous Animal Behaviors (SAABs) Jpn. J. Appl. Phys. 35: 4587-4594

Imamura T, Ogawa T (1995) Radiative damping of gravity waves in the terrestrial planetary atmospheres. Geophys. Res. Lett. 22: 267-270

Imankulov AC, Struminski VI, Nepein NA (1987) Measurement of the atmospheric electric field during the earthquake of 27 October 1987 in Tadzhik SSR . In: Atmospheric electricity. Proceedings of III all-union symposium. Gidrometeoizdat. Leningrad pp. 33-37

Imankulov AC, Struminski VI, Tatarinov SP (1990) Results of observation of anomalous variations of electric field intensity before earthquakes. Proceedings of the IV all-union symposium on atmospheric electricity. Gidrometeoizdat. Leningrad pp. 33-37

Inan US, Bell TF, Helliwell RA (1978) Nonlinear pitch angle scattering of energetic electrons by coherent VLF waves in the magnetosphere. J. Geophys. Res. 83:3235-3248

Ismaguilov VS, Kopytenko YA, Hattori K, Voronov PM, Molchanov OA, Hayakawa M (2001) ULF magnetic emissions connected with under sea bottom earthquakes. Natural Hazards and Earth System Sci. 1: 23-31

Iudin DI, Korovkin NV, Molchanov OA, Surkov VV, and Hayakawa M (2002) Model of earthquake triggering due to gas-fluid "bubble" upward migration. I. Physical rationale, in Atmospheric and Ionospheric electromagnetic phenomena associated with earthquakes. Eds. M.Hayakawa and O.A.Molchanov, TERRAPUB, Tokyo: 177-185

Isräel H (1973) Atmospheric Electricity/ Vols 1 and 2. Israel Program for Scientific Translation

Isräel YA (1996) Radioactive precipitations after nuclear explosions and emergencies. Progress-Pogoda, Sankt-Petersburg 355 p.

Jackson JE (1969) The reduction of Topside Ionograms to Electron-Density Profiles. Proc. IEEE. 57: 960-976

Jaffe G (1913) Elementary processes in particle tracks. Ann. Phys. Leipzig 42: 303-307

Jakowski N (1996) TEC Monitoring by Using Satellite Positioning Systems. In: Modern Ionospheric Science. EGS Publ. Katlenburg-Lindau. 371-390

Jason S, Pulinets S, da Silva Curiel A, Sweeting M (2003) Earthquake Forecast Science Research with a Small Satellite. Royal Soc. Philosoph. Trans.: Math., Phys.&Eng. Sci., **361**, No. 1802: 169-173

Jianguo H (1989) Near earth surface anomalies of the atmospheric electric field and earthquakes. Acta Seismol. Sin. 2: 289–298

Kanamori H (1977) The energy release in great earthquakes. J. Geophys. Res. 82: 2981-2987

Kanamori H (1994) Mechanics of Earthquakes, In: Annual Review of Earth and Planetary Science, Annual Reviews Inc., Palo Alto, Ca, USA, 22: 207-237

Karpachev AT, Deminova GF, Pulinets SA (1995) Ionospheric changes in response to IMF variations, J. Atm. Terr. Phys. 57: 1415-1432

Kasahara K (1981) Earthquake mechanics, Cambridge University Press, Cambridge

Kawamoto H, Ogawa T (1984) A steady state model of negative ions in the lower stratosphere. Planet. Space Sci. 32: 1223–1233

Kawamoto H, Ogawa T (1986) First model of negative ion composition in the troposphere. Planet. Space Sci. 34: 1229-1239

Kazimirovski ES (2002) Coupling from below as a source of ionospheric variability: a review. Annals of Geophysics 45: 1-29

Kazimirovsky E, Herraiz M, De la Morena BA, (2003) Effects on the ionosphere due to phenomena below it. Surveys in Geophysics. 24: 139–184

Kebarle P, Searles SK, Zolla ZuA, Seabrogh J, Arshadi M (1967) The solvatation of the hydrogen ion by water molecules in the gas phase. J. Am. Chem. Soc. 89: 6393-6403

Keesee RG, Lee N, Castleman Jr. AW (1979) Properties of clusters in the gas phase. 3. Hydration complexes of carbonate(1-) and bicarbonate(1-) ions. J. Am. Chem. Soc. 101, 2599-2604

Keilis-Borok VI, Kossobokov VG (1990) Premonitory activation of earthquake flow, Algorithm, M8. Phys. Earth Planet. Inter. 61: 73-83

Kelley MC (1989) The earth's ionosphere: plasma physics and electrodynamics. Academic Press Inc.

Kikuchi H (2001) Electrodynamics in Dusty and Dirty Plasmas, Kluver Academic Publishers, Dordrecht, Netherlands, 2001

Kim VP, Hegai VV, Illich-Svitych PV (1994, Russian edition Sept-Oct 1993) On the Possibility of a Metallic Ion Layer Forming in the E-Region of the Night Midlatitude Ionosphere Before Great Earthquakes. Geomagn. and Aeronomy. 33: 658-662

Kim VP, Hegai VV, Illich-Svitych PV (1994) On one possible ionospheric precursor of earthquakes Physics of the Solid Earth 30: 223-226

Kim VP, Hegai VV, Nikiforova LI (1995) On the possible disturbance of the Night E-region of the ionosphere over the large scale tectonic fault. Physics of the Earth. No. 7: 35-39

Kim VP, Hegai VV (1997) On Possible Changes in the Midlatitude Upper Ionosphere before Strong Earthquakes. J. Earthq. Predict. Res. 6: 275-280

Kim VP, Hegai VV (1999) A Possible Presage of Strong Earthquakes in the Night-Time Mid-Latitude F_2 Region Ionosphere. in Atmospheric and Ionospheric Electromagnetic Phenomena Associated with Earthquakes. Ed. Hayakawa M, Terra Scientific Publishing Company, Tokyo, pp. 619-627

Kim VP, Pulinets SA, Hegai VV (2002) The theoretical model of the possible changes in the night-time midlatitude D-region of the ionosphere over the zone of strong earthquake preparation, Radiophys. Quantum Radiophys. 45: 289-296

King C-Y (1996) Gas geochemistry applied to earthquake prediction. An overview. J. Geophys. Res. 91: 12,269-12,281

King Freund (1984)

King C-Y, Zhang W, and King B-S (1993) Radon Anomalies on Three Kinds of Faults in California. Pageoph. 141: 111-124

Klusman RW (1993) Soil Gas and Related Methods for Natural Resource Exploration. Wiley, Chichester

Kodama T (1994) The Outline of Electric and Magnetic Field Observation Mission - ELMOS: Electric and Magnetic field Observation Satellite. Proceedings of the 17th Japanese Conference on Remote Sensing, Dec. 1991

Kodama T, Molchanow OA, Hayakawa M (2000) NASDA EARTHQUAKE REMOTE SENSING FRONTIER RESEARCH - Feasibility of Satellite Observation of Seismoelectromagnetics. Adv Space Res. 26: 1281-1284

Kolokolov LE, Shalagina NA (1978) Sporadic E-layer formation after Karymski volcano eruption, in Studies on geomagnetism and aeronomy problems in high latitudes. Magadan, USSR Acad. Sci., p. 76

Kondo G (1968) The variation of the atmosphere electric field at the time of earthquake. Memoirs of the Kakioka magnetic observatory. 13(1): 11 - 23

Korovkin NV, Iudin DI, Molchanov OA, Hayakawa M, and Surkov VV (2002) Model of earthquake triggering due to gas-fluid "bubble" upward migration II. Finite-automaton model, in Atmospheric and Ionospheric electromagnetic phenomena associated with earthquakes. Eds. M.Hayakawa and O.A.Molchanov, TERRAPUB, Tokyo: 187-194

Kossobokov VG, Keilis-Borok VI, Turcotte DL, and Malamud BD (2000) Implications of a Statistical Physics Approach for Earthquake Hazard Assessment and Forecasting. Pure appl. Geophys. 157: 2323-2349

Kossyi IA, Kostinsky AY, Matveev AA, Silakov VP (1994) Plasmachemical processes in the non-equilibrium nitrogen-oxygen mixture. Inst. General Phys. Trans. V. 47. Ed. Kovrizhnykh LM. Nauka, Moscow pp. 37-57

Kostoglodov V, Bandy W, Domínguez J, Mena M. (1996) Gravity and seismicity over the Guerrero seismic gap, Mexico, Geophys. Res. Lett. 23: 3385-3388

Kostoglodov V, Larson K, Singh SK, Lowry A, Santiago JA, Franco SI, and Bilham R (2003) A large silent earthquake in the Guerrero seismic gap, Mexico, accepted for Geophys. Res. Lett.

Kouris SS, Spalla P, Zolesi B (2001) Could ionospheric variations be precursors of a seismic event? A short discussion. Annali di Geofisica 44: 395-402

Kozyreva LI (1993) Sizes (Diameters) of Zones in which Biological Earthquke Precursors Occur. Transactions (Doklady) of the Russian Acad. Sci. Geophysics 333A(9): 1-7

Krall NA, Trivelpiece AW (1973) Principles of Plasma Physics. McGraw-Hill Book Company

Krasikov NN (2001) The Characteristic of Electricity in Lower Layers of the Atmosphere. Doklady. Earth Sciences. 377(2): 263-265

Krider EP and Roble RW Eds (1986) The Earth's Electrical Environment. National Academy Press, Washington D.C.

Kushida Y, Kushida R (2002) Possibility of earthquake forecast by radio observations in the VHF band, J. Atm. Electr. 22: 239-255

Kuzhevskij BM, Nechaev OY, Shavrin PI, Volodichev NN (1996) Variations of neutron concentration in the thermal neutron field of the Earth atmosphere. Preprint SINP 96-7/414, 1996

Larkina VI, Migulin VV, Nalivaiko AV, Gershenzon NI, Gokhberg MB, Liperovsky VA, Shalimov SL (1983) Observation of VLF emissions, related with seismic activity, on the Intercosmos-19 satellite. Geomagn. Aeronom. 23: 684-687

Larkina VI, Migulin VV, Molchanov OA (1988) Seismicity and seismic hazard forecast by the results of observation of the low frequency noise-like emissions at the altitudes of upper ionosphere. Vulkanologiya i seismilogiya (Vulcanology and seismology) No 6: 94-102

Larkina VI, Migulin VV, Molchanov OA, Khar'kov IP, Inchin AS, Schvetcova VB (1989) Some statistical results on very low frequency radiowave emissions in the upper ionosphere over earthquake zones. Phys.Earth Planet.Inter. 57: 100-109

Larkina VI, Migulin VV, Sergeeva NG, Senin BV (2001) A Comparative Analysis of Electromagnetic Effects above Tectonic Fracture Zones. Doklady Earh Sci. 376: 127-133

Lee N, Keesee RG, Castleman Jr. AW (1980) The properties of clusters in the gas phase. IV. Complexes of H2O and HNOx clustering on NOx. J. Chem. Phys. 72, 1089-1094

Leonard RS, Barnes RA (1965) Observation of ionospheric disturbances following the Alaska earthquake. J.Geophys. Res. 70: 1250-1253

Lin'kov EM, Petrova LN, Osipiv KS (1990) Seismic Gravitational Pulsations of the Earth and Atmospheric Disturbances as Possible Precursors of Strong Earthquakes. Doklady USSR Acad. Sci. 313: 1095-1098

Liperovskaya EV, Silina AS, Saidshoev A, Liperovsky VA, Meister C.-V., Vasil'eva NE (2000) On the Effect of Spread- E_s of Night Sporadic Layers. Geomagnetism and Aeronomy 40: 112-114

Liperovsky VA, Alimov OA, Shalimov SA, Gokhberg MB, Liperovskaya RH, Saidshoev A (1990) Ionosphere F-region studies before earthquakes. Izvestiya Ussr Acad. Sci. Physics of the Solid Earth No. 12: 77-86

Liperovsky VA, Gladyshev VA, Shalimov SL (1991) Lithospheric-Ionospheric Relationships Prior to Earthquakes. Izvestiya, Earth Physics 27: 190-197

Liperovsky VA, Pokhotelov OA, Shalimov SA (1992) Ionospheric precursors of the earthquakes. Nauka, Moscow, 304 p (in Russian)

Liperovsky VA, Pokhotelov OA, Liperovskaya EV, Parrot M, Meister C.-V., Alimov OA (2000) Modification of sporadic E-layers caused by seismic activity. Surveys in Geophysics. 21: 449-486

Liu JY, Tsai HF, Jung TK (1996) Total electron content obtained by using the global positioning system, Terr. Atmos. Oceanic Sci. 7: 107-117

Liu JY; Chen YI, Pulinets SA, Tsai YB, Chuo YJ (2000) Seismo-ionospheric signatures prior to M ≥ 6.0 Taiwan earthquakes. Geophys. Res. Lett. 27: 3113-3116

Liu JY; Chen YI, Chuo YJ, Tsai HF (2001) Variations of ionospheric total content during the Chi-Chi earthquake. Geophys. Res. Lett. 28: 1381-1386

Liu JY, Chuo YJ, Pulinets SA, Tsai HF, Zeng X (2002) A study on the TEC perturbations prior to the Rei-Li, Chi-Chi and Chia-Yi earthquakes. In "Seismo-Electromagnetics: Lithosphere-Atmosphere-Ionosphere Coupling", Eds. Hayakawa M and Molchanov OA, TERRAPUB, Tokyo, pp. 297-301

Liu JY, Chen YI, Chuo YJ, Tsai HF, Liu CH, Pulinets SA (2003) A Statistical Study of Ionospheric Precursors of M≥5.0 Earthquakes in Taiwan Area (Private communication)

Liu JY, Chuo YJ, Shan SJ, Tsai YB, Pulinets SA, Yu SB (2004) Pre-earthquake anomalies registered by continuous GPS TEC measurements. Annales Geophysicae, 22: 1585-1593

Lomnitz C (1994) Fundamentals of earthquake prediction. Wiley, New York

MacGorman DR, Rust WD (1998) The Electrical Nature of Storms. Oxford University Press, Oxford

Maeda K (1999) Pulsed Emission Observed at 22.2 MHz at the Time of the Kobe Earthquake in 1995. In: Atmospheric and Ionospheric Electromagnetic Phenomena Associated with Earthquakes, ed. M. Hayakawa, TERRAPUB, Tokyo, 1999, pp. 543 – 549

Makino M, Ogawa T (1985) Quantitative Estimation of Global Circuit. J. Geophys. Res. 90: 5961-5966

Manson AH., Meek CE, Koshyk J, Franke S, Fritts DC, Riggin D, Hall CM, Hocking WK, MacDougall J, Igarashi K, Vincent RA (2002) Gravity wave activity and dynamical effects in the middle atmosphere (60–90km): observations from an MF=MLT radar network, and results from the Canadian Middle Atmosphere Model (CMAM). J. Atm. Sol. Terr. Phys. 64: 65 – 90

Mareev EA, Iudin DI, Molchanov OA (2002) Mosaic source of internal gravity waves associated with seismic activity. In "Seismo-Electromagnetics: Lithosphere-Atmosphere-Ionosphere Coupling", Eds. Hayakawa M and Molchanov OA, TERRAPUB, Tokyo, pp. 335-342

Martel EA (1985) Enhanced Ion Production in Convective Storms by Transpired Radon Isotopes and Their Decay Products. J. Geophys. Res. 90: 5909-5916

Matveyev IV, Morgunov VA, Statiev AV (1991) Atmospheric electric field observation in areas with tectonic activity. In: Methods of seismoelectromagnetic processes analysis. Nauka, Moscow, pp. 54-66

Mayorov SA, Tkachev AN, Yakovlenko SI (1994) The metastable state of overcooled plasma. In: Ball lightning in the laboratory, Moscow, Khimiya publ., 228–247

McCormick RJ, Rodger CJ, Thomson NR (2002) Reconsidering the effectiveness of quasi-static thunderstorm electric fields for whistler duct formation. J Geophys. Res. 107: art.no.1396 NOV 2002

Medvedev YA, Stepanov BM, Fedorovich GV (1980) Physics of radiative excitation of electromagnetic fields Atomizdat, Moscow, 104 p.

Mendillo M, Rishbeth H, Roble RG, Wroten J (2002) Modelling F2-layer seasonal trends and day-to-day variability driven by coupling with the lower atmosphere. J. Atm. Sol.-Ter. Phys. 64: 1911-1931

Meyerott RE, Reagan JB, Evans JE (1985) Tropospheric Aerosol Perturbations Due to Volcanism as Measured by Electrical Conductivity of Ocean Air. J Geophys Res. 90: 5925-5932

Migulin VV, Larkina VI, Molchanov OA et al. (1982) Detection of the earthquake effects on the VLF-ELF noises in the upper ionosphere. Preprint IZMIRAN No 25(390) Moscow, 28 p.

Mikhailov AV, Förster M (1997) Day-to-day thermosphere parameter variations as deduced from Millstone Hill incoherent scatter radar observations during March 16 22, 1990 magnetic storm period. Ann. Geophys. 15: 1429-1435

Mikhailov YM, Mikhailova GA, Kapustina OV, Depueva AK, Buzevich AV, Druzhin GI, Smirnov SE, Firstov PP (2002) Variations in Different Atmospheric and Ionospheric Parameters in the Earthquake Prepartion Periods at Kamchatka: the Preliminary Results. Geomagn. Aer. 42: (769-776)

Mikhailov YM, Mikhailova GA, Kapustina OV, Buzevich AV, Smirnov SE (2003) Variations in the Power Spectra of the Electric Fields in the Near-Earth Atmosphere at Kamchatka. Geomagn. Aer. 43: (395-401)

Mitra AP (1974) Ionospheric Effects of Solar Flares D. Reidel Publishing Company Dordrecht - Holland/Boston - USA, 1974

Mjachkin V, Brace W, Sobolev G, and Dietrich J (1975) Two models of earthquake forerunners. Pageoph. 113: 169-181

Mogi K (1985) Earthquake Prediction. Tokyo: Academic Press

Molchanov OA, Mazhaeva OA, Goliavin AN, Hayakawa M (1993) Observation by the Intercosmos-24 satellite of ELF-VLF electromagnetic emissions associated with earthquakes. Ann. Geophysicae., No 11: 431-440

Molchanov OA, Hayakawa M, Ondoh T, Kawai E (1998a) Precursory effects in the subionospheric VLF signals for the Kobe earthquake. Phys. Earth Planet. Interr. 105: 239-248

Molchanov OA, Hayakawa M (1998b) Subionospheric VLF signal perturbations possibly related to earthquakes. J. Geophys. Res. 103: 17,489-17,504

Molchanov OA, Hayakawa M (2001) VLF Monitoring of Atmosphere-Ionosphere Boundary as a Tool to Study Planetary Waves Evolution and Seismic Influence. Phys. Chem. Earth. 26: 453-458

Molchanov OA, Hayakawa M, Afonin VV, Akentieva OA, Mareev EA (2002 a) Possible influence of seismicity by gravity waves on ionospheric equatrial anomaly from data of IK-24 satellite. 1. Search for idea of seismo- ionosphere coupling. In: Hayakawa M, Molchanov OA (Eds.) Seismo Electromagnetics: Lithosphere-Atmosphere-Ionosphere Coupling. Terrapub, Tokyo, pp. 275-285

Molchanov OA, Hayakawa M, Afonin VV, Akentieva OA, Mareev EA, Trakhtengerts VY (2002 b) Possible influence of seismicity by gravity waves on ionospheric equatrial anomaly from data of IK-24 satellite. 2. Equatorial anomaly and small-scale ionospheric turbulence. In: Hayakawa M, Molchanov OA (Eds.) Seismo Electromagnetics: Lithosphere-Atmosphere-Ionosphere Coupling. Terrapub, Tokyo, pp. 287-296

Moore GW (1964) Magnetic Disturbances preceding the 1964 Alaska Earthquake. Nature, 203: 508-512

Morgunov VA (1998) Электрические явления, предшествующие Шикотанскому землетрясению и его афтершокам, - Докл. РАН, 1998, т. 359, № 1. с. 102 – 105.

Morgunov VA, Matveyev IV (1991) Электрические и электромагнитные эффекты в эпицентральной зоне афтершоков Спитакского землетрясения, - Изв. АН СССР, Физика Земли, 1991, № 11, с. 124 – 128.

Morozova AL, Pudovkin MI, Barliaeva TV (2000) Variations of the Cosmic Ray Fluxes as a Possible Earthquake Precursor. Phys. Chem Earth (A) 25: 321-324

Mühleisen R (1958) The Influence of Water on the Atmospheric Electrical Field. In: Recent Advances in Atmospheric Electricity. Proceed. of the Second conf. on Atm., Electricity, Mar 20-23, 1958, Portsmouth, New Hampshire, Pergamon Press, pp. 213 – 221

Mühleisen R (1977) The global circuit and its parameters. In Electrical Processes in Atmospheres. H. Dolezalek and R. Reiter eds. Dr. Dietrich Steinkopff, Darmstadt. Pp. 467-476

Musya K (1931) On the luminous phenomenon that attended the Idu earthquake, November 26th, 1930, Bull. Earthquake Res. Inst., Tokyo Univ. 9: 214-215

Musya K (1932) Investigations into the luminous phenomena accompanying earthquakes, Bull. Earthquake Res. Inst., Tokyo Univ. 10: 666—673

Musya, K. (1934). On the luminous phenomena that accompanied the great Sanriku tsunami in 1933.(part I). Bull. Earthquake Res. Inst., Tokyo Univ., Suppl. 1, 87-111

Nagao T, Enomoto Y, Fujinawa Y, Hata M, Hayakawa M, Huang Q, Izutsu J, Kushida Y, Maeda K, Oike K, Uyeda S, Yoshino T (2002) Electromagnetic anomalies associated with 1995 Kobe earthquake. Journal of Geodynamics 33: 401–411

Nasyrov GA (1978) Correlation between night sky emissions and seismic activity. Izvestiya Acad. Sci. Tadzh. SSR, No. 2, pp. 119-122

Nava B, Radicella SM, Pulinets S, Depuev V (2001) Modelling bottom and topside electron density and TEC with profile data from topside ionograms, Adv.Space Res., 27: 31-34

Nikiforova NN, Michnowski S (1995): Atmospheric electric field anomalies analysis during great Carpatian Earthquake at Polish Observatory Swider. IUGG XXI General Assembly Abstracts. Boulder, Colorado. VA11D-16

Nishimura S, Katsura I (1990) Radon in soil gas: applications in exploration and earthquake prediction, - Geochemistry of gaseous elements and compounds. Theophrastus Publ., Athens, pp. 497 – 533

Nomikos K, Vallianatos F (1998) Electromagnetic variations associated with the seismicity of the frontal Hellenic arc. Geologica Carpatica 49: 57-60

Nunn D (1971) A theory VLF emissions. Planet. Space Sci. 19: 1141-1167

O'Keefe SG, Thiel DV (1999) Electromagnetic Emissions from Underground Rock Fracture Events. In: Atmospheric and Ionospheric Electromagnetic Phenomena Associated with Earthquakes, ed. by M. Hayakawa, TERRAPUB, Tokyo, pp. 233-244

Oliver MA, Webster R (1990) Kriging: a Method of Interpolation for Geographical Information Systems. Int. J. Geographical Information Systems, 4: 313-318

Ondoh T (1998) ionospheric disturbances associated with great earthquake of Hokkaido southwest coast, Japan of July 12, 1993. Phys. Earth Planet. Interr. 105: 261-269

Ondoh T (1999) Seismo-Ionospheric Effects. In: Atmospheric and Ionospheric Electromagnetic Phenomena Associated with Earthquakes. Terra Scientific Publishing Company, Tokyo, pp. 789-803

Ondoh T (2000) Seismo-Ionospheric Phenomena. Adv. Space Res. 26: 1267-1272

Ondoh T, Hayakawa M (1999) Anomalous Occurrence of Sporadic E-layers before the Hyogo-ken-Nanbu Earthquake, M 7.2 of January 17, 1995. In: Atmospheric and Ionospheric Electromagnetic Phenomena Associated with Earthquakes. Terra Scientific Publishing Company, Tokyo, pp. 629-640

Ondoh T, Marubashi K (eds) (2001) Science of Space Environment. Ohmsha, IOS Press

Onsager L (1938) Phys. Rev. 54: 554 – 557

Oraevsky VN, Boyarchuk KA, Pulinets SA (2002) The problems of satellite monitoring of earthquake precursors and Russian satellite constellation "Vulkan". 2002 International Workshop on earthquake precursors, June 5-6, 2002, NCU, Taiwan, Extended abstracts, 2002, pp. 15-19

Oraevsky VN, Boyarchuk KA, Dokukin VS (2003) Scientific instrumentation for the small satellite platform “Vulkan”. 4th IAA Symposium on Small Satellites for Earth Observation, Berlin, Germany, 7 - 11 April 2003, Abstr. 1109P

Oraevsky VN, Danilkin VA, Boyarchuk KA, Dokukin VS (2003) Small satellite COMPASS-2. 4th IAA Symposium on Small Satellites for Earth Observation, Berlin, Germany, 7 - 11 April 2003, Abstr. 1110P

Oreskes N ed. (2003) Plate Tectonics, Perseus Publishing, 424 p

Orville RE, Spenser DW Global lightning flash frequency. Mon. Weather Rev. 107: 934-943

Ouzounov D (2003) Earth-atmospheric coupling during strong earthquakes by analyzing IR remote sensing data. Proceeding of International workshop on earth system processes related to Gujarat earthquake using space technology Jan. 27-29 2003. Indian Institute of technology, Kanpur, p. 65

Papadopoulus GA (1999) Luminous and Fiery Phenomena Associated with Earthquakes in the East Mediterranean. In: Atmospheric and Ionospheric Electromagnetic Phenomena Associated with Earthquakes. Terra Scientific Publishing Company, Tokyo, pp. 559-575

Park CG, Dejnakarintra M (1973) Penetration of thundercloud electric fields into the ionosphere and magnetosphere, 1. Middle and auroral latitudes. J. Geophys. Res. 84: 960-964

Park CG, Dejnakarintra M. (1977) Thundercloud electric fields in the ionosphere, in Electrical Processes in Atmospheres, Dolezhalek, H.and Reiter, R., eds., Steinkopff, Darmstadt, pp. 544-551

Parrot M (2002) The micro-satellite DEMETER. J. Geodynamics 33: 535-541

Parrot M, Lefeuvre F (1985) Correlation between GEOS VLF emissions and earthquakes, Ann. Geophys., No 3: 737-748

Parrot M, Lefeuvre F, Corcuff Y, Godefroy P (1985) Observations of VLF emissions at the time of earthquakes in the Kerguelen Islands. Ann. Geophys. No 3: 731-736

Parrot M, Mogilevsky MM (1989) VLF emissions associated with earthquakes and observed in the ionosphere and the magnetosphere. Phys. Earth. Planet. Interr. 57: 86-99

Parrot M, Achache J, Berthelier JJ, Blanc E, Deschamps A, Lefeuvre F, Menvielle M, Plantet JL, Tarits P, Villain JP (1993) High frequency seismo-electromagnetic effects. Physics of Earth and Planetary Interiors 77: 65-83

Parrot M (1994) Statistical study of ELF/VLF emissions recorded by a low altitude satellite during seismic events. J. Geophys. Res.

Parrot M (1994) Seismo-electromagnetic waves detected by low altitude satellites. In: Electromagnetic Phenomena Related to Earthquake Prediction, Hayakawa and Fujinawa (eds.), TERRAPUB, Tokyo, pp 361-372

Paulikas GA (1975) Precipitation of particles at low and middle latitudes. Rev. Geophys. Space. Phys. 13: 709-734

Pertsev NN, Shalimov SL (1996) Excitation of atmospheric gravity waves in a seismically active region and their ionospheric effect. Geomagn. Aeronom. 31: 801-807

Piggott WR, Rawer K (eds) (1972a) U.R.S.I. Handbook of ionogram interpretation and reduction, Second Edition, November 1972, 324 pp.

Piggott WR, Rawer K (eds) (1972b) U.R.S.I. Handbook of ionogram interpretation and reduction, Second Edition, Revision of Chapters 1-4, November 1972, 135 pp.

Pikaev AK (1986) The modern radiative chemistry. Gas and liquids radiolysis. Nauka Publ. Moscow 440 p (in Russian)

Pilipenko V, Shalimov S, Uyeda S, Tanaka H (2001) Possible mechanism of the over-horizon reception of FM radio waves during earthquake preparation period, Proc. Japan Academy, 77 (B), No. 7: 125-130

Pokhmelnykh LA (2003) The gas crystal and ball lightning. In: The problems of environment and natural resources. ISSN 0235-5019, VINITI Publishing, Moscow: 20-31

Przerzecki SY, Dmitriev MT (1978) Radiative physico-chemical processes in the air. Atomizdat Publ. 182 p., (in Russian)

Pulinets SA (1989) Prospects of Topside Sounding. In: WITS handbook Vol. 2, Liu CH ed., Chapter 3, SCOSTEP Publishing, Urbana, Illinois: 99-127

Pulinets SA (1998a) Strong earthquakes prediction possibility with the help of topside sounding from satellites. Adv. Space Res. 21: 455-458

Pulinets SA (1998b) Seismic activity as a source of the ionospheric variability, Adv. Space Res. 22: 903-906

Pulinets SA, Fligel MD, Karpachev AT, Kochenova NA (1989) Topside sounder results for the 87 SUNDIAL Campaign, SUNDIAL Workshop VI, McLean, VA, USA, Brief reports, vol. 1: 1-10

Pulinets SA, Legen'ka AD, Karpachev AT, Kochenova NA, Fligel MD, Migulin VV, Oraevsky VN (1991) The earthquakes prediction possibility on the base of topside sounding data, IZMIRAN preprint N 34a(981), 25 p. (paper was presented at VII IAGA Scientific Assembly, Buenos-Aires, Argentina, 1993)

Pulinets SA, Legen'ka AD, Alekseev VA (1994) Pre-earthquakes effects and their possible mechanisms in "Dusty and Dirty Plasmas, Noise and Chaos in Space and in the Laboratory". Plenum Publishing, New York, pp. 545-557

Pulinets SA, Yudakhin KF, Evans D, Lester M (1996) The Study of the Ionospheric Variability within the Euro-Asian Sector During the ATLAS-1/SUNDIAL Mission, Journal Geophysical Research, 101: 26759-26767

Pulinets SA, Legen'ka AD (1997) First simultaneous observations of the topside density variations and VLF emissions before the Irpinia earthquake, November, 23, 1980 in magnetically conjugated regions. In: Proceedings of International Workshop on Seismo Electromagnetics, 1997, University of Electro-Communications Publ, Chofu, Japan, pp. 56-59

Pulinets SA, Alekseev VA, Legen'ka AD, Khegai VV (1997) Radon and metallic aerosols emanation before strong earthquakes and their role in atmosphere and ionosphere modification, Adv. Space Res., 20: 2173-2176

Pulinets SA, Khegai VV, Boyarchuk KA, Lomonosov AM (1998a) Atmospheric Electric field as a Source of Ionospheric Variability, Physics-Uspekhi, 41: 515-522

Pulinets SA, Legen'ka AD, Zelenova TI (1998b) Local-Time Dependence of Seismo-Ionospheric Variations at the *F*-Layer Maximum. Geomagnetism and Aeronomy. 38: 400-402

Pulinets SA, Kim VP; Hegai VV, Depuev VK, Radicella SM (1998c) Unusual longitude modification of the nighttime midlatitude F2 region ionosphere in July 1980 over the array of tectonic faults in the Andes area: Observations and interpretation. Geophys. Res. Let., 25: 4143-4136

Pulinets SA, Benson RF (1999) Radio-Frequency Sounders in Space. In: Review of Radio Science. Stone WR, Ed. Oxford University Press, Chapter 28: 711-733

Pulinets SA, Alekseev VA, Boyarchuk KA, Hegai VV, Depuev VKh (1999) Radon and ionosphere monitoring as a means for strong earthquakes forecast. Il Nuovo Cimento. 22C: 621-626

Pulinets SA, Boyarchuk KA, Khegai VV, Kim VP, Lomonosov AM (2000) Quasielectrostatic Model of Atmosphere-Thermosphere-Ionosphere Coupling. Adv. Space Res. 26: 1209-1218

Pulinets SA, Legen'ka AD (2002) Dynamics of the Near-Equatorial Ionosphere Prior to Strong Earthquakes. Geomagnetism and Aeronomy 42: 239-244

Pulinets SA, Boyarchuk KA, Hegai VV, Karelin AV (2002a) Conception and model of seismo-ionosphere-magnetosphere coupling, In: Seismo-Electromagnetics: Lithosphere-Atmosphere-Ionosphere Coupling, Eds. M.Hayakawa and O.A.Molchanov, TERRAPUB, Tokyo, 2002, pp. 353-361

Pulinets SA, Boyarchuk KA, Lomonosov AM, Khegai VV, Liu JY (2002b) Ionospheric Precursors to Earthquakes: A Preliminary Analysis of the *foF*2 Critical Frequencies at Chung-Li Ground-Based Station for Vertical Sounding of the Ionosphere (Taiwan Island). Geomagnetism and Aeronomy 42: 508-513

Pulinets SA, Depuev VH, Karpachev AT, Radicella SM, Danilkin NP (2002 c) Recent advances in topside profile modeling. Adv. Space Res. 29: 815-823

Pulinets SA, Legen'ka AD (2003) Spatial-temporal characteristics of the large scale disturbances of electron concentration observed in the F-region of the ionosphere before strong earthquakes. Kosmicheskie issledovaniya (Cosmic Research) 41(3): 1-10

Pulinets SA, Depuev VK (2003) Global Scale Ionospheric Irregularities Associated with Thunderstorm Activity, Proceeding of the IRI Task Force Activity 2002, IC/IR/2003/3, ICTP, Trieste, pp. 71-80

Pulinets SA, Hernández-Pajares M, Liu JY, Legen'ka AD (2003 a) Fountain effect depression before the strong earthquakes in low latitude areas. Proceeding of International workshop on earth system processes related to Gujarat earthquake using space technology Jan. 27-29 2003. Indian Institute of technology, Kanpur, p. 93

Pulinets SA, Legen'ka AD, Gaivoronskaya TV, Depuev VK (2003 b), Main phenomenological features of ionospheric precursors of strong earthquakes. J. Atm. Solar Terr. Phys. 65: 1337-1347

Pulinets SA, Leyva Contreras A, Bisiacchi G, Ciraolo L., Singh R (2003 c) Ionospheric and thermal precursors of Colima earthquake of 22 January 2003. Annual Meeting of Mexican Geophysical Union. GEOS 23: 170

Pulinets SA, Liu JY, Safronova IA (2004 a) One more interpretation of the statistical data analysis of the critical frequency foF2 variations before earthquakes by the data of Chung-Li (Taiwan) ionospheric station. Geomagn. Aeronom. 43: 102-106

Pulinets S, Leyva Contreras A, Kostoglodov V, Perez de Tejada H, Urrutia-Fucugauchi J (2004 b) PREVENTION project: the complex geophysical observatory in Mexico as a test facility for lithosphere-atmosphere-ionosphere coupling models. Natural Hazards and Earth System Sci. 4 to be published

Pustovetov VP, Malyshev AV (1993) Spatial-temporal correlation of the earthquakes and variations of high energy particle flux in the inner radiation belt. Cosmic Res. 31(5): 84-90

Rees MH (1963) Auroral ionization and excitation by incident energetic electrons. Planet. Space Sci. 11: 1209-1218

Reid HF (1910) The mechanism of the earthquake, In: The California Earthquake of April 18, 1906, Report of the State Earthquake Investigation Commission, 2, Washington D.C., Carnegie Institution, pp. 1-192

Reiser YP (1992) Physics of the gas discharge Nauka Moscow 536 p.

Reiter R (1985) Fields, Currents and Aerosols in the Lower Troposphere. Amerind, New Deli, India

Renisch BW (1996) Modern ionosondes. In: Modern Ionospheric Science. EGS Publ. Katlenburg-Lindau. 440-458

Rishbeth H, Mendillo M (2001) Patterns of ionospheric variability. J. Atm. Sol.-Ter. Phys. 63: 1661-1680

Rikitake T (1976) Earthquake Prediction, Amsterdam: Elsevier

Roberts CS (1969) Pitch-angle diffusion of electrons in the magnetosphere. Rev. Geophys. Space Phys. 7: 305-337

Roble RG (1991) On modeling component processes in the Earth's global electric circuit. J. Atmos. Terr. Phys. 53: 831-847

Roble RG, Tzur I (1986) The Global Atmospheric-Electrical Circuit. In The Earth's Electrical Environment, Studies in Geophysics series, National Academy Press, Washington D.C., 206-231

Rodger CJ, Thomson NR, Dowden RL (1998) Testing the formulation of Park and Dejnakarintra to calculate thunderstorm dc electric fields. J. Geophys. Res. 103: 2171-2178

Rodger CJ, Clilverd MA, Thomson NR (1999) Modeling of subionospheric VLF signal perturbations associated with earthquakes. Radio Science 34: 1177-1185

Roffman A (1972) Short-Lived Daughter Ions of Radon 222 in Relation to Some Atmospheric Processes. J. Geophys. Res. 77: 5883 – 5899

Rosenberg M, Mendis DA (1995) UV-induced Coulomb Crystalization in a Dusty Gas. IEEE Trans. Plasma Sci. 23: 177-180

Row RV (1966) Evidence of long-period acoustic gravity waves launched into the F - region by the Alaskan earthquake of March 28.1964. J.Geophys.Res., 71: 343-345

Rulenko OP (2000) Operative precursors of earthquakes in the near-ground atmosphere electricity. Vulcanology and Seismology No 4: 57–68

Rulenko OP, Ivanov AV, Shumeiko AV (1992) The short-term atmosphere-electric precursor of the Kamshatks earthquake 06.03.1992, M=6.1. Doklady RAS 326, № 6, c. 980-982.

Rulenko OP, Druzhinin GI, Vershinin EF (1996) Measurements of the atmospheric electric field and natural electromagnetic emission before the Kamchatka earthquake 13.11.93 r., M=6.1. Doklady RAS 326: 814 - 816

Rundle JB, Turcotte DL, Klein W (eds) (2000) GeoComplexity and the Physics of Earthquakes. Geophysical Monographs Series, American Geophysical Union, Washington D.C.

Ruzhin YY, Larkina VA (1996) Magnetic conjugation and a time coherency of seismoionosphere VLF bursts and energetic particles. In: Proceedings of International Wroclaw Symposium on Electromagnetic Compatibility. Poland. Wroclaw. 1996. pp. 645-648

Ruzhin YY, Larkina VI, Depueva, AK (1998) Earthquake Precursors in Magnetically Conjugated Ionosphere Regions. Adv. Space Res. 21: 525-528

Ruzhin Y, Nomicos C, Vallianatos F, (2000) High frequency seismoprecursor emissions. In: Proceedings of XV International Wroclaw Symposium on Electromagnetic Compatibility (EMC 2000), Poland, 2000, Part-II, pp.512 - 516

Rycroft MJ (1991) Interactions between whistler mode waves and energetic electrons in the coupled system formed by the magnetosphere, ionosphere and atmosphere. J. Atmos.Terr. Phys. 53: (849-857)

Sadovski MA (1982) Electromagnetic precursors of earthquakes. Nauka Moscow 88 p.

Salikhov RS, Vladimirov AV, Sennik NA, Oraevsky VN, Dokukin VS, Boyarchuk KA (2003) Small satellite platform "Vulkan". 4th IAA Symposium on Small Satellites for Earth Observation, Berlin, Germany, 7 - 11 April 2003, Abstr. 1108P

Salm YY, Loots AM (1988) Kinetics of negative light air-ions formation in troposphere. Sci. Memoir. Tartu State Univ. V. 809, pp. 64-70

Scholz CH (1990) The mechanics of earthquakes and faulting. Cambridge University Press, Cambridge

Scholz CH, Sykes, LR, and Aggarwal YP (1973) Earthquake prediction: A physical basis, Science, 181: 803-809

Schulz J, Meywerk J, Ewald S, Schlussel P (1997) Evaluation of satellite-derived Latent Heat Fluxes, J. Climate, 10: 2782-2795

Schunk RW (1988) A Mathematical Model of the Middle and High Latitude Ionosphere. Pure Appl. Geophys. 127: 255-303

Schunk RW, Nagy AF (2000) Ionospheres: physics, plasma physics and chemistry. Cambridge University Press, Cambridge

Segovia N, De la Cruz RS, Mena M, Ramos E, Monnin M, Siedel JL (1989) Radon in soil anomaly observed at Los Azufres Geothermal field, Michoacan: a possible precursor of the 1985 Mexico earthquake ($M_s = 8.1$). Natural Hazards 1: 319-329

Segovia N, Ponciano G, Ruiz W, Godínez L (2002) Indoor radon and airborne particles. Geofisica Internacional 42: (255-259)

Sentman D.D., Weskott E.M. et al. (1995) Preliminary results from the Sprites94 campaign: Red sprites. Geophys. Res. Lett. 22: 1205-1208

Serebryakova ON, Bilichenko SV, Chmyrev VM, Parrot M, Rauch JL, Lefeuvre F, Pokhotelov OA, (1992) Electromagnetic ELF radiation from earthquake regions as observed by low-altitude satellites. Geophys. Res. Lett. 19: 91-94

Sgrigna V, Console R, Conti L, Galper AM, Malvezzi V, Parrot M, Picozza P, Scrimaglio R, Spillantini P, Zilpimiani D (2003) The ESPERIA Mission, Fall AGU Meeting 2003, T52D-01

Shalimov SL (1992) Lithosphere-ionosphere relationship: a new way to predict earthquakes? Episodes. Intern. Geophys. Newsmag. 15: 252-254

Shalimov SL, Gokhberg MB (1998) Lithosphere-ionosphere coupling mechanism and its application to the earthquake in Iran on June 20, 1990. A review of ionospheric measurements and basic assumptions. Phys. Earth Planet. Interr. 105: 211-218

Shapley AH, Croehl HW, Allen JH (1975) Solar-Terrestrial Physics and Meteorology: A Working document. Washington, SCOSTEP Secretariat, July 1975

Sharadze ZS, Dzhaparidze GA, Zhvaniya ZK, Kvavdze ND, Kikvilashvili GB, Liadze ZL, Matiashvili TG, Mosashvili NV, Nikoliashvili NS (1991) Disturbances in the Ionosphere and the Geomagnetic Field Associated with the Spitak Earthquake. Izvestiya, Earth Physics 27: 989-996

Shearer PM (1999) Introduction to Seismology. Cambridge University Press, Cambridge

Shklyar DR, Nagano I (1998) On VLF wave scattering in plasma with density irregularities. J. Gophys. Res. 103: 29,515-29,526

Shklyar DR, Truhlik V (1998) On the modification of light ion concentration profiles above seismically active regions: a qualitative consideration. J. Atm. Solar Terr. Phys. 60: 1025-1033

Sidorin AY (1992) Earthquake precursors. Nauka, Moscow 192 p.

Silina AS, Liperovskaya EV, Liperovsky VA, Meister C.-V. (2001) Ionospheric phenomena before strong earthquakes. Natural Hazards 1: 113-118

Singh R, Simon B, Joshi PC (2001) Estimation of surface latent heat fluxes from IRSP4/MSMR satellite data, Proc. Indian Acad. Sci. (Earth Planet Science), 110: 231-238

Singh RP, Bhoi S, Sahoo AK (2002) Changes observed on land and ocean after Gujarat earthquake 26 January 2001 using IRS data. Int. J. Rem. Sens. 23: 3123-3128

Smirnov BM (1978) Negative ions. Atomizdat, Moscow 176 p. (p. 61)

Smirnov VV (1980) Condensation nuclei formation in the areas of increased ionization IEM Trans. Iss. 24 (89), pp. 80–98

Smirnov VV (1983) The electric factors of air purity. IEM Trans. Iss. 30(104) pp. 74-106

Smirnov VV (1987) Experimental studies of the ion cluster formation in air. Abstracts of the V all-union conference "Aerosols and their application in national economy". Moscow V II, pp.88 – 89

Smirnov VV (1980) Condensation nuclei formation in the areas of increased ionization IEM Trans. Iss. 24 (89), pp. 80–98

Smirnov VV Ionization in troposphere (1992) Gidrometeoizdat Publ., Sankt-Petersburg, 312 p. (in Russian)

Smith PA, Kaiser BA (1967) Estimates of ionospheric composition and temperature derived from topside sounder electron scale height data. JATP 29: 1345-1353

Sobolev GA (1993) The basics of earthquake prediction. Moscow, Nauka Publ.

Sobolev GA, Husamiddinov SS (1985) Pulsed electromagnetic Earth and ionospheric field disturbances accompanying strong earthquakes, Earthquake Prediction Res. No. 3: 33-45

Sobolev GA, Shestopalov IP, Kharin EP (1998) Implications of Solar Flares for the Seismic Activity of the Earth. Izvestiya, Phys. Solid Earth 34: 603-607

Sobolev GA, Zakrzevskaya NA, Kharin EP (2001) On the Relation between Seismicity and Magnetic Storms. Izvestiya. Phys. Solid Earth 37: 917-927

Solovyov VA (1941) Experiments on the atmosphere ionization by the X-rays. Meteorology and hydrology No. 3: 19 - 30

Sorokin VM, Chmyrev VM, Yashchenko AK (2001) Electrodynamic model of the lower atmosphere and the ionosphere coupling. J. Atm. Sol. Terr. Phys. 63: 1681-1691

Stakhanov IP (1996) On the physical nature of the ball-lightning. Nauchnyi Mir, Moscow, 263 p.

Stein S, Wysession M (2003) An introduction to seismology, earthquakes, and Earth structure. Blackwell Publishing

Stozhkov YI, Basilevskaya GA, Ermakov VI, Pokrevski PE (1997) Ion formation in the atmosphere. Izvestiya- Physics 61: 1155 – 1158

Suarez, G, Monfret T, Wittlinger G, David C (1990) Geometry of subduction and depth of the seismogenic zone in the Guerrero gap, Mexico. Nature. 345: 336-338

Sugisaki R, Ann H, Aedachi M, Ui H (1980) Geochemical features of gases and rocks along active faults. Geochem J. 14: 101-112

Sytinskiy AD (1987) Sviaz' seismichnosti Zemli s solnechnoy aktivnost'yu I atmosfernymi processami (Correlation of the Earth's Seismicity with Solar Activity and Atmospheri Processes). Leningrad, Gidrometeoizdat.

Sytinskiy AD (1989) Correlation of earthquakes with Solar Activity. Izvestiya, earth Physics 25: 86-98

Szuszczewicz EP, Torr D, Roble R, Wilkinson P, Richards P, Emery B, Lu G, Abdu M, Evans D, Pulinets SA. (1996) F-Region Climatology During the SUNDIAL/ATLAS-1 Campaign of March 1992: Model-Measurement Comparisons and Cause-Effect Relationships.Journal Geophysical Research. 101: 26741-26758

Szuszczewicz EP, Lester M, Wilkinson P, Blanchard P, Abdu M, Hanbaba R, Igarashi K, Pulinets S, Reddy BM (1998) Global ionospheric storm characteristics during solar maximum equinox. J. Geophys. Res. 103: 11,665 – 11,684

Taran YA, Bernard A, Gavilanes J-C, Lunezheva E, Cortes A, Armienta MA (2001) Chemistry and mineralogy of high-temperature gas discharges from Colima volcano, Mexico. Implications for magmatic gas-atmosphere interaction. J. Volc. Geotherm. Res. 108: 245-264

Tarasov NT, Tarasova NV (2003) Seismicity changes under electromagnetic impact in seismoactive zones. Abstracts of 100-years anniversary of acad. Gamburtsev conference, Institute of Physics of the Earth, RAS, Moscow

Terada T (1931). On luminous phenomena accompanying earthquakes. Bull. Earthquake Res. Inst.,Tokyo Univ. 9: 225-255

The Earth's Electrical Environment (1986) Studies in Geophysics series, National Academy Press, Washington D.C. 263 p.

Thomas L (1971) The lower ionosphere, — J. Atm. Terr. Phys. 33: 157 – 195

Tinsley BA (2000) Influence of Solar Wind on the Global Electric Circuit, and Inferred effects on Cloud Microphysics, Temperature, and Dynamics in the Troposphere. Space Sci. Rev. 94: 231-258

Titheridge JE (1986) Ionogram analusis with the generalized program POLAN. World data Center A for STP. Report UAG-93

Titheridge JE (1988) The real height analysis of ionograms. Generalized formulation. Radio Sci. 23: 831-849

Toroshelidze TI, Fishkova LM (1988) Analysis of variations of nighttime emission of middle and upper atmosphere preceding earthquakes. Doklady USSR Acad. Sci. 302(2): 313-316

Toutain J-P, Baubron J-C (1998) Gas geochemistry and seismotectonics: a review. Tectonophysics 304: 1-27

Tramutoli V, Di Bello G., Pergola N., Piscitelli S. (2001) Robust satellite techniques for remote sensing of seismically active areas. Annali de Geofisica 44: 295-312

Tronin, AA (2000) Thermal IR satellite sensor data application for earthquake research in China. International Journal of Remote Sensing, 21: 3169-3177

Tronin AA, Hayakawa M, Molchanov OA (2002) Thermal IR satellite data for earthquake research in Japan and China. Journal of Geodynamics 33: 519-534

Turcotte DL (1991) Earthquake prediction. Annu. Rev. Earth Planet. Sci. 19: 263-281

Turcotte DL, and Spence DA (1974) An analysis of strain accumulation on a strike slip fault, J. Geophys Res., 79: 4407-4412

Tulinov VF (1967) On the corpuscular radiation role in the lower atmosphere formation. Cosmic Res. 5: 240-244

Turman BN, Edgar BC (1982) Global lightning distributions at down and dusk. J. Geophys. Res. 87: 1191-1206

Tzur I, Roble RG (1985) The Interaction of a Dipolar Thunderstorm With Its Global Electrical Environment. J. Geophys. Res. 90: 5989-5999

Understanding Plasma Interactions with the Atmosphere: The Geospace Electrodynamic Connections (GEC) Mission (2001) Report of the NASA Science and Technology Definition Team for the Geospace Electrodynamic Connections (GEC) Mission, NASA/TM—2001–209980

Utkin VI (1997) The gaseous breath of the Earth. Soros Educationsl Journal No 1, 1997, pp. 57-64 (http://www.nature.ru/db/msg.html?mid=1158809&uri=index.htm)

Utkin VI, Bulashevich YP, Yurkov AK, Nikolaev VV (1993) Forecast of mine blasts in deep mines. Proc. Intern. Conf. SEG-93. Moscow pp. 15-17

Utkin VI, Yurkov AK (1998) The Dynamics of Radon Liberation from a Rock Massif as a Short-Term Earthquake Forerunner. Doklady Earth Sci., 359: 262-266

Uyeda S, Nagao T, Orihara Y, Yamaguchi T, Takahashi I (2000) Geoelectric potential changes: Possible precursors to earthquakes in Japan, PNAS, 97: 4561-4566

Uyeda S, Nagao T, Hattori K, Hayakawa M, Miyaki K, Molchanov O, Gladychev V, Baransky L, Shtchekotov A, Fedorov E, Pokhotelov O, Andreevsky S, Rozhnoi A, Khabazin Y, Gorbatikov A, Gordeev E, Chebrov V, Sinitzin V, Lutikov, Yunga AS, Kosarev G, Surkov V, Belyaev G (2001) Geophysical Observatory in Kamchatka region for monitoring of phenomena connected with seismic activity. Natural Hazards and Earth System Sciences 1: 3–7

Uyeda S, Hayakawa M, Nagao T, Molchanov O, Hattori K, Orihara Y, Gotoh K, Akinaga Y, Tanaka H (2002) Electric and magnetic phenomena observed before the volcano-seismic activity in 2000 in the Izu Island Region, Japan, PNAS 99: 7352-7355

Valensise G (1993) Summary of Contributions on the 23 November 1980, Irpinia earthquake, Annali di Geofisica, 36: 345-351

Vallianatos F, Nomicos K (1998) Seismogenic Radioemissions as Earthquake Precursors in Greece. Phys. Chem Earth. 23: 953-957

Vampola AL, Kuck GA (1978) Induced precipitation of inner zone electrons 1. Observations. J. Geophys. Res. 83: 2543-2551

Vampola AL, Gorney DJ (1983) Electron energy deposition in the middle atmosphere J. Geophys. Res. 88: 6267-6274

Vershinin EF, Buzevich AV, Yumoto K, Saita K, Tanaka Y (1999) Correlations of seismic activity with electromagnetic emissions and variations in Kamchatka region. in Atmospheric and Ionospheric Electromagnetic Phenomena Associated with Earthquakes, Ed. by M. Hayakawa. Terra Scientific Publishing Company, Tokyo: 513-517

Voinov VV, Gufeld IL, Kruglikov VV (1992) Effects in the ionosphere and atmosphere before the Spitak earthquake. News of USSR Academy, Fizika Zemli (in Russian), No.3, pp. 96-101

Voitov GI, Dobrovolsky IP (1994) Chemical and isotopic-carbon instabilities of the native gas flows in seismically active regions. Izvestiya Earth Science 3: 20-31

Volland H ed. (1995) Handbook of Atmospheric Electrodynamics Vol.1 CRC Press

Volodichev NN, Kuzhevskii BM, Nechaev OY, Panasyuk MI, Podorol'skii AN, Shavrin PI (2000) Solar-lunar-terrestrial links: Peaks of neutron emission and seismic activity. Solar Syst. Res. 34: 173-175

Volodichev NN, Podorolskii AN, Levin BV, Kuzhevskii BM (2001) Correlation of Appearance of Large Series of Earthquakes with the Phases of New and Full Moons. Vulkanol. Sesimol. No.1: 60-67

Vorobiov AA, Remizov VP (1984) Atmospheric electric field changes as a possible earthquke precursors. In: Electromagnetic fields in biosphere. V. 1. Nauka, Moscow pp. 311–315

Voronov SA, Galper AM, Kirillov-Ugriumov VG, Koldashov SV, Mikhailov VV, Popov AV, Chesnokov VY (1989) Registration of high energy particle fluxes increase in the area of Brazilian anomaly on 25 September 1985. Cosmic Res. 27: 629-631

Voronov SA, Galper AM, Koldashev SV, Maslennikov LV, Mikhailov VV, Nikitina NV, Popov AV (1990) The increase of charged particles of high energy fluxes in the area of Brazilian anomaly and the Earth' seismicity. Cosmic Res. 28: 789-791

Voss HD, Smith LG (1980) Global zones of energetic particle precipitation. J. Atmos. Terr. Phys. 42: 227-239

Wakita H (1988) Short-term and intermediate-term geochemical precursors. Pageoph. 126: 267-278

Wang C.-Y., Shin T.-C., Taiwan Seismicity: an Overview (2002) in 2002 International Workshop on earthquake precursors, June 5-6, 2002, NCU, Taiwan, Extended abstracts, 2002, pp. 26-29

Warwick JW, Stoker C, Meyer TR (1982) Radio Emission Associated With Rock Fracture: Possible Application to the Great Chilean Earthquake of May 22, 1960. J. Geophys. Res. 87: 2851-2859

Wayne RP (2000) Chemistry of Atmospheres. Oxford University Press. 775 p.

Weaver PP, Yuen PO, Prölss GW, Furumoto AS (1970) Acoustic coupling into the ionosphere from seismic waves of the earthquake of Kurile islands on August 11, 1969. Nature, 226: 1239-1241

Werner S, Prölss GW (1997) The position of the ionospheric trough as a function of local time and magnetic activity. Adv. Space Res. 20: 1717-1722

Whitehead JD (1970) Production and prediction of sporadic E. Rev. Geophys. and Space Phys. 8(4): 65-114

Wicks A, Jason S, Harrison J (2001) An EO constellation based on the TopSat microsatellite: global daily revisit at 2.5 metres. In Proc. 15th Annual Small Satellites Conf., Utah State University, UT, USA, 13–16 August 2001, paper no. SSC01-I-6. (Available from http://www.aria.cec.wustl.edu/SSC01/papers/1-6.pdf)

Willis C, Boyd AW (1976) Excitation in radiation-chemistry of inorganic gases. Int. J. Rad.Phys. 8: 71-76

Wilkinson PJ, Richards P, Igarashi K, Szuszczewicz EP (1996) Ionospheric climatology and weather in the Australian-Japanese sector during the SUNDIAL/ATLAS 1 campaign. J. Geophys. Res. 101: 26,769-26,782

Wyss M, Booth DC (1997) The IASPEI procedure for the evaluation of earthquake precursors. Geophys. J. Int. 131: 423-424

Yakovlenko SA (1995) Metastable plasma of hydrated ions. Izvestiya VUZov Fizika, No 4, pp. 3–10

Yasui Y (1968) A study on the luminous phenomena accompanied with earthquake (part 1), Mem.Kakioka Mag. Obs., 13: 25-61

Yasui Y (1971) A study on the luminous phenomena accompanied with earthquake (part II), Mem.Kakioka Mag. Obs. 14: 67-68

Yasui Y (1972) Seismo-luminous phenomena at Santa Rosa, Mem. Kakioka Mag. Obs. 15: 81-86

Yoshihisa I, Yoji K, Takashi T (2002) Large earthquakes initiate by the acceleration of slips on the downward extensions of seismogenic faults. Earth Planet. Sci. Let. 202: 337-343

Yuen FC, Weaver PF, Suzuki RK, Furumoto AS (1969) Continuous traveling coupling between seismic waves and the ionosphere evident in May 1968 Japan earthquake data. J. Geophys.Res.,74: 2256-2264

Zadorozhny AM (2001) Effects of charged dust on mesospheric electrical structure. Adv. Space Res., 28: 1059-1064

Zelenova TI, Legen'ka AI (1989) Ionospheric Effects Related to the Moneron Earthquake of September 5(6), 1971. Izvestiya, Earth Physics 25: 848-853

Zhang Tie-zheng (1998) Prediction of the Tengshan Ms 7.8 earthquake with the "Magnetic Storm Double Method". Northwest Seismology 20(2): 29-32

Zinn J, Sutherland CD, Gangul S (1990) The Solar Flare of August 18, 1979: Incoherent Scatter Radar Data and Photochemical Model Comparisons. J. Geoph. Res. 95: 16,706-16,718

Subject Index